Late Stalinist Russia

The late Stalinist period, long neglected by researchers more interested in the high-profile events of the 1930s, has recently become the focus of new and innovative research. This book examines late Stalinist society's interaction with ideology, state policy and national and international politics. It dispels the notion of late Stalinism as the apogee of Stalin's rule. Rather than being cowed by terror and control, Soviet post-war society emerges as highly diverse and often contradictory. The analysis of issues such as the impact of demographic changes, shifts in popular opinion, the rise of new generations and the construction of post-war spaces demonstrates that alongside the needs of reconstruction, a strong desire for reinvention existed. The late Stalinist years are thus shown to be a crucial turning point between the Soviet Union's revolutionary origins and its later appearance as a mature and consolidated empire beset by problems of stagnation and corruption. It was in this time that the Soviet Union acquired many of the characteristics that gave it the face it was to have until its demise.

Juliane Fürst is a Junior Research Fellow at St. John's College, Oxford. She has published widely on the life, culture and thought of young people in the late Stalinist era. Her book titled *Stalin's Last Generation: Youth, Identity and Culture* is forthcoming.

BASEES/Routledge series on Russian and East European studies

Series editor:
Richard Sakwa, Department of Politics and International Relations, University of Kent

Editorial Committee:
Julian Cooper, Centre for Russian and East European Studies, University of Birmingham
Terry Cox, Department of Central and East European Studies, University of Glasgow
Rosalind Marsh, Department of European Studies and Modern Languages, University of Bath
David Moon, Department of History, University of Durham
Hilary Pilkington, Department of Sociology, University of Warwick
Stephen White, Department of Politics, University of Glasgow

Founding Editorial Committee Member:
George Blazyca, Centre for Contemporary European Studies, University of Paisley

This series is published on behalf of BASEES (the British Association for Slavonic and East European Studies). The series comprises original, high-quality, research-level work by both new and established scholars on all aspects of Russian, Soviet, post-Soviet and East European Studies in humanities and social science subjects.

Late Stalinist Russia

Society between reconstruction
and reinvention

Edited by Juliane Fürst

LONDON AND NEW YORK

First published 2006
by Routledge
2 Park Square, Milton Park, Abingdon, Oxon OX14 4RN

Simultaneously published in the USA and Canada
by Routledge
270 Madison Ave, New York, NY 10016

Routledge is an imprint of the Taylor & Francis Group, an informa business

© 2006 Selection and editorial matter, Juliane Fürst; individual chapters,
the contributors

Typeset in Times by Wearset Ltd, Boldon, Tyne and Wear
Printed and bound in Great Britain by TJI Digital, Padstow, Cornwall

British Library Cataloguing in Publication Data
A catalogue record for this book is available from the British Library

Library of Congress Cataloging in Publication Data
A catalog record for this book has been requested

ISBN10: 0-415-37476-6 (hbk)
ISBN10: 0-203-96968-5 (ebk)

ISBN13: 978-0-415-37476-7 (hbk)
ISBN13: 978-0-203-96968-7 (ebk)

Contents

Notes on contributors

Mark Edele is Lecturer of Modern History at the University of Western Australia in Perth. He completed his PhD on Soviet veterans at the University of Chicago.

Beate Fieseler is Professor of Russian History at the University of Bochum, Germany. She has published widely on questions of veterans, women and street children.

Donald Filtzer is Professor of Russian History at the University of East London. He has published several monographs on labour history including, most recently, *Soviet Workers and Late Stalinism: Labour and the Restoration of the Stalinist System after World War II*.

Sheila Fitzpatrick is Bernadotte E. Schmitt Professor of History at the University of Chicago. Among her numerous publications are *Stalin's Peasants* (Oxford, 1993), *Everyday Stalinism* (Oxford, 1999), *Tear off the Mask* (Princeton, 2005) and *Stalinism: New Directions* (London, 2002).

Juliane Fürst is a Junior Research Fellow at St. John's College, Oxford. She has published widely on the life, culture and thought of young people in the late Stalinist era. Her book titled *Stalin's Last Generation: Youth, Identity and Culture* is forthcoming.

James Heinzen is Associate Professor at Rowan University in New Jersey. He recently published *Inventing a Socialist Countryside: The Soviet State and the transformation of Rural Russia Before Collectivization*. He is currently working on a new project on post-war corruption.

Cynthia Hooper is Assistant Professor at Holy Cross College, Massachusetts. She completed a PhD on the mechanisms of the Great Purges at Princeton, for which she received the Frankel Prize of the Wiener Library, London. She has published widely on questions of populist participation and comparative aspects of Soviet history with Nazi Germany.

Timothy Johnston is a DPhil candidate at the University of Oxford. He is working on popular responses to international politics in Stalin's Russia.

Jean Lévesque is Assistant Professor at Texas Tech University. He completed his PhD at the University of Toronto and has published widely on peasants and rural life in the post-war period.

Ann Livschiz is Assistant Professor at the University of Indiana and Purdue at Fort Wayne. She finished her PhD on Soviet children and childhood at Stanford University.

Rebecca Manley is Assistant Professor in the History department of Queen's University in Kingston, Canada. She is currently completing a book manuscript on the Soviet experience of the Second World War, focusing on the politics and culture of internal population displacement.

Mie Nakachi is a PhD candidate at the University of Chicago. She is currently completing her work on post-war motherhood and reproductive policies.

Monica Rüthers is teaching at the University of Basel in Switzerland. Her book on Moscow's public spaces is forthcoming. Currently she is investigating the topographies of Soviet childhoods between 1950 and 1985.

Acknowledgements

The editor would like to thank all the contributors of this volume for their hard work, patience and good humour. Further, she is grateful to Peter Sowden at Routledge, editor of the BASEES series, who has been unwaveringly supportive and helpful throughout. She would also like to thank her colleagues at St. John's College and the History Faculty of the University of Oxford, particularly Dr Ross McKibbin whose encouragement and interest in this book were invaluable.

Introduction

Late Stalinist society: history, policies and people

Juliane Fürst

The period known as late Stalinism has long been a neglected area in Soviet studies. Squeezed between the better-documented 1930s and the subsequent epic Great Fatherland War on the one side and the more transparent and exciting events of the Thaw on the other, late Stalinism was until recently treated as the bizarre appendix to Stalin's rule – stale, at times malicious and, ultimately, superfluous to the understanding of Soviet history. Other terms and descriptions applied to the years between the end of the war and the death of Stalin indicate that, rather than considering it a period in its own right, Sovietologists understood it primarily in relation to pre-war Stalinism. Labels such as 'high Stalinism' demonstrate the widespread belief that this period was seen as the apogee of Stalin's power and the culmination of many facets of Stalin's rule. Strict restrictions on visitors to the Soviet Union and an ever-growing opacity of the workings of Soviet government and society contributed to the belief that postwar Stalinism was essentially an intensified version of the 1930s minus the upheavals of collectivization and the extremes of the Great Terror. Yet, ultimately, such a view clouds, rather than illuminates, the unique nature of these years. Late Stalinism had its own dynamics and internal logic. It was undoubtedly indebted and linked to events that had taken place and norms that had been set in the 1930s. Yet it was also a post-war era. It showed the scars of a long and protracted, but victorious, war and displayed the societal dynamics typical of periods of recovery. While the 1930s represented a system and a society in consolidation, the post-war years brought unprecedented – and mostly unintended – upheaval with large swathes of the Soviet population on the move and much of the Soviet physical and ideological infrastructure in tatters. Yet Stalin's last years also heralded many developments associated with so-called mature socialism. The emergence of a fledgling consumer society, the birth of youth countercultures and the rise of a Soviet middle class question not only our understanding of 'high Stalinism', but also force us to rethink the roots of the Thaw and our commonly accepted periodisation of Soviet history.

This volume of articles proposes to look at late Stalinism not as a period of time, but as a phenomenon in its own right. It dismisses the notion that late Stalinism is neatly framed by 9 May 1945, Soviet victory day, and 5 March 1953, the day of Stalin's death. When the war ended late Stalinism was well underway.

Reconstruction had been a topic in the corridors of power ever since the Germans' defeat outside Moscow. On the ground people had started to repair not only their houses and belongings, but also their relationship with the Soviet system, which for many had either been temporarily suspended or severely questioned. Yet with reconstruction came the question what exactly was 'normal'? What constituted the post-war Soviet ideal? What did it mean to be 'Soviet' in the post-war era? The experience of war generated many answers, ranging from very inclusive and elastic versions of Sovietness to demands for even stricter discipline and ideological purity. The period was by no means static in its character and nature. Rather, 1947 seems to have been a significant turning point both economically and ideologically. Major political campaigns provided a constant roller-coaster of drives and demands, while new realities shaped new policies. The Stalinist 1950s were quite different to the immediate post-war years – Monica Rüther's article in the last chapter of this book deals with very different questions than Mie Nakachi's and Beate Fieseler's, who open the section 'When the war was over'. Yet change can also be overemphasized. The transformation of Gorky Street, described by Rüther, was an ongoing process, while neither the demographic problem of Nakachi's article nor Fieseler's veterans disappeared quickly. Stalin's death was a significant and indeed, as memoirs testify, cataclysmic event for many late Stalinist subjects. Yet change was neither immediate nor complete or entirely due to the disappearance of the persona of Stalin. Stalin's post-war years, like any period of transition, were Janus-faced with one side looking back to the immediate and long-term past and the other looking forward to the future. Late Stalinism was just as much about reinvention as it was about reconstruction.

Capturing an elusive entity: late Stalinism as seen by contemporaries and historians

The very first people to pay scholarly attention to late Stalinism were Western contemporaries, who, in their role as diplomats or journalists, observed post-war Soviet society at first hand – albeit, severely hampered by travel restrictions and a growing fear among the Soviet public of association with foreigners.[1] Ironically, many of their reports seem to have been ignored in later scholarly treatments, which, under the spell of non-witnesses such as Hannah Arendt and George Orwell, preferred an interpretation of late Stalinist society as cowed and atomized.[2] After Stalin's death many more people were allowed to enter the Soviet Union. Their more colourful and exciting reports very quickly eclipsed the world described by their predecessors.[3] It became very quiet around late Stalinism. There was a smattering of publications devoted to domestic power politics and a flood of works concerning foreign policy and the emergence of the Cold War. Very few of either treated late Stalinism as a period in its own right.[4] Chris Ward argued that it was not a lack of sources or events that caused this dearth of scholarly interest, but the rigidity of the totalitarian school; 'Once the lineaments of the party-state and the nomenklatura system had been described and categorized, "mature" or "late" Stalinism was packaged as the classic representation of Orwell's nightmare – a world without history, a place where nothing changed and nothing signified outside the paranoid mind of the *vozhd'*.'[5]

It was only in the 1970s that the silence surrounding the social and cultural developments of late Stalinism was broken. Vera Dunham's study of the post-war Soviet novel and her proposal of a 'Big Deal' between the Stalinist regime and the middle classes echoed many of the assumptions of the 'Great Retreat' defined by Timosheff several decades earlier, yet it had very different connotations. Instead of a sense of return or retreat, Dunham interpreted late Stalinist society as developing into something in its own right, which neither aped Western habits nor followed a pure, revolutionary model. Here was a world that was neither in opposition nor in accordance with Bolshevik ideology.[6] When in 1985 a volume of collected articles on the impact of the Second World War on Soviet society appeared, the post-war years were en route to rehabilitation.[7] For the first time questions, which had long been raised relating to other Soviet periods, were posed in the context of late Stalinism and the cataclysmic event which ushered in its existence. Essays on the consequences of war on the party, the city of Leningrad and non-Russian speaking minorities gave a first taste of how far-reaching and complex the impact of war was.[8] Sheila Fitzpatrick's thesis that late Stalinist society never really demobilized, hinted to the enormous problems that beset the system at the time. In her contribution 'The Return to Normalcy' she also implicitly raised the question of what normalcy actually meant for the Soviet period.[9] Were the 1930s really the norm? What does 'normal' mean in a Soviet context? Years later Amir Weiner would argue that indeed the post-war period was normal in the Bolshevik sense, since it continued the revolutionary project of purging and cleansing in order to achieve the promised socialist utopia.[10] David Shearer has put forward the argument that mobilization was the natural state of Stalinism and that 'the idea of militarization and war' could serve as 'an explanatory model' for Stalinism overall.[11] Such interpretations shift the angle from which we usually examine post-war periods away from comparisons with pre-war periods and raise multiple questions about the precise nature of such commonly accepted terms such as reconstruction, normalcy and recovery.

Despite new and innovative publications on the topic of late Stalinism, these questions have by no means been answered. Yet with the plurality of approaches increasing late Stalinism is rapidly becoming less of a black hole. In the early 1990s Elena Zubkova took advantage of the newly opened Soviet archives and painstakingly constructed a picture that was far removed from interpretations of late Stalinism as the apogee of control and power. Indeed, Zubkova's post-war Soviet state is embattled on all possible fronts and can only hang onto control through the return to and intensification of repressive measures.[12] Her investigation was path-breaking and illuminating, putting the post-war period almost single-handedly back on the scholarly map. Yet it is also a witness to its time. Zubkova's work is strongly rooted in the spirit of perestroika and its desire to turn the Manichean world view of Soviet times on its head without losing the belief in a dichotomy of good and bad. Post-war society emerges as people, who try to struggle free from oppression and counter 'paper-cut norms' with their own ideas and identities.[13] Youth, peasants and the intelligentsia all appear as the champions of a new, freer and better way of life. The grey zones of post-war Soviet society remain grey, the entanglement of actors, victims and bystanders unexplored. This interpretation of late Stalinism as merely a herald of a 'hint of freedom', which was ultimately squashed and superseded by the subsequent Thaw, is also at the heart of the many influential memoirs on late Stalinism, such as those by the writers Konstantin Simonov and Ilia Ehrenburg, which emerged in the twilight days of the Soviet Union.[14]

The most recent work on late Stalinism has tried to counter such notions and put late Stalinism into new contexts. Yoram Gorlizki and Oleg Khlevniuk, concentrating on the mechanism of high politics in late Stalinism, have shown that the separation between the international front of the Cold War and domestic politics is an unhelpful one. The period can only be successfully understood when events of international importance are juxtaposed with domestic campaigns. In their thorough analysis of Politburo documents late Stalinism emerges as a series of logical interactions between several players both within and outside the Soviet Union.[15] David Brandenburger widens the picture even further. He argues that the primary influence in the formation of post-war policy and identity remained a Soviet notion of Russian history, which he terms 'national Bolshevism'. He disputes the centrality of the war, placing the decisive turning point of Soviet history firmly in the pre-war period.[16] In contrast Amir Weiner considers the myth of the Great Fatherland War as essential to the fate of the Soviet/Bolshevik project in the post-war period. He argues for an interpretation of the war as, on the one hand, a watershed, which fundamentally and irrevocably altered the mechanisms of Soviet identity-building, and, on the other, a catalyst of reinforcement, which continued the history of the revolutionary project. He points to the fact that the war by no means signified the end of revolution, but rather its continuation with a new set of norms and rules.[17] His interest in the myth of the war and the social and cultural consequences of its dominance is echoed in Jeffrey Brooks' work on public culture, which understands the post-war years as a period when people were robbed of their own memory of war. Rather than continuity with the revolutionary mission Brooks sees a continuation of Stalinist performance culture.[18] The extent to which late Stalinism heralded the failure of the Soviet project has been made visible by work done on the social history of the period. Neither entirely devoid of ideology nor entirely devoted to it, late Stalinism was a harbour for contradictory policies and realities. Still anchored in the sincerity of the revolutionary project, these years also perfected the appearance of pretty facades, the petrification of ritual and the omnipresence of hypocrisy and dissimulation.[19] Julie Hessler has painted a convincing picture of war and post-war trade, which, at least in the opinion of Soviet policy-makers attempting to foster 'cultured', e.g. controlled trade, had taken several steps back by giving prominence to primitive and unofficial trade forms such as bazaars and flea markets.[20] Donald Filtzer in his monograph on workers in late Stalinism stressed the chaos and poverty prevailing in these years and the recalcitrant nature of working class behaviour – the very section of society on which the regime – at least rhetorically – based its power. Filtzer sees here the seeds of disillusion and discontent, which ultimately lead to the downfall of the Soviet system.[21] New work on post-war subjectivities shows that, while the research on Soviet self-perception relating to the 1930s is invaluable in identifying mechanisms for analysis of subjective realities, the researcher of the post-war years has to unearth new norms and codici of Soviet thinking.[22] Post-war subjectivities were the results not only of the cataclysmic events of the war, but also of a new Soviet generation coming of age – indeed of the whole Soviet project maturing both for better and for worse.

Transition and negotiation: approaches to understanding late Stalinist society

How thus to navigate the complexities of late Stalinism? How to make sense of its contradictory evidence, its confusing plurality of voices and its double-natured character as a period tightly connected to the revolutionary past on the one hand and heralding all the hallmarks of mature socialism on the other? Late Stalinist society can be understood in a multiplicity of ways. The following approaches, drawing mostly from the evidence provided in the articles of this volume, represent just four of many attempts to unravel the complexities of the period and develop methodologies which will help to shed light on the interaction between policies and people, subjects and rulers, authorities and citizens.

A society shaped by war

Late Stalinism was a post-war society. It is this facet of its character that has recently come to the forefront of the attention of scholars, yet still remains largely unexplored and shrouded in mystery. At the end of the war all European societies had good reason to suppress the real consequences of war with a blanket of official and commonly accepted stereotypes. Collaboration, betrayal or the sheer pain of remembrance led to the war years being referred to in platitudes and a lowered voice. The late Stalinist state was particularly determined not to be associated with trauma, ambiguity and societal division – phenomena typical for societies emerging from war and occupation.[23] Certainly, the extensive damage done to buildings, infrastructure and agriculture could not be ignored. Yet the message was that all injuries would heal as soon as Soviet power returned – including the many psychological injuries inflicted on the victims of fighting and displacement. Reconstruction only meant the reconstruction of houses, streets and factories, the establishment of Soviet educational and cultural services and the re-employment of demobilized soldiers. Cripples were viewed as merely physically impaired, orphans were nothing more than children without parents and veterans were just soldiers who had come home. Restoration of the outer trappings of normalcy was to cover and eradicate inner scars and traumas. The difficulties of reintegration of those who had been at the front with those who had stayed behind or been on enemy territory were studiously ignored. Over time popular memory of the war as a people's war was suppressed and replaced by the collective memory of a victorious war led by Stalin, the generalissimo.[24] Nonetheless, silence and reinvention could not hide the fact that the experience of war shaped late Stalinist society both individually and collectively. War and the myth of war informed political decisions on all levels – internationally and domestically, in the *raion* committees as well as in the Politburo. War was never far from people's minds, both as a memory and as a future threat. Consequences of the physical damage of war informed the everyday reality of all but the most privileged citizens. War shaped how people thought, acted and interacted. The experience of war and its inevitable by-product of displacement introduced new ideas and knowledge into Soviet societal discourse. War formed generations, war created new focal points of identification and war provided the mechanism for a new, societal stratification. Ultimately, war could not be ignored.

First and foremost war forced the hands of post-war policy-makers and administrators. Any issue discussed – may it relate to agriculture or industry, propaganda or crime – had to take into account the effects of destruction and upheaval. As soon as the first shock of the invasion had waned off, the Soviet administrative machinery started rolling to deal with the most immediate damage and initiate a programme of reconstruction. Mie Nakachi demonstrates in her article on the 1944 Family Law in this volume that Soviet leaders were well aware of the problem of immense population losses long before the war had concluded. The newly drafted family law was unabashedly pro-natalist and, with its provision for state alimony for single mothers, encouraged not only the one-parent family (unimaginably shameful according to 1930s Soviet norms), but also a shift in gender relations, since it encouraged pre- and extra-marital and uncommitted sex. Nakachi's work indicates not only that directives for the post-war years were set long before the Red Army reached Berlin, but that Khrushchev's heavy involvement in the drafting of the Law made this policy live beyond Stalin's death. It is no coincidence that the heroine of one of the most popular Soviet films of the 1970s – *Moscow does not believe in tears* – is the living embodiment of this policy: a young woman left with a child by an irresponsible man who refused to marry her. It is also no coincidence that, despite the fact that the policy was created in the 1940s, it took Soviet society until the 1970s to embrace its consequences.

More often than not, however, the consequences of war made themselves felt before policy-makers could turn their attention to the looming problems. Jean Lévesque details the developments taking place in the countryside, where the consequences of war, which forced the hands of policy-makers, were more indirect. While severely affected by war per se – through the loss of manpower, physical destruction and diversion of much-needed funds – life in the village posed even more of a challenge to post-war leaders, because of its inhabitants' ingenuity in engaging in non-conformist survival and avoidance strategies.[25] The multiple ways in which peasants made use of the 'grey zone' of rural life – which to no small extent had come into existence as a result of diminished attention and growing chaos during the war – generated sharp responses, designed to regulate the unruly countryside. The drive to bring order to the post-war village culminated in 1948 in an expulsion order for parasitic elements from kolkhozes and their forced exile into eastern Russia. *Kolkhnozniki* were forced to select the victims from their very midst in open assemblies. The campaign, which ran only in a few trial *oblasti*, was, however, not judged a success and ultimately abandoned – presumably, because the rural population did not display the expected reaction.

Cynthia Hooper picks up the theme of rural disobedience in her article on corruption and black-marketeering. Her corrupt kolkhoz chairmen, who carved out considerable personal freedoms during and after the war, were not persecuted by the state. Rather they came to a form of truce with the higher authorities, despite – or maybe because – of the harsh legislation passed from above: lower and mid-level functionaries supported the general line, including the implementation of unpopular policies, in exchange for a few kickbacks. Avoidance of participation in official life, emergence of entire alternative economic systems, exploitation of available loop holes in Soviet law and, indeed, outright violent crime were not problems confined to the countryside or to any specific part of the country.[26] They could be found in towns and industrial centres as well as on kolkhozes and sovkhozes. They were on the agenda of party meetings in the Central Committee in Moscow as well as in remote regions of the Soviet periphery. Some areas such as the Baltics and Western Ukraine had problems far beyond corruption and ordinary crime. Here the Soviet system was literally fighting for its survival.[27] Looming behind the regime's failure to keep law and order was the simple fact that the Soviet system was not functioning very well after it had been drained of 27 million people, lost millions of roubles' worth of infrastructure and found its ideology under attack. Pursuing the same level of service, control and coverage as in pre-war times, the post-1945 Soviet system found itself in institutional overstretch.

The fact that, above all, war meant poverty and lack of resources is highlighted in Don Filtzer's contribution to this volume. He proves that in the post-war years the question of whether the late Stalinist system was benign or hostile to its own population was often quite irrelevant. In many instances, it simply could not cope with the challenges it faced. While the hunger he describes was mainly a consequence of the harvest failure of 1946, famine was exacerbated by the inability of the Soviet state to muster an adequate response and by the generally abysmal state of housing, sanitary installations, health service and food supply.[28] The reality was that, while left-wing observers in Western Europe believed that the Soviet victory had sanctioned the socialist system, war had almost broken the backbone of the Soviet order. Recalcitrant workers with little labour discipline, a wayward and defiant peasantry, homeless and loitering children and youngsters and a general rise in crime, corruption and violence challenged the system on a daily basis.[29] The state took refuge in harsh legislation, which helped to criminalize large swathes of people, including many children and adolescents.[30] The ethnic cleansing campaigns carried out in the later stages of the war against the Tartars, Ingushetians, Chechens and other minorities enlarged the number of those excluded from the Soviet project. The implications went much further than the first three hungry and violent years of the post-war period. The failure of the collective welfare state initiated the rise of individual survival strategies and the decline of the collective as commonly of value. The legacy of practices of exclusion and persecution was to haunt the Soviet Union long after Stalin's death.

Analysis of late Stalinism through the prism of the experience of war quickly indicates that it was not merely destruction, trauma and chaos that followed victory. Much of the war's impact was more indirect and subtle. Some consequences were engineered from above, while some were unexpected, even for those who thought themselves to be in control. The regime's attempt to transform the war into a new founding myth is revisited in the work of Ann Livschiz. While concurring on the immense impact the war had on the identity and self-perception of the late Stalinist subject, she turns her attention to the ambiguities of this impact, the authorities' desperate attempt to control its consequences, and the anxieties about the sheer force the new myth generated. To market 'the experience of war' after it had ended turned out to be a more difficult task than expected. Children not only learnt to value sacrifice, they also took away ideas of spontaneity and independence. Heroic stories of young defenders of the motherland served not only to inspire more work in school and factory, but also encouraged youngsters to turn their attention to injustices closer at home. Well-meaning publications guiding children's values and norms turned out to have subversive undertones. The fact that active and dutiful participation in the war effort did not automatically mean future devotion and conformism is borne out in Mark Edele's chapter on the veterans. Here it is the experience rather than the propaganda of war that leaves soldiers reeling. Despite their victorious campaign over Germany, not all veterans return convinced of the superiority of the Soviet system. Expectations raised during the course of the war were often disappointed, while the cautious propaganda on behalf of the Western allies was taken up with rapturous enthusiasm. Yet, as both Mark Edele and Timothy Johnston, who analyses the workings of rumours and war scares, stress, such diverse views do not signify the existence of resistance. Rather, they testify to the variety of discourses possible even in a controlled and isolated society. This could include narratives of unfavourable comparison or opposition.

An obsession with control

The paradigm of late Stalinism as an era of absolute and unbending control has been successfully challenged in recent years. Nonetheless, it is no coincidence that this interpretation enjoyed popularity for as long as it did, and it would be foolish to underestimate the extent to which late Stalinism can be understood as a period of intensive moulding and reshaping from above. A simple look at the mere events in the Stalinist post-war years reveals that indeed this time was characterized by a rapid succession of various ideological and legal campaigns, which were wound up to fever pitch for a few weeks or months and then left to simmer. The late Stalinist leaders, most notably Stalin himself, displayed an ever growing desire to control every corner of society and to take hold of even the most ephemeral of thoughts by its members. Campaigns ranged from the rooting out of petty crime and corruption to the implementation of correct historical consciousness and Stalinist values. While the pre-war Soviet Union was characterized by a few, very distinctive campaigns, the late Stalinist years displayed a frenzy of different drives, many of which were ridiculous in their utopianism and twisting of facts. Nonetheless, for many late Stalinist subjects the expectations and demands expressed in these campaigns were very real – both as ideological beliefs and as potential factors, impacting on their personal lives.

Campaigns clustered around a vague notion of anti-Westernness, anti-intelligentsia and anti-Semitism. Post-war campaigns, in contrast to pre-war campaigns, were highly ambiguous, couched in incomprehensible and intransparent terms and left wide open to interpretation. The 1946 attack on the scientists Roskin and Kliueva represented the starting signal for a series of elaborate crusades against the intelligentsia, its Western ways and its moral and ideological corruption. Party and Komsomol protocols of secret meetings called to discuss the case indicate that behind the condemnation of the scientists sat a deep fear that people were not ideologically prepared for the conflict with the West that loomed on the political horizon. The so-called Zhdanovshchina sought to purify the output of the cultural establishment. In different places the Zhdanovshchina showed different faces. In Ukraine the campaign's attention was focused on the battle of nationalism.[31] In the field of opera and theatre, inappropriate historical interpretation were under attack, while in music composers were accused of formalism.[32] It was clear that the attack was a broader attempt to rein in the arts. Finally, the anti-cosmopolitan campaign started off with attacks on a number of Jewish theatre critics. While never endorsed as an anti-Semitic campaign, the populist version was a witch-hunt of Jews, especially those in privileged positions. All of these campaigns had several subtexts, ranging from the channelling of generational conflict to the re-enforcement of collective values. Yet looming behind every single one of them lay the Soviet regime's desire to bring order to the post-war ideological world, which had acquired a worrying amount of parallel discourse. Ultimately, the campaigns were designed to gain or regain control of this most elusive entity of Soviet life – the Soviet mind.

The regime was no less active in the domain of law and order. The tightening of the criminal law system reached new heights during late Stalinism, at times even bordering on the hysterical. Harsh labour laws, introduced shortly before and after the beginning of war, remained in force for the entire late Stalinist period with only some minor modifications made after 1947.[33] Yet the very same year saw the ratification of a number of *ukazy* concerning theft and misappropriation of state and private property, which featured the same draconian punishments for even minor offences. According to Peter Solomon, criminal law became Stalin's preferred tool of control in the post-war period.[34] While the regime employed the practice of political purges in the Leningrad or Mingrelian affairs, the late Stalinist years saw a definite shift towards criminal convictions. Hooliganism, shirking work and misappropriation became the traps in which 'socially harmful' elements were caught.[35] The criminal laws became ever more critical tools for completion of the socialist project. The fact that a criminal conviction meant almost automatic banishment into prison or labour camps indicates that late Stalinist social control was based on an exclusionary model, which neatly divided society between those who were allowed and expected to participate, and those who were exiled to, what Monica Rüthers terms in her article, Soviet 'non-spaces'. Notions of redemption and reinvention, which had prevailed during the 1920s and again during the war, were disposed of. Late Stalinist rule had no time for correcting imperfections.

The desire to shape Soviet subjectivities and cleanse them of the pollution of war and Western contact was not only apparent in the big campaigns and major decrees. It was perfected in the practice of the cult of Stalin's personality, convincing Soviet subjects of the unimaginability of a world without Stalin. It was apparent in the strict guidelines that governed the production of cultural life. And it made its ideological visions and demands visible through the construction of new spaces, which were both expressions and generators of Soviet power. A number of articles in this book explore this new approach to late Stalinist control, viewing it less through the lens of totalitarianism than within the context of ideology, identity and modernity. Monica Rüthers discusses the reshaping of the old main road of Tsarist Moscow into the central artery of the Soviet capital. By no means a late Stalinist idea, the reconstruction of *Tverskaia*, subsequently *Gorkogo*, reached its climax – and in many ways its completion – in the post-war years. Walking down Gorky street, especially on public holidays, entailed an implicit confirmation of the regime and its official values. Ann Livschiz in her contribution points to the same desire to control people's notion of their Soviet identity. The system's desire to control the workings of childhood minds and re-establish the concept of order physically and ideologically was hampered by the duality of the memory of war – pride and anxiety. The very act of control created spaces in which unexpected consequences could arise. Wayward writers misunderstood the task they had been set; children interpreted war-time heroism as a call to spontaneity or even opposition.

The impossibility of 'total control' is also highlighted in the contributions by Cynthia Hooper and James Heinzen. Corruption was particularly harmful in the eyes of the authorities, since it was an activity that not only contradicted high-minded Soviet values of collectivism and honesty, but also represented business taking place outside the regime's control. It thus undermined the regime's control in multiple ways and was consequently fought with all severity – thus ironically opening even more possibilities for patron–client relationships, as the red tape and punishments made protection from above an ever more common feature of any managerial activity. Hooper concludes that indeed at a certain point the desire to control shifted from the desire to have clear, state-controlled business transactions to the ratification of a 'dirty Big Deal', which allowed certain personal gratification on the part of functionaries in return for implementing law and order. Control ceases to be absolute in order to be effective, thus establishing a completely new meaning to the axiom of state power and its execution.

A society of individuals

While the previous section has demonstrated that the concept of control from above is a powerful paradigm to analyse late Stalinism, it has also become apparent that any kind of supervision was permanently challenged from below. Late Stalinism was not only a system of campaigns and demands which were more or less successfully executed. Rather, late Stalinism is better understood as a debate between official norms and the people affected by them. Neither side held a static position or, indeed, a clearly defined stance. Hundreds and thousands of permutations made up society's relationship with those in power. The equation becomes even more complicated when considering the viability of a state/society dichotomy in the face of multi-dimensional power relationships. Indeed, the line between these two entities was entirely fluid, often confused and dependent on perspective. Pockets of power were enjoyed by party and Komsomol activists, trade union representatives or professional bosses. Most Soviet citizens found themselves in a double role. At times they were representatives of the state and at other times they became its subjects and victims of its rule. How they positioned themselves vis-à-vis Soviet power was defined by a complex set of interactions which, like a spider's web, defined their mental and physical habitat. It was durable and fragile at the same time. Individual parts of the web could be altered and destroyed, yet ultimately, very few citizens left the general framework of Soviet thoughts and norms.

Soviet society with its emphasis on the imagined tomorrow, its utopian dreams and its belief that reality is as much what it ought to be as what it is, lends itself to approaches focusing on the subjective view of individual agents. It is therefore not surprising that a number of articles in this volume have chosen to pay particular attention to the viewpoint of the late Stalinist subject and explore late Stalinist society through the eyes and minds of individuals. The results provide a very different, yet not necessarily contradictory picture, to that achieved in a reading, which emphasizes the power of forces from above. Rather than absolute control in every corner, challenges and imperfections seem to signal an ideological malaise, which, as we know with hindsight, was to haunt the Soviet Union to its very end. Yet, at the same time, a closer examination of individual subjectivities reveals that challenging directives and norms from above does not signify automatic opposition, let alone resistance. While critical thoughts and actions were a feature of late Stalinist society, very few people were consistent in their rejection. Partial support was joined with resigned acceptance. Discontent often had a marriage of convenience with pragmatic participation. Critical anger was often rooted in enthusiastic support.[36] The shades between the white of conformity and the black of resistance produced many greys indeed.

The fact that individual survival strategies could significantly undermine the regime's blueprint for normative behaviour has already become apparent. Jean Lévesque and Cynthia Hooper detail in their articles on peasant life and corruption how individuals could easily distort campaigns or avoid integration into the late Stalinist fold by finding alternative paths for survival. Indeed, the world of racketeers or non-collectivized peasants could either function entirely parallel to the official world of socialist trade and strictly managed kolkhozes or blend into it, taking advantage of some of the benefits on offer, but choosing to shirk socialist obligations. Mark Edele makes the point that veterans displayed a whole range of attitudes towards the regime, which drew on a plethora of ideas and influences, the central argument of his contribution. Even a group of citizens who at least rhetorically were supported and fostered by the regime were by no means only victims of 'brainwashing', but used their experiences, past events and rumours as building blocks for their own specific world views. Similarly, as Timothy Johnston describes, the very existence of post-war rumours demonstrates that individual practices were not only capable of challenging the official discourse, but could form new collectives, thus reaching far into a domain the Soviet state considered its monopoly. Monica Rüthers and I show in our contributions that such alternative discourse, created by individuals or subgroups within society, did not necessarily have to be verbal, but rather could find expression in a multitude of markers, which included fashion, entertainment and street behaviour.[37] Even with small gestures, individuals were able to enter into a debate with the regime that allowed them to switch in and out of conformity in different settings at different times. The simple fact that young, stylish people referred to Gorky Street as *Broadway* indicates that individuals were able to reinvent the whole meaning of public spaces designed and interpreted from above with a flick of the tongue. Rather than the expression of a reconstructed Moscow, that had reinvented itself as a modern city, the term *Broadway* implied ownership of the street by the young and evoked parallels with American society. Yet only the widening and refashioning of the old Tsarist shopping mile under Stalin could provide the background for such a claim. Thus, precisely in the regime's highly developed sense of control rests the very possibility to avoid, undermine and reinterpret. From this perspective the campaigns of the post-war years seem less a mechanism of control and more an engagement in a battle that was taking place in every little corner of the Soviet Union. They were responses rather than initiatives, enacted by a regime that felt under threat.

All contributors in this volume, however, are adamant that to conclude that late Stalinist society was a hotbed of opposition would be premature. Rather, the attempt to see the late Stalinist world through the eyes of its inhabitants reveals that conscious opposition was extremely rare. People considered their actions necessary for survival, compatible with the norms of the regime or insignificantly outside the prescribed norms. Mark Edele distinguishes between three different forms of dissent among veterans, of which only one scratches on the foundations of the Soviet system. Jean Lévesque and Cynthia Hooper describe sections of society which took advantage of loopholes – they did not challenge the legitimacy of the Soviet system as a whole, but rather sidestepped its implications or came to an arrangement that is ideologically dubious, but economically beneficial. Young people challenged official culture in their quest for identity, not being content with the options the regime provided through its official ideological transmission belts. Yet very few of them can be considered to have been engaged in resistance. Their subjective motives came from a universe far removed from the politics of the Party.

The multi-faceted nature of the relationship between system and individual in late Stalinism is best understood as a dialogue between authorities and individuals, where the very same people could represent different sides at different times. The war had widened and intensified the range of topics available for discussion. Both people and leaders realized that post-war realities would never permit a complete recreation of the 'normalcy' of the 1930s. The field of possible reinventions was left wide open and ranged from reversing collectivization to experimenting with the usage of public spaces. The right or wrong of the Soviet system was rarely under discussion. Rather, the question was what exactly Sovietness meant. Fervent patriotism could go hand in hand with criticism of living conditions, while Komsomol membership did not preclude a liking of jazz. Just as the line between authorities and subjects was not clearly drawn, the line of debate did not necessarily have to run between impersonal system versus individual citizen. Debate could also take place between different sections within society or among individuals, as the sense of living through a time of transition and change intensified. One of the most contested areas was the memory of the war. Rebecca Manley describes in her contribution how something as mundane as the post-war housing question was prone to reveal rifts between different memories of war and different perceptions of who was entitled to reap the fruits of victory.

A society of the future

Last, but not least, late Stalinism heralded many developments that were to become characteristic of the Soviet post-war period and so-called mature socialism. Just as much as late Stalinist society was indebted to events of the past – most notably the Great Fatherland War – it was linked to the years that were to follow. Late Stalinist society, as will become apparent in many articles in this volume, was a transition society, which belonged both within the realm of Stalin's rule and provided the preparatory ground for the years of the Thaw, Stagnation and perestroika.[38] Often excluded from examination of the post-war years, the late Stalinist period marks the beginning of a new phase in Soviet history – one that was characterized less by terror, force and ideological fervour, and more by competition with the West, ideological crisis and the rise of a consumer society. When Stalin died, Soviet society did not make an immediate break with the past.[39] Rather, the developments that followed the shock of the death of the great leader had their roots in the preceding years, where often a veil of silence imposed on the Soviet media had hidden their existence.

It is no coincidence that a number of articles in this volume are concerned with the material aspects of Soviet life. While always a natural preoccupation of the ordinary citizen, the confrontation with the wealthier West during the war had given the topic great urgency, while at the same time a loosening of state control had allowed the return of certain 'bourgeois' habits of ownership and consumption.[40] Jean Lévesque and Rebecca Manley detail the complicated questions arising from the debates about ownership in the post-war period, when the question of who owned what determined a citizen's relationship with, and standing within, the Soviet system. After years of loss and deprivation, personal belongings had not become less, but more, important – a fact that was very apparent among the young generation of the time, who chose to identify themselves via external markers rather than via a shared ideological belief. The fact that material circumstances were at the centre of most people's mental worlds is also demonstrated in Donald Filtzer's and Beate Fieseler's contributions, which draw attention to the extreme poverty suffered by the urban population at large and by the veterans in particular. Mark Edele shows that this deprivation indeed informed the thinking of some veterans, who drew on visions of a utopia, located in a different time or a different place, in order to criticize the existing conditions. Last, but not least, corruption which, as detailed in Cynthia Hooper's and James Heinzen's contributions, was rampant in this period was one of the purest expressions of the dominance of materialism over ideology. Hooper's identification of a 'darker Big Deal' already indicates that the system recognized the inevitable rise of a material culture. While the phenomenon was ignored and swept under the carpet by Stalin, Khrushchev tried to tackle the challenge to the primacy of ideology head on, by devising programmes that would foster material *and* ideological gains. His desire to combine the practical with the ideological is already apparent in his reproductive policy. Mie Nakachi demonstrates that the need to replenish the Soviet population squared well with Khrushchev's belief that more people meant more power and thus a better survival guarantee for Soviet socialism. It was not only the policy itself that survived well into Khrushchev's own rule, his general view of society as in need of material concessions within a tight ideological framework continued to inform his politics until his own demise.

Late Stalinism set the tone for mature socialism in many other ways. The mythology of war, so carefully constructed in the immediate post-war years, continued to inform Soviet hierarchies and perception of identity and status. If anything, the further war receded into the past, the more important became its myth. The debate over the exact nature and content of this myth raged not only in late Stalinism, but also became a topos of the Khrushchev years. The commemoration of war, so central in many of the articles in this volume, continued to be shaped in the decades after Stalin's death. Films such as *Ivan's Childhood* and *Fate of a Man* prove that the question as to who could claim the war was far from over in the 1960s. Brezhnev's rejuvenation of victory celebrations demonstrated how important and central the subject was to Soviet internal politics and state–society relations.[41] War had forever substituted class as the ultimate marker in Soviet society. The rise of war-related patriotism went hand in hand with multiple attempts to rejuvenate the system and a decline in socialist radicalism. Again, a closer study of late Stalinism reveals the post-war period to harbour the roots of this development. Not only did concerted and formalistic procedures stifle enthusiasm and spontaneity, one of the core illnesses of mature socialism, the first attempts at rejuvenation can also be found in these years. The Zhdanovshchina and anti-cosmopolitan campaign were not only designed to frighten writers and intellectuals into conformity, they were also meant to rouse the young and radical against ideological complacency. In general, the press was full of self-flagellating articles and cartoons against formalism, bureaucratization and stultification of Soviet practices. While more muted than the Khrushchevite campaigns, they were orientated along the same line of belief in the supremacy of the revolutionary and war-time period. Nonetheless, late Stalinism also saw the beginning – or at least intensification – of a certain process of depoliticization and an increasing split between the public and the private in Soviet life. People began to separate their lives increasingly into official life, taking place at work and public organizations, and private life, which withdrew more and more from public places into the realm of kitchens, dachas and private flats.[42] Such developments necessarily caused critique and consternation among some people. While a large number of people withdrew from the Soviet project, another section of society deplored the state of socialism and began to hope and work for reforms from within the socialist framework. The modern dissident was thus also born in the post-war years – partly as a necessary sub-consequence of the trend towards apoliticization and partly out of the experience of having encountered other societies and values.[43] The confidence the war bestowed on people was an important precursor to the confidence that led people later to the Pushkin monument on Gorky street or to the courtroom of the Siniavskii/Daniel trial.

Conclusion

The list of subjects rooted in late Stalinism could continue ad infinitum. Whatever phenomenon the historian of the later Soviet Union investigates, ultimately Stalin's last years are likely to prove crucial. The articles in this volume make a strong case for the rehabilitation, re-examination and reinterpretation of a neglected, and often misunderstood, era. Rather than attempting to provide a textbook-like overview of the period, the following essays present in-depth research on certain topics, based on new archival sources. The articles have been grouped into five parts, each of which provides a different angle on the analysis of late Stalinism. Part I, 'When the war was over', gives central place to the impact of war on late Stalinist society. Mie Nakachi, Beate Fieseler and Timothy Johnston illuminate how the Soviet state tried (or failed) to cope with the devastation and destruction of years of fighting. These three articles demonstrate on how many levels war-related problems arose and how difficult the task to overcome not only the material, but also the psychological consequences of war was for both system and people. Responses ranged from the practical to the ideological, from the rational to the superstitious. The fact that war was never far from people's minds is also evident in the two articles of Part II. Donald Filtzer draws attention to the extreme poverty which ruled in Soviet cities, while Jean Lévesque details survival strategies employed in the countryside. Stalin's decision to continue his politics of fostering heavy industries over consumer production meant that life in both countryside and town was difficult in the extreme. The sheer pressure placed on population, officials and leaders resulted in all kinds of wheeling and dealing outside the official sphere – a phenomenon that is the subject of Part III. Corruption, tolerance of corruption and official collaboration with corruption were all trends that intensified in the chaotic and insecure post-war years. Part III indicates that in many ways the worst consequences of war for late Stalinist society were not the destruction and devastation brought upon people, but the secondary consequences of war, which created a system that was deeply rotten, yet covered by a veil of silence and ideological control.

It is not surprising that such an extraordinary time created new generational cohorts, which were bound by age or common experience. Veterans represented a new sector of society, whose experience of the front and other places gave them the confidence and ability to voice beliefs different to the official discourse. Yet, as Mark Edele in Part IV argues, despite their shared background, they failed to constitute a coherent group with one voice. Rather their disparate views soon became common currency among the population at large and symptomatic of the fragmented nature of late Stalinist society. Ann Livschiz and I look at specific age cohorts, who lived through the post-war years as children or youngsters. The fact that neither regime nor young people themselves were able to find a consensus on the nature of post-war Soviet identity indicates the instability of the regime's ideological control and the diversity of ideas permeating the late Stalinist discourse. Finally, Part V on post-war spaces explores late Stalinist society by focusing on specific sites which were instrumental in shaping identity and memory in these years. While Rebecca Manley looks at the question of the private apartment, which became a contested battlefield when evacuees or returnees challenged new inhabitants over their rights, Monica Rüthers explores the space of Gorky Street, which became a place that both regime and ordinary people claimed for different, sometimes overlapping, sometimes contradictory, purposes. Both demonstrate that these sites became much more than spaces to be filled or to be owned – rather their chosen subjects reveal many of the debates and fault-lines that ran within and across late Stalinist society.

It is predominantly a young generation of historians who have turned to the study of late Stalinism and the articles thus present not only new themes and topics, but also indicate a new direction in Soviet history, which attempts to bridge the gap between social and cultural history. Neither entirely discursive nor entirely empirical, Soviet history has taken an 'anthropological' turn. The close observation of real phenomena does not stand in the way of a cultural and contextual interpretation. At the same time, while the 'archival turn' initially resulted in a fascination with documents, ten years later, diversity is the hallmark of the field.[44] The plethora of sources used in the articles in this volume is representative of the new style of scholarship which allows a combination of sources and does not champion the primacy of one type of source over another. Authors draw from sources, ranging from trade union and prosecution documents through information provided by Party and Komsomol to eye-witness accounts, interviews and pictures. The breadth of topics and the variety of material presented by the authors allow the emergence of a general picture which transcends the specific subjects investigated in this volume. Late Stalinism has gradually ceased to be the 'black hole' of Soviet history – not least because of the pioneering work done by the contributors to this volume. Many facets of post-war life have been illuminated, many problems have been investigated and many interpretations offered. Yet, as will also become apparent to the attentive reader of the following articles, late Stalinism with its double-faced nature and contradictory characteristics still poses more questions than answers.

Notes

1 Walter Bedell Smith, *My Three Years In Moscow*, Philadelphia: Lippincott, 1950. Alexander Werth, *Russia: The Post-war Years*, New York: Taplinger, 1971. Don Dallas, *Dateline Moscow*, London: William Heinemann, 1951. Michel Gordey, *Visa to Moscow*, New York: Alfred Knopf, 1952. Alexander Werth, *Russia after the War*, London: Robert Hale, 1971. See also Harry Nerwood, *To Russia and Return: An Annotated Bibliography of Travellers*, Ohio: Ohio State University Press, 1968.

2 See especially Hannah Arendt, *The Origins of Totalitarianism*, New York: Meridian Books, 1958. George Orwell's novel *Animal Farm* and *1984*, first published in 1946 and 1948, respectively.

3 See for example Edward Cranckshaw, *Russia without Stalin*, London: Michael Joseph, 1956. Helen and Pierre Lazareff, *The Soviet Union after Stalin*, London: Odhams Press, 1955. Maurice Hindus, *House without a Roof*, London: Doubleday, 1961.

4 The exceptions were some political histories of the Stalinist post-war years. Robert Pethbridge, *A History of Post-War Russia*, London: Allen and Unwin, 1966. W. McCagg, *Stalin Embattled*, Wayne State University Press, 1978. Werner Hahn, *Postwar Soviet Politics: The Fall of Zhdanov and the Defeat of Moderation 1946–1953*, Ithaca: Cornell University Press, 1982.

5 Chris Ward, *Stalin's Russia*, London: Arnold, 1993, p. 187.

6 Vera Dunham, *In Stalin's Time: Middle Class Values in Soviet Fiction*, Cambridge: Cambridge University Press, 1976. Timasheff, *The Great Retreat: The Growth and Decline of Communism in Russia*, New York: Dutton, 1946.

7 Susan Linz (ed.), *The Impact of World War II on the Soviet Union*, Totowa: Rowman and Allenhead, 1985.

8 Cynthia Kaplan, 'The Impact of World War II on the Party', Linz (ed.), *Impact*, pp. 157–188. Edward Bubis and Blair Ruble, The Impact of World War II on Leningrad, Linz (ed), *Impact*, pp. 189–206. Barbara Anderson and Brian Silver, Demographic Consequences of World War II on the Non-Russian Nationalities of the USSR, Linz (ed.), *Impact*, pp. 207–243.

9 Sheila Fitzpatrick, 'Postwar Soviet Society: the "Return to Normalcy", 1945–1953', Linz, *Impact*, pp. 129–156.

10 Amir Weiner, *Making Sense of War: The Second World War and the Fate of the Bol-shevik Revolution*, Princeton: Princeton University Press, 2002.

11 David Shearer, 'Elements Near and Alien: Passportization, Policing, and Identity in the Stalinist State, 1932–1952, *Journal of Modern History* 76 (December 2004), pp. 835–881.

12 Elena Zubkova, *Reformy i Obshchestvo*; Elena Zubkova, *Poslevoennoe Obshchestvo*, Moscow: Rosspen, 1999.

13 Zubova, *Obshchestvo*, p. 137.

14 Konstantin Simonov, *Glazami Cheloveka moego Pokoleniia*, Moscow: Novosti, 1988. Ilia Ehrenburg, *Liudi, Gody, Zhizn'*, Moscow: Sovetskii Pisatel', 1990.

15 Yoram Gorlizki, Oleg Khlevniuk, *Cold Peace: Stalin and the Soviet Ruling Circle, 1945–1953*, Oxford: Oxford University Press, 2004.

16 David Brandenburger, *National Bolsheism: Stalinist Mass Culture and the Formation of Modern Russian National identity 1931–1956*, Cambridge: Harvard University Press, 2002, pp. 183–248.

17 Amir Weiner, 'The Making of a Dominant Myth: The Second World War and the Construction of Political Identity with the Soviet Polity', *Russian Review* 55(4), Oct. 1996, 638–660.

18 Jeffrey Brooks, *Thank You, Comrade Stalin! Soviet Pubic Culture from Revolution to Cold War*, Princeton: Princeton University Press, 2000, pp. 195–232.

19 See for example Timothy Colton's discussion on 'Governance in the Stalinist Manner' in Colton, *Moscow: Governing the Socialist Metropolis*, Cambridge: Harvard University Press, 1995, pp. 299–305.

20 Julie Hessler, *A Social History of Trade: Trade Policy, Retail Practices and Consumption, 1917–1953*, Princeton: Princeton University Press, 2004, pp. 273–295.

21 Donald Filtzer, *Soviet Workers and Late Stalinism: Labour and the Restoration of the Stalinist System after World War II*, Cambridge: Cambridge University Press, 2002.

22 On research of the 1930s see Hellbeck Jochen, 'Laboratories of the Soviet Self: Diaries from the Stalin Era', PhD Dissertation, Columbia University, 1998. Igal Halfin, *From Darkness to Light: Class, Consciousness, and Salvation in Revolutionary Russia*, Pittsburgh: University of Pittsburgh Press, 2000. On research of post-war subjectivities see Jeffrey Jones, 'People without a "Definite Occupation": the Illegal Economy and Speculators in Rostov-on-Don 1943–48', in Don Raleigh (ed.), *Provincial Landscapes: Local Dimensions of Soviet Power, 1917–1953*, Pittsburgh: University of Pittsburgh Press, 2001, pp. 236–254. Elena Iarskaia-Smirnova, Pavel Romanov, 'At the Margins of Memory: Provincial Identity and Soviet Power in Oral Histories, 1940–1953', in Raleigh (ed.), *Landscapes*, pp. 299–329.

23 For an attempt to capture the multitude of damages done see the chapter 'The Pantheon', in Catherine Merridale, *Night of Stone: Death and Memory in Russia*, London: Granta, 2000, pp. 307–344.

24 See Brooks, *Thank You*, pp. 195–232. 'Kluften der Erinnerung: Russland und Deutschland 60 Jahre nach dem Krieg', *Osteuropa/Nepriosnovennyi Zapas*, 55 (4–6), Juni 2005. Nina Tumarkin, *The Living and the Dead: The Rise and Fall of the Cult of World War II in Russia*, New York: Basic Books, 1994.

25 On unruly peasants see 'Iz prigovorov narodnykh sudov o privlechenii k ugolovnoi otvetstvennosti kolkhoznikov po ukazu 1947 goda', *Sovetskie arkhivy*, No. 3, 1990, pp. 55–60. Zubkova, *Obshchestvo*, pp. 61–68.

26 On this theme see Hessler, *Trade*, pp. 277–279. Uwe Gartenschläger, 'Living and Surviving in Occupied Minsk', Robert Thurston, Bernd Bonwetsch (ed.), *The People's War*, Urbana: University of Illinois Press, 2000, pp. 13–28. Richard Bidlack, 'Survival Strategies in Leningrad during the First Year of the Soviet-German War', Thurston/Bonwetsch, *People's War*, pp. 84–107.

27 On the Baltics see Elena Zubkova, 'Problemnaia zona: osobennosti sovetizatsii respublik Baltii v poslevoennye gody 1944–1952 gg', in Isai Nariz, Tomita Tokza

(ed.), *Novyi Mir Istorii Rossii*, Moscow: AIRS, 2001. On Western Ukraine see Jeffrey Burds, *The Early Cold War in Soviet West Ukraine, 1944–1948*, Pittsburgh: The Carl Beck Papers No. 1505, 2001.

28 V. Zima, *Golod v SSSR 1946–194 godov: Proiskhozhdenie i Posledstviia*, Moscow: RAN, 1996.

29 See Filtzer, *Soviet Workers*, pp. 158–200. Zubkova, *Obshchestvo*, pp. 89–101. Vladimir Kolzlov, *Massovye Besporiadki v SSSR pri Khrushcheve i Brezhneve*, Novosibirsk: Sibirskii Khronograph, 1990, pp. 60–82.

30 Peter Solomon, *Soviet Criminal Justice under Stalin*, Cambridge: Cambridge University Press, 1996, pp. 404–446.

31 Serhy Yekelchuk, 'Celebrating the Soviet Present: The Zhdanovshchina Campaign in Ukrainian Literature and the Arts', in Raleigh (ed.), *Landscapes*, pp. 255–275.

32 Elizabeth Wilson, *Shostakovich: A Life Remembered*, London: Faber, 1994.

33 Filtzer, *Soviet Workers*, pp. 161–164

34 Solomon, *Criminal Justice*, p. 405.

35 This is different from the 1930s when such people were often swooped up in political persecutions. Paul Hagenloh, 'Socially "Harmful Elements" and the Great Terror', in Sheila Fitzpatrick (ed.), *Stalinism: New Directions*, London: Routledge, 2000, pp. 286–308.

36 For a more detailed discussion of opposition and resistance in late Stalinism see Juliane Fürst, 'Prisoners of the Soviet Self? Political Youth Opposition in late Stalinism', *Europe-Asia Studies* 54(3), 2002, 353–375. 'Re-examining Opposition under Stalin: Evidence and Context – A reply to Kuromiya, *Europe-Asia Studies* 55(5), 2003, 789–803.

37 See also Mark Edele, 'Strange Young Men in Stalin's Moscow: The Birth and Life of the Stiliagi, 1945–53, *Jahrbücher für Geschichte Osteuropas*, 50, 2002, 41. For other nascent subcultures see M. Menshikov, 'Zolotaia koronkoa', *Komsomol'skaia pravda* 18 May 1946, 3. S. Gorbusov, 'Vecher v Gigante', *Komsomol'skaia pravda* 16 April 1946, 2.

38 For critiques on Stalin's death as defining the end of an époque see Yoram Gorlizki, 'Party Revivalism and the Death of Stalin', *Slavic Review* 54(1), Spring 1995, 1–22. Polly Jones, 'From Stalinism to Post-Stalinism: De-Mythologising Stalin 1953–56', Harold Shukman (ed.), *Redefining Stalinism*, London: Franz Cass, 2003, pp. 127–148.

39 Polly Jones (ed.), *The Dilemmas of De-Stalinization*, London: Routledge, 2006.

40 On the emergence of new economic and social relations see also Eric Duskin, *Stalinist Reconstruction and the Confirmation of a New Elite, 1945–1953*, Basingstoke: Palgrave, 2001. Julie Hessler, 'A Postwar Perestroika? Toward a History of Private Enterprise in the UUSR', *Slavic Review* 57(3), Fall 1998, 516–541.

41 Tumarkin, *The Living and the Dead*.

42 Lewis Siegelbaum, *Borders of Socialism*, Basingstoke: Palgrave, 2006. Vladimir Shlapentokh, *Public and Private Life of the Soviet People: Changing Values in Post-Stalin Russia*, Oxford: OUP, 1989.

43 See Juliane Fürst, 'Prisoners of the Soviet Self? Political Youth Opposition in Late Stalinism?', *Europe-Asia Studies* 54(3), May 2002, 353–376.

44 On a discussion of archival sources see 'Ten Years After', *Kritika: Explorations in Russian and Eurasian History* 2(2), Spring 2001.

Part I

When the war was over

1 Population, politics and reproduction

Late Stalinism and its legacy[1]

Mie Nakachi

> ... the population of the Soviet Union is 170 million, and the population of England is no more than 46 million. Economic capacity/power (*moshchnost'*) is indicated not by volume of industrial product as a whole, without regard to the country's population, but by the size of the demand for this product per capita ... the more population there is in a country, the more there is demand for consumption goods, which means all the more industrial production for such a country.
>
> I. V. Stalin (1939)[2]

> The more people we have, the stronger our country will be. Bourgeois ideology has invented many cannibalistic (*liudoedskie*) theories such as the theory of over-population. They think about how to reduce birth-rate and decrease population growth. Comrades, we are different. If we added another 100 million to 200 million, it would still be too little!
>
> N. S. Khrushchev (1955)[3]

The Soviet Union won a victory in the Second World War, but what awaited the country at fighting's end was an unprecedented demographic crisis. Not only did the Soviet Union lose 27 million soldiers and civilians, but also a large part of its surviving population was injured, ill and emaciated after years of fighting, malnutrition, unsanitary conditions and fatigue.[4] Moreover, at the end of the war a large percentage of the Soviet population was dislocated by repeated mobilization, evacuation, deportation and occupation. In rural areas, for the age category 18–49 there were only 28 men for every 100 women.[5] Soviet leaders recognized that re-establishing a healthy, expanding population was essential to post-war economic and social reconstruction and developed various policies to increase the birth-rate. But no one really knew the appalling extent of the catastrophe.

The 1944 Family Law replaced its 1936 counterpart to become post-war reproductive policy. Aid was increased to women with many children (*mnogodetnye materi*) and they began to receive medals for their efforts. Given the imbalanced sex ratio, it became important to encourage births among unmarried women, who also became eligible for financial support. Laws on marriage, divorce and birth registration were amended so that single mothers (*odinokie materi*) could not seek child support from their male partners. Through these changes, for the first time in Soviet history, a separate legal category of father-less children was created. Newly declassified documents reveal that although pieces of this policy package were present in other proposals, the key draft con-taining all the elements above was written by N. S. Khrushchev.

All the Soviet leaders, in particular Stalin (see epigraph above), thought that population was a measure of and limit on national strength. As long as they believed there were not enough Soviet people, they would support pro-natalist policies. For them the problem was demographic, but it was Khrushchev who devised a comprehensive approach to reproductive politics that affected preval-ent conceptions of legitimate gender relations and led to universal mobilization of the population to the post-war task of replacing the dead.[6] Every individual was included in its scope. With very few exceptions, all would be required to reproduce or pay for those who were more fertile. Ironically, just as population increase became the dominant concern of Soviet family policy, demographic data were becoming increasingly unreliable. In the aftermath of the 1937 census, Stalin had decimated the ranks of Soviet demographers, causing unwillingness in the profession to conduct further censuses to generate full and accurate popu-lation data. In the post-war period, lack of empirical information on the war's effects led to the adoption of a simplistic, erroneous and politically expedient assumption: the Soviet Union would return to pre-war demographic trends by 1950. This took into full account none of the important social changes that had taken place during the war. In this way, inadequate calculations were used to help pass Khrushchev's proposal into law, soon to be followed by unfortunate unforeseen consequences affecting the lives of millions. Despite indications to the contrary, the Central Statistical Administration (TsSU) continued to focus on the policy's successes. These evaluations together with Khrushchev's rise to power guaranteed that the 1944 Family Law would remain in force until his fall. The All-Union Basic Family Code finally came out in 1968.

By focusing on social history, significant shifts from pre-war policies to late Stalinism become apparent. In fact, it was the new policy initiatives that affected the lives of ordinary Soviet citizens as fundamentally as did the Great Patriotic War. Furthermore, the study of areas which underwent great transition in the years 1944–1945 reveals features of late Stalinism that continued after Stalin's death. Key elements of late Stalinist reproductive and demographic policy cer-tainly continued into the Khrushchev years, for the epigraph above makes clear that Khrushchev had certainly not changed his mind by 1955.

The Law came out on 8 July 1944 under the heading: "On increasing govern-ment support for pregnant women, mothers with many children, single mothers, and strengthening protection of motherhood and childhood; on the establishment of the honorary title 'Mother Heroine,' the foundation of the order 'Motherhood Glory,' and the medal 'Motherhood Medal.'" The first paragraph identified "strengthening the family (*ukreplenie sem'i*)" as "one of the most important tasks of Soviet government."[7] Although the 1944 Family Law's public goals promised protection of mothers and children, the secret documents that preceded it made clear that the primary aim was not the population's welfare, but population itself.

Building blocks of Soviet pro-natalism: the prehistory of the 1944 Family Law

The Soviet government first articulated pro-natalist policies in the mid-1930s in response to the declining birth-rate, which had been in evidence since the beginning of the decade. Stalin's rapid industrialization and urbanization in the late 1920s and early 1930s drew many women from homes to factories. The liberal abortion policy of the revolutionary government caused a rise in the number of abortions. Collectivization, famine and repression also took their toll. Together these factors explain the falling birth-rate.[8] Soviet leaders did not simply accept the declining birth-rate as a natural consequence of their economic and social policies. Ideologically, Soviet leaders rejected the very idea that workers limited the number of children in the family because good Marxists believed that under socialism, as living conditions improved, workers would have many children, thus increasing population. Strategic concerns were also important. Similarly to several interwar Western European countries, such as France, Germany and Italy, in the 1930s the Soviet government considered that depopulation would have a negative impact on labour force and military strength, the very measures of national power.[9]

To fight the declining birth-rate, the Soviet government introduced a new family law and ban on abortion in 1936. This pro-natalist policy aimed at promoting reproduction in several ways. First, the ban on abortion was to save the lives of millions who would have been aborted under the liberal abortion policy. The text of the anti-abortion law suggested that this policy was not a repression against women's choice, but a result of successful Soviet support for women whose nature is to reproduce. It emphasized that abortion was no longer necessary for working women because of their improved material and cultural life under socialism. Second, mothers who had seven or more children were granted large government subsidies for several years. This policy promoted the ideal of a large family and gave incentives to mothers with several children to have additional children. Third, the law expanded the network of maternity homes and childcare facilities. This policy was developed to reduce the burdens of childcare for working mothers and reduce infant mortality. Finally, the law created more complex requirements for divorce registration and increased responsibility for child support payment.[10]

Overall, the pro-natalist policy of the mid-1930s was structured around the idea of creating incentives for Soviet women of reproductive ages to carry additional children to term. This marked a shift from the revolutionary emphasis on female reproductive health to an emphasis on increasing fertility and population.[11] The important implication of this shift is that the new family law redefined the reproductive roles of men and women in terms of the state pro-natalist goals. Reproduction became an important task of Soviet women in their relationship to the state, while the state was to assure that their male partners shared the responsibility of raising children. The state would also provide a dependable social, legal and economic environment for childrearing. What was constructed as a result of this legislation was a loose hierarchy of women ranked by their reproductive contribution to the state, as well as a system of rewards for those at the top and punishments for those who contravened. With the introduction of the decree "On taxing bachelor, single and childless citizens of the USSR" of 21 November 1941, the system of categorizing citizens by reproductive contribution to the state was expanded to male citizens. The law was drafted following Khrushchev's suggestion to Stalin, establishing the former's credentials in reproductive policy-making.[12] The law stipulated that the childless pay taxes subtracted from monthly salaries and that the amount be class dependent.[13]

In the post-war period, the Soviet government faced demographic threats on an unprecedented scale. The reproductive conditions which caused the low birth-rate differed from those of the 1930s, and accordingly, different reproductive policies were needed to meet the new challenges. The defining problems of post-war reproduction were the significantly reduced number of citizens of reproductive age and the distorted male–female sex ratio. Although the overall average in rural areas was 28:100, there was wide fluctuation from region to region, ranging between 19:100 (Smolenskaia *oblast'*) and 70:100 (Tadzhikistan).[14] Soviet leaders used the information on sex ratio to discuss the problem of gender-specific labour shortage in the post-war Soviet Union.[15] They also recognized that the sex imbalance was a reproductive problem and developed a new pro-natalist policy to counter it.

In late 1943 and early 1944, several Soviet leaders foresaw post-war problems with low birth-rate and drafted new decrees for improving government support for mothers. Two factors determined the timing of these projects. First, the most difficult part of the war with the Nazis was over after the Soviet Union's victories at Stalingrad and Kursk in the winter of 1942–1943. Soviet leaders were able to consider allocating resources to priorities that were not directly for military use or post-war reconstruction. Second, in 1943 the wartime collapse of the birth-rate became clear. As N. A. Voznesenskii reported to Stalin, V. M. Molotov, A. I. Mikoian, G. M. Malenkov and L. P. Beria on 15 March 1942, during the first seven months of war between July 1941 and January 1942, the "inevitable reduction of birth-rate under wartime conditions is not yet visible. Reduction of the birth-rate, caused by the departure of men to the army, will probably become manifest in the spring–summer of 1942."[16] As predicted, in 1942 and 1943 the birth-rate fell dramatically (Table 1.1).

This data adumbrated the approaching crisis in reproduction and drove Soviet leaders to discuss three draft laws to increase the birth-rate, each with its own particular demographic focus. The least novel of the new initiatives was a draft *ukaz* "on government subsidies for mothers with many children," simply intended to increase the provisions of the 1936 Law. The following points were discussed: (1) the number of children required to qualify for governmental subsidies, (2) the amounts accorded, and (3) the number of years of support. On 9 October 1943 three versions of the draft *ukaz* were submitted to Molotov,

Table 1.1 Birthrate in USSR, 1940–1943

Year	Births per 1,000 citizens
1940	35.3
1941	31.6
1942	20.5
1943	11.2

Source: RGASPI f. 82, op. 2, d. 387, ll. 35, 90.

together with the People's Commissariat of Finance (Narkomfin) projected costs based on demographic statistics provided by TsSU. Estimated budgets ranged between 1,405.9 million and 1,985.2 million roubles.[17] These financial calculations were important for an accurate assessment of the most cost-effective way to raise birth-rate. Already on the following day (10 October) Molotov received a revised version of the draft (*pererabotannyi proekt*), which proposed to give 1,000 roubles for the fifth and sixth children for two years, 2,000 roubles for the seventh and eighth children for five years, 3,000 roubles for the ninth and tenth children for five years, and 5,000 roubles for five years for the eleventh child and up. The total project cost was estimated to be 1,837.2 million roubles. Thus, the initial threshold to become a "many-child" mother was lowered to five (from seven in 1936) and a bonus higher than the 1936 Law was offered for mothers with ten or more children.[18] A few days later the pot was sweetened to 10,000 roubles for eleventh children.[19] Thus, over the course of four days, at least five versions of the draft *ukaz* circulated at the highest level of the Soviet government. Discussion and fine-tuning would continue until April 1944, when Khrushchev submitted his trail-blazing proposal that included and therefore superseded legislation aimed exclusively at the prolific.

Almost simultaneously, in October 1943, Molotov examined a draft decree submitted by G. A. Miterev, the People's Commissar of Public Health (Narkomzdrav). This proposal also aimed at improving living conditions for mothers as a way to raise the birth-rate, but the basic approach was different with Miterev arguing that government support for mothers should be expanded to include all pregnant and nursing women, regardless of family size. Narkomzdrav was alarmed, not only by falling birth-rate, but also by worsening numbers for premature births. Miterev argued that to reverse these tendencies the government needed to "carry out urgent measures to establish maximally favourable conditions for pregnant and nursing women."[20] Using analysis by in-house statisticians, he argued that population increase depended not only on family size, but also on overall female reproductive health. In order to decrease premature births, Miterev proposed to shift the focus of support from women in childbirth (*rozhenitsy*) to an earlier stage, namely pregnancy. Miterev proposed three areas for reform: (1) improving working conditions for pregnant and nursing women, (2) enriching the diet of pregnant women and newborns, and (3) founding special refuges (*Dom materi i rebenka*) for mothers who were soldiers, partisans or otherwise disadvantaged. Molotov forwarded a copy of this proposal to Voznesenskii, the head of Gosplan, and Mikoian, the Deputy Chair of Sovnarkom, but before either variant was finalized, events took an unexpected turn.[21]

On 13 April 1944, Khrushchev, at the time the First Secretary of the Communist Party of the liberated Ukraine, sent Molotov a draft *ukaz* "on increasing government support to women in childbirth (*rozhenitsy*) and mothers with many children, and the reinforcement of the protection of motherhood and childhood." To this he attached an informational note (*spravka*) called "on measures for increasing the population of the USSR," in which the first paragraph spelled out the demographic logic underpinning the policies in the draft *ukaz*:[22]

> The Great Patriotic War demanded of the peoples of the USSR an unprecedented exertion (*napriazhenie*) to secure victory over the Nazi-German invaders. Naturally, this led to significant population loss and the consequent important task of replacing as quickly as possible the lost population (*vozmeshchenie*) to assure future demographic acceleration.

Khrushchev's initiative regarding the 1944 Family Law has been suggested in the past, but newly declassified documents reveal that the issue of primary economic, political and social importance was not the titular "support to women and mothers," but reproduction of the population. This latter topic was never addressed publicly, but was clearly the central issue of Khrushchev's confidential note to other top Soviet leaders.[23] With Khrushchev's project, the conceptualization of post-war pro-natalist policy diverged definitively from pre-war efforts. The draft of the autumn of 1943 under consideration by Molotov was in line with the antebellum emphasis on fertile women. Narkomzdrav's autumn 1943 project widened the scope of pro-natalist policy to the pool of all pregnant and nursing mothers.[24] Khrushchev, while keeping elements of previous drafts, included all women and men of reproductive ages in the scope of pro-natalist policies and created a new hierarchy of citizenship determined by the level of participation in reproducing the Soviet population. The project stated that non-fulfilment of this obligatory participation was subject to penalties, even punishment, regardless of a given citizen's reproductive capabilities, "because this tax is also a form of participation for the childless citizens in the state expenditure for raising the new generation."[25]

This extreme emphasis on fertility and population increase led to a redefinition of legitimate sites of reproduction. In previous Soviet family laws, all children were identified by their mother and father, suggesting that the site of legitimate reproduction was a sexual union between a woman and a man, who would subsequently take on mutual responsibility for childrearing.[26] In Khrushchev's project, two legitimate sites for reproduction were created. One of them was, as previously, an officially registered conjugal relationship. The new site was single motherhood, where only the mother, with state aid, was responsible for childrearing.[27] The single mother's sexual partner had no legal responsibility or obligations. The legitimization of single motherhood as a site of reproduction was one of the most significant novelties of post-war reproductive politics. Yet, the Soviet state failed to provide the full complement of necessary economic and social support causing single-parent homes to become a site of significant suffering and hardship for mothers and children alike.

The success of Khrushchev's pro-natalist strategy: inclusive, yet gender-specific

Although the tax base for the 1944 Family Law drew on almost the whole Soviet population, more specific measures differentiated the stimuli to which men and women were expected to react. Below I will discuss the demographic logic underlying both. Encouragement of single motherhood and bonuses for increased fertility made women their exclusive objects. New barriers to divorce were primarily aimed at men, for studies conducted by the People's Commissariat of Justice made it clear that males filed 70–80 per cent of divorce requests.[28] The relationship between child support and state aid proved to be the key link, as the fate of tens of millions of children became dependent on the financial specifics of the new measures. By including key aspects of pre-existent drafts in his project, Khrushchev's proposal quickly found supporters in Moscow, becoming law less than three months after arriving from Ukraine, even before many of its implications could be fully analysed. Nonetheless, the kinds of support and amendments offered provide insight into the variety of bureaucratic concerns regarding reproductive issues.

Khrushchev's pro-natalist project considered reproduction a civic responsibility. His proposal explained how many children a citizen must produce, as well as what the advantages and disadvantages would be for those who did and those who didn't. First, Khrushchev set the quota at two children. In order to encourage women to have more than two, the government aid for mothers would begin when a mother bore a third child.[29] For the third child one-time government subsidies would be provided at birth. In addition, monthly subsidies would be provided for five years, beginning from the age of two.[30] Thus, the award for exceeding the reproductive quota would begin when a citizen replaced herself, her spouse, and then made an addition towards population growth.

There were some important qualitative differences between the Khrushchev proposal and the previous projects for increasing governmental aid to mothers. First, Khrushchev redefined a mother with many children not from the perspective of promoting large families, as in the 1936 Family Law, but to rewarding those who participated in reproducing population.[31] Khrushchev thought that the propaganda effect of glorifying large families was not sufficiently effective (*nedostatochno effektivno*). The newly lowered threshold directly reflected the demographic thinking of the policy-makers who considered that regeneration of the Soviet population was only possible when the average number of children exceeded two.[32] Penalties for those who did not fulfil the reproductive quota took two forms, taxes and prison sentences. Khrushchev's proposal was clear on how to finance additional subsidies for mothers and children following the logic of the tax on insufficiently fertile citizens decreed in 1941, yet these sources would not have been sufficient to support the new plan. Therefore, he reduced the number of excepted categories. The need for additional resources and the justification of new taxes were combined to generate the idea that all citizens must participate in the state project to replace the dead. From now on, the main exempt categories would be pensioners without additional sources of income and soldiers on duty at the front. The new idea, that reproduction was a responsibility for all Soviet citizens, would have different meanings for women and men. Since there were more male than female military personnel eligible for exemption, it was mainly men who benefited from the rules of exception. Students were the only segment of the female population left untaxed.

The new reproductive quota of two children was reflected in the new tax for insufficiently fertile citizens, not only the childless. Khrushchev's note stated that those who had only one child would also be taxed. Low birth-rate was not only about the childless (*bezdetnost'*), but also about family planning practices, which were now labelled "incompatible with government interests, especially during the war and in the immediate post-war years."[33] The new focus on one-child families created material disadvantages for the insufficiently reproductive (*malodetnykh suprugov*), while stressing the negative views "of the government and society (*obshchestvennost'*) toward the one-child family (*ogranicheniiu sem'i odnim rebenkom*)."[34] Additional penalties for unlicensed production of contraceptive devices and substances were also recommended.

Beyond penalty came punishment. According to Khrushchev, incarceration was the answer to women's anti-reproductive behaviour. The proposal refers directly to "crimes against the health and life of mothers and children," code words for abortion and infanticide, in asking for increased jail terms for abortionists, as well as intensified medical surveillance over suspected abortion cases. Khrushchev also called for prison sentences for women who underwent unauthorized abortions, a measure not present at all in the 1936 Law.[35] This draconian measure was necessitated, explained Khrushchev, by the growing number of underground abortions, particularly alarming "not only because the birth-rate has significantly decreased, but also because we have lost many women who were physically healthy and capable of procreation."[36]

Khrushchev even foresaw the need for preventative incarceration. In addition to a general expansion of childcare and maternity facilities, the project proposed to organize special rest homes for single mothers and weakened nursing mothers. Such a measure was considered necessary, particularly because it was believed that single women were often unable to give birth to a child in their place of residence because of hostile family relations or bad living conditions and that they should temporarily leave for a different place during pregnancy, childbirth and post-partum recovery. Leaving single mothers alone in questionable conditions was harmful because they often "place[d] their children with relatives and abandon[ed] them, and sometime even kill[ed] them." From the description of the project, it is clear that the major function of the home for single mothers was not glorification of single motherhood. Rather, it was to

facilitate a system of monitoring single mothers in order to "remove such phenomena (*ustranit' podobnye iavleniia*)" as abortion, infanticide and/or abandonment. This is why the medical control commission, rather than maternal request,
would determine who stayed in these facilities and for how long.[37]

One of the most novel divergences of the post-war from the pre-war pro-
natalist project was encouragement for women without husbands to reproduce.
Khrushchev stated:

> The question of stimulation of procreation among women who are not
> married for one reason or another (widows of those who died in the war and
> unmarried girls) has special significance at the present moment and in the
> approaching post-war period.[38]

The collapse of the distinction between war widows, respected members of
the community entitled to government support as bereaved wives, and unmarried mothers, often stigmatized, was possible only in the context of reproductive
politics.[39] These two groups, the polar extremes of female respectability, now
formed a single category of women for policy-makers focused on the state's
great reproductive task. Khrushchev's optimistic materialism suggested that any
woman without a husband would give birth, given the means.

But the Soviet government had already experienced the limitations of the pre-
war system in which all too many fathers refused to pay child support.[40] From
the perspective of raising the birth-rate, this was harmful because mothers
without child support were unlikely to have additional children. Equally unre-
productive, fathers who provided child support were likely to limit the number
of children in a new family. Thus, the existing system did not encourage procreation, both among women receiving child support unreliably and men paying
regularly.

To overcome these demographic drawbacks in incentive structure,
Khrushchev's project proposed a system where both single mothers and their
reproductive male partners would not be discouraged from procreation after the
end of their relationship and where the welfare of their existing and future children could be assured. This was done by providing government aid to single
mothers.[41] Because of the state's full involvement in raising out-of-wedlock children, at least on the financial level, women would not have to be afraid of
getting pregnant, and male partners would not have to be afraid of impregnating
their sexual partners. The possibility that a single mother might want to receive
child support from her partner rather than the government for reasons other than
economic was never considered. Interestingly, because of these specific formulations of post-war aid to single mothers, the new project was designed to
encourage both men and women to have non-conjugal sexual relationships that
would result in procreation. Birth control was discouraged and prophylactic aids
were effectively unavailable.

In the post-war period government aid to single mothers and fatherless
children, but not for two-parent households, created a clear boundary between

the two types of family. In addition, marriages needed to be permanent to prevent subsequent reallocations of childcare resources and responsibilities, which would result in runaway costs at state expense. The fluidity of pre-war conjugal relations illustrated this potential danger. To this end, the ambiguities concerning marital status were abolished through tight legislation.[42] After 1944, only registered marriage would be legally binding. As a corollary, only legal wives could register their children under the patronymic and family name of the father. With this definition of legal marriage the government could easily identify single mothers and provide childcare support, while their male partners would be relieved of any and all responsibility. Similarly, the government could easily identify two-parent families, fully responsible for their own childcare. Khrushchev proposed stricter divorce procedures in order to prevent dissolution of this type of family.

The proposal condemned those who desired divorce as being "frivolous (*legkomyslennoe*) toward family and family responsibilities," while mandating "increased material expenditures for the break-up of conjugal relations" and a series of embarrassing, public "formalities." The new procedure for divorce would become: (1) submit a petition to the court about intention to divorce accompanied by a payment of 100 roubles; (2) summon spouse and witnesses to the court; (3) publish the announcement of filing a divorce in a local newspaper at the expense of the plaintiff; (4) go through mediation and, if unsuccessful, a court investigation which would assign post-divorce childcare responsibilities; (5) pay 300–1,000 roubles to register the divorce at ZAGS; and (6) make a note about the divorce in both spouses' passports.[43] On arrival in Moscow, Khrushchev's draft law immediately became the centre of attention, superseding discussion of earlier legislation. In the three months before promulgation on 8 July, several versions circulated at the highest levels of the Communist Party and the Soviet bureaucracy, with each entity adding or subtracting its pound of flesh. Narkomfin cut into the aid to be provided to single mothers, while a small tax (0.5 per cent of income) for those with "only" two children was added.[44] Narkomzdrav successfully opposed the mandatory prison terms for abortion. Additional tax exemptions appeared for war widows and invalids, possibly at the behest of the army. Also in the military spirit, having five or more children would now earn you a Motherhood title, medal or order, as well as cash. Below I examine a few of the positive evaluations that accompanied the final preparation of the 1944 Family Law to explore the concerns, assumptions and data on which they were based.

On 3 May 1944 Miterev reported to Molotov about Narkomzdrav's examination of Khrushchev's draft.[45] In general, Narkomzdrav expressed strong support for the proposed *ukaz*:

> [T]he question about the measures for stimulating the rise of birth-rate (*stimuliruiushchikh povyshenie rozhdaemosti*) and population growth in the USSR, raised by Comrade Khrushchev is timely and measures proposed by him are appropriate.[46]

In order to show that Khrushchev's concerns that came from Ukraine were relevant to the USSR, Miterev attached the analysis of the natural population movement for 1943 by Narkomzdrav's top statistician, A. M. Merkov.[47] He assessed the losses of the Second World War as "much graver than the demographic consequences of the war of 1914," where the reduction of the birth-rate was only 22 per cent. Merkov warned that during the present war "even without counting those who perished in the war, population decline continues to grow."[48]

Referring to Merkov's demographic analysis, Narkomzdrav harmonized with Khrushchev's basic message, arguing that the Soviet Union needed to take measures towards increasing the birth-rate:

> Our preliminary analysis of the natural population movement in the USSR for 1943, illustrated in the attached informational note, shows the rapid fall of the birth-rate. This necessitates measures toward stimulation of the birth-rate.[49]

Nonetheless, Miterev recommended a few qualitative changes to Khrushchev's proposal, anticipating negative impacts on the overall physical and psychological health of women and children. Some were incorporated, others rejected. Narkomzdrav's earlier proposal to broaden coverage for pregnant women and nursing mothers, while increasing maternity leave and improving working conditions for all pre- and post-partum women, were incorporated into the revised versions of Khrushchev's draft and the final *ukaz*. Narkomzdrav opposition also helped remove the stiffer penalties Khrushchev had proposed for abortion-related crimes. However, its suggestion to give genetic fathers the option of registering out-of-wedlock children under their last names was rejected, because it might increase responsibility and therefore inhibit further procreation both in and out of wedlock, the driving concept of post-war reproductive politics.

As one would expect, Narkomfin's contribution to this discussion took the form of calculations of the cost of the increased government support for mothers. For example, the number of out-of-wedlock children for 1945 was estimated at 500,000, given the TsSU projection of two million births for 1945, of which up to 25 per cent would be out of wedlock. The number of out-of-wedlock children for 1946–1950 was calculated in the same way.[50] Narkomfin simply multiplied the estimated numbers of single mothers and mothers with many children by the proposed level of one-time and monthly governmental subsidies. The total cost of the project for 1945–1950 was calculated to be 58.8 billion roubles. Lacking data on the number of single mothers who would qualify as mothers with many children (three or more), these projected costs did not include government subsidies for single mothers with many children, suggesting that total costs would be somewhat higher.[51] Nonetheless, Khrushchev's draft included new income as new categories of insufficiently fertile citizens were taxed. This could only draw approval from Narkomfin as it sought to minimize costs and maximize revenue.

Table 1.2 Births in USSR

Year	Number of births (in millions)	Yearly increase (+) or decrease (−) (in millions)
1913	6.49	−
1926	6.47	−
1936	5.35	−
1937	6.41	+1.06
1938	6.32	−0.09
1939	6.29	−0.03
1940	5.75	−0.54
1941	4.63	−1.12
1942	2.09	−2.54
1943	1.36	−0.73

Source: RGASPI f. 82, op. 2, d. 387, l. 92.

As shown above, Narkomfin made use of materials provided by TsSU, the unit of Gosplan responsible for the collection and analysis of demographic statistics. As the sole source for comprehensive data, TsSU was indispensable for all spheres of Soviet economic, social and cultural planning. Post-war reproductive policy-making also required demographic data regarding children, single mothers and mothers with many children. On 31 May 1944, V. N. Starovskii, the head of TsSU, reported these three key projections to Molotov. The report first provided historical data (Table 1.2).

Starovskii discussed the sharp drop in birth-rate between 1941 and 1942 and identified the primary cause as the "drawing (*otvlechenie*) of a significant part of the male population into the army." Following the same logic, TsSU projected that the birth-rate would grow after the end of 1945 because the demobilization of soldiers would accelerate birth-rate, partially compensating for the wartime loss of male population and the general extermination of population in the occupied regions.[52] According to this projection, pre-war levels would resume by the beginning of the 1950s (Table 1.3). Details of TsSU's calculation of the level of births for 1945–1950 are not in the report. It seems that the projection did not consider the structural and qualitative changes in the post-war demography that could affect the post-war birth-rate pattern. It also appears that the projection was based on the assumption that the normal level of births for the post-war period would be slightly less than the 1939 level due to the decrease in male population that would persist into the post-war period. In short, without additional empirical study, it was hard for TsSU to quantify the demographic disaster of the Second World War and Starovskii's estimates were far from the actual results (Table 1.4).

Although the yearly birth-rate rose quickly from wartime lows to fluctuate stably around five million, it would not regain the pre-war magnitude of Starovskii's prediction. It was this incorrect data, based on the unsubstantiated

Table 1.3 1944 TsSU estimate of USSR births to 1950

Year	Millions of births (estimated)
1944	1.5
1945	2.0
1946	3.0
1947	4.0
1948	5.0
1949	6.0
1950	6.0

Source: RGASPI f. 82, op. 2, d. 387, l. 92.

Table 1.4 Actual USSR birth totals, 1945–1952

Year	Millions of births
1945	2.5
1946	4
1947	4.5
1948	4.2
1949	5
1950	4.8
1951	5
1952	4.9

Source: RGAE f. 1562, op. 33, d. 2638, l. 76.

assumption of a return to the *status quo ante*, that was central to post-war leaders' belief that Khrushchev's plan could indeed reproduce the lost population. For the second data set, Starovskii projected the number of single mothers in the post-war years based on the following pre-war census results. First, according to the 1926 census, 11–12 per cent of all children in urban areas lived only with their mothers.[53] Second, the 1939 census reported that there were 40.1 million women between the ages of 18 and 49, of which 11.3 million women were not married, and 28.8 million women were in either registered or common-law marriages. At that time, the absolute number of women in this age group exceeded the number of men of those ages by 3.1 million. Third, in the same census, it became clear that approximately 15–20 per cent of young married couples did not register their marriage.

Starovskii's report concluded that "taking into account this data, and also the significant loss of male population during the war, it is possible, as an approximate calculation, to assume that the number of children, born to unmarried mothers (*zhenshchiny, ne sostoiavshikh v zaregistrirovannom brake*) will consist of about one-fourth of all births."[54] Again, it is not clear exactly how the

Table 1.5 Breakdown of USSR family sizes in 1944

Number of children	Adult women (in thousands)
Childless (*bezdetnye*)	11,952
1	10,010
2	8,081
3	3,469
4	2,031
5	1,111
6	531
7 and 8	253
9	36
10	10
11 and up	2
Total	37,486

Source: RGASPI f. 82, op. 2, d. 387, l. 93.

percentage of illegitimate children was drawn from the data provided above. However, what is apparent is that the projection was made based on the assumption that 1920s and 1930s marriage trends would resume once abnormal wartime conditions were normalized. But this was not to be the case, for both Soviet men and women had developed new patterns of sociability in the cauldron of war. Nor did Starovskii imagine the difficult economic conditions and inhibiting prejudices that single mothers and their offspring would face, leading many to avoid or abort pregnancies. In fact, the state never fully delivered on its 1944 promise of financial support and the sharp line drawn by Khrushchev's legislation between the married and unmarried only darkened the stigma placed on the latter. As a result, after hovering between 15 and 20 per cent from 1945 to 1953, the percentage of single mothers settled near the lower end of this range.[55]

Finally, Starovskii presented an estimate of adult women broken down by the number of their children at the beginning of 1944 (Table 1.5). The three sets of data provided by TsSU on birth-rate trends and the number of illegitimate children were transmitted to Narkomfin, which used them to estimate the costs of promulgating a revised version of Khrushchev's draft *ukaz*. It should come as no surprise that demographic statistics based on pre-war data, flawed assumptions and ignorance of actual social conditions would leave the validity of Narkomfin calculations in doubt as well.

The centrality of 1944: population and reproduction

With the outcome of the Great Patriotic War all but settled, 1944 became a watershed year as post-war planning began in earnest. In economics, the State Defence Committee decided to request six billion dollars in American lend–lease in order to jumpstart the five-year plan (1946–1950), still in the making. At the People's Commissariat of Foreign Affairs, diplomats A. A. Gromyko, M. M. Litvinov and I. M. Maiskii put forth their visions of a new world order in which the USSR enjoyed superpower status.[56] In the social sphere, no single measure had greater and longer-lasting influence on gender relations and family life than Khrushchev's Family Law. The effects of the 1944 Family Law ran deep and wide, touching millions for decades, shaping decisions and desires about post-war marriage and family. Many demobilized men who had pre-war families, but had formed new liaisons, often did not want to return to their old households.[57] If the pre-war marriage was common-law, they were free to remarry, although the financial responsibility for earlier children remained. On the other hand, if they had registered pre-war relationships, they were now subject to the new and stricter law, which dampened the post-war divorce rate. The artificiality of this judicial/administrative measure is clearly indicated by the fact that divorces continued to increase in number, reaching the pre-war level by the late 1950s. When the 1965 amendment to the divorce provisions of 1944 went into effect, the divorce rate jumped over 70 per cent in 1966 alone.[58] Although millions of divorces had been blocked in the course of two decades, no one could claim that strengthening the Soviet family, an explicit goal of the 1944 law, had been achieved.

As men abandoned their former families and/or formed unofficial post-war "families," they created "single mother" families. The surplus of post-war women made it easy to find new partners, but even a superficial reading of the 1944 Law sufficed to discourage marrying intentions in many males. In a world without contraceptives, this was a simple recipe for more "single mothers," a recipe that could be repeated many times over. The production of "single mothers" and the reproduction of "bastards" were both growth industries in the post-war Soviet Union.[59] The varied life-courses described above produced millions of unmarried mothers and illegitimate children. TsSU data for 1945–1955 show 8.7 million "fatherless" children in the Law's first ten years.[60] Partial data for later years suggest similar rates until 1968, when changes in registration practices made identification of illegitimate children more problematic.[61] Abortion rates also skyrocketed. Reliable totals are impossible to assemble for the years when the operation was outlawed, but assuming a steady police effort, the number of botched abortions that ended up in the hospital should help us to calculate the general upward trends.[62] For example, 661,000 cases in 1946 had risen to 1.9 million in 1954. Even worse, without alternative contraceptives, abortion became the method of (no) choice. This became all too clear once the ban was lifted. In 1956, 4.7 million abortions were performed, with this upward trend only peaking in 1965 at 8.5 million.[63]

Narkomzdrav's successful push to prevent incarceration of women who had undergone abortion pointed forward to the key role that ministry would play in its legalization. The failure of criminalized abortion was recognized by its repeal in 1955, after nearly 20 years as a key plank of pro-natalism.[64] Narkomzdrav's failed recommendation to allow paternal registration of out-of-wedlock children, along with the other central components of post-war demography, difficult divorce and the single-mother system, would not be modified until they had done another decade of damage. The unaltered sway of Khrushchev's Family Law continued for TsSU's early evaluation of its effectiveness, so early in fact as to have almost no empirical relevance, left a lasting impression as to the Law's success.[65] Although this was certainly appreciated by the high-ranking cadres who had worked together to promulgate the Law, including Starovskii himself, this boded ill for the serious development of applied demographic statistics as a policy-relevant academic field. The year 1944 saw a further reflection of TsSU's disinterest in demographic rigour, when an initiative from a much-respected professor and authority on census evaluation, A. I. Gozulov, for the re-opening of an Academy of Sciences demographic institute was summarily dismissed by Starovskii.[66]

Gozulov's idea, like Khrushchev's, drew immediate inspiration from "war's demographic catastrophe," but unlike Starovskii's simplistic calculations, predicting automatic return to the pre-war situation within five years, Gozulov pointed out the many factors that would affect the labour supply "for several decades," while calling for a special section in the proposed institute to "study the elements of population replacement." Another department would specialize in "the influence of war on population." But the very first research group on Gozulov's list would be responsible for censuses and it was to the avoidance of any comprehensive counting of the USSR population that Starovskii's next 15 years would be devoted.

Gozulov had already enlisted the support of several other prominent academics,[67] but Starovskii wrote back politely, yet bluntly on 16 February 1944:

> In the near term, an institute or other scientific institution handling demographic issues should limit itself to bibliographic work, the examination of methodological issues of population statistics, and the study of foreign governments' statistics.
>
> It would be inappropriate (*netselesoobrazno*) at the present time to assign the study of statistical materials related to the population of the USSR to any scientific institution.[68]

Clearly, by early 1944, Starovskii had decided what the contribution of science to his statistical and political responsibilities should be. He may have chosen wisely for he would maintain his illustrious position in charge of Soviet statistics (including demography) until retirement in 1975, while garnering three Orders of Lenin.[69]

To understand Starovskii's exclusion of demographers, we need to look back to the then recent experience of the 1937 census, when his immediate predecessors were purged for failing to balance basic science and political expedience.[70] Although I. A. Kraval, who headed TsSU's predecessor organization, TsUNKhU, was well aware that a price would have to be paid for letting the census prove wrong Stalin's 1933 claim to the 17th Party Congress that there were 168 million Soviet citizens, he nonetheless tried hard to conduct a scientifically impeccable one-day count on 5–6 January 1937. After reporting the unwelcome results in March, he was swiftly arrested and soon shot.[71] O. A. Kvitkin, the head of the census department, suffered the same fate, as did numerous top statisticians and demographers. Others were sent to the Gulag. By the end of the year, I. D. Vermenichev, the man who had replaced Kraval to carry out the purge, had also been arrested.

The following year, N. Osinskii, Kraval's predecessor, was also shot. It was he who, as head of TsUNKhU, had reportedly questioned Stalin on the origin of the 168-million population statement. According to memoir literature, probably circulating as gossip at the time, Stalin replied that he knew himself what figures to indicate. This anecdote was published by M. V. Kurman, assistant to both Kraval and Osinskii, after he was amnestied from the Gulag in the 1950s. Starovskii appears to have quickly learned this lesson in political and physical survival. Already in 1938, while analyzing literacy statistics, Starovskii presented then TsUNKhU director, I. V. Sautin, with a memorandum that included Stalin's statements on the subject as one of the decisive parameters.[72] The 1939 census, for which Starovskii already bore responsibility as the new Deputy Director of TsUNKhU, may have been less accurate, but at least it produced politically palatable results, in particular an official "upper range" total of 170.5 million.[73] According to Blum and Mespoulet, Starovskii's final conclusions on the 1939 census, made in a 5 April 1940 memorandum to Molotov, Voznesenskii and Saburov, "offer fine examples of the formulations that would characterize the communication of sensitive data to politicians after the great purges of the late 1930s." Blum and Mespoulet also add that "all of his [Starovskii's] future conduct would be marked by this experience."[74]

In 1946, Stalin was interviewed by *Pravda* and announced that the Soviet Union had lost seven million in the Great Patriotic War.[75] Although this number may be more or less accurate as a measure of the number of active-duty servicemen who lost their lives in the war, the number of Soviet citizens who died during the conflict was far greater.[76] So many indicators were affected by the wartime losses that almost any comprehensive statistics would immediately prove Stalin's statement false. Possibly, it is in this light that the law of 1 March 1948, assigning all population data to the category "state secret," should be interpreted. It was certainly for this reason that Starovskii would not even consider a census until after Stalin's death.[77] Starovskii made similar efforts not to undercut Khrushchev's statements. In a February 1962 speech to the Presidium of the Academy of Sciences, he focused on the relationship between population and productivity. Not only did he echo Khrushchev's pro-natalist analysis of 1955 (see epigraph) in promising 400 per cent increases in production for a 30 per cent rise in population, but he also condemned the "misanthropic neo-Malthusians." The favourable probabilities of achieving these growth rates were attributed to Khrushchev's initiative of 1944, in particular its emphasis on reducing divorce and concomitant child support, while encouraging single mothers to breed.[78]

Starovskii's loyalty was reciprocated. For example, on 10 November 1963 during a presentation to the USSR Presidium[79] by the head of Gosplan, P. F. Lomako, a discussion about statistics drew forth a Khrushchev comment: "We should have invited Starovskii." A year later, a similar conversation drew forth a longer confession.

> Khrushchev: We don't have any other numbers. Comrade Starovskii is a very careful man. I get my numbers from him. I don't have my own numbers. This is our Union-wide accounting (*soiuznaia bukhgalteriia*). It is the only source, which I can use and rely on. I believe in him/it. (*Ia emu veriu.*) He/it has my complete confidence.[80]

In fact, so closely did they work together, that on the day of Khrushchev's removal from power, Starovskii was accused by A. I. Mikoian of "confusing (*putaet)*" Khrushchev.[81] Minutes later, Khrushchev would resign, but Starovskii remained for another decade.

Keeping the experience of 1937 in mind, a tragic event that had passed all too close to Starovskii, it is easy to understand his unwillingness to let an academic institution meddle in his responsibilities in 1944. In the 1960s, when proposals for an Institute of Demography were once again made, Western analysts found it "interesting" that the call had "primarily gone without a response."[82] Starovskii was still at his post and it is not difficult to imagine that attitudes hardened over the decades persisted, while accumulated power made it all too easy for him to have his way, even as the USSR faced new demographic challenges without the tools to analyse and answer them.[83] Just as Khrushchev's Family Law of 1944 went unamended until the late 1960s due to the continuing power of its author and widespread belief in pro-natalism, Starovskii's 1944 stunting of policy-applicable demographic studies would only be reconsidered in the Brezhnev era.[84] The political lives of Khrushchev and Starovskii, which took shape in the crucible of the 1930s before coming to grips with the new and daunting tasks of the war and post-war, illustrate well the shift to late Stalinism. The purges made Starovskii's career, most directly, and unmade Soviet demography, somewhat indirectly, for many years to come. More immediately, with the end of the Second World War, key policy-makers emerged who would shortly become post-Stalinist leaders with a stake in the continuation of the legislation they had created. Khrushchev is the most prominent case. The social engineering tasks envisioned in his 1944 Family Law were very much in tune with Stalinist practices, both in its production of beneficiaries and victims. Stalin approved the Law, but unlike earlier measures, the bulk of the victims, millions of women and children, would suffer on into the post-Stalinist era. In this way, the study of population, politics and reproduction helps the student of the Soviet era identify the changes and continuities from Stalin to Khrushchev and beyond.

Notes

1 The author would like to thank Alain Blum and Mark Edele for expert commentary on an earlier draft of this article.

2 Stalin's speech at the 18th Party Congress on 10 March 1939. I. V. Stalin, *Sochineniia* t.1 [XIV] 1934–1940, ed. by Robert H. McNeal (Stanford, California: Hoover Institution on War, Revolution, and Peace, 1967), 350.

3 *Izvestiia*, 8 January 1955. Khrushchev addressing young people about to leave for the virgin lands.

4 The historiography of wartime losses in the Second World War is almost as contentious as the discussion of "excess deaths" in the 1930s, because together these two issues provide the empirical basis for computing the demographic consequences of Stalinism. A full examination of this literature in English would include many articles by Barbara Anderson, Robert Conquest, Michael Ellman, Steven Rosefielde, Brian Silver and Stephen Wheatcroft, all of them commenting and criticizing the pioneering work of Frank Lorimer, *The Population of the Soviet Union: History and Prospects* (Geneva: League of Nations, 1946). A good introduction to the quantitative intricacies and political minefields is Michael Ellman and S. Maksudov, "Soviet Deaths in the Great Patriotic War: A Note," *Europe-Asia Studies* 46:4 (1994), 671–680.

In brief, Stalin used a seven million figure, Khrushchev spoke of 20 million and the Gorbachev-era committee endorsed 27 million "excess deaths" by subtracting best estimates of the December 1945 population from the 1941 population and adjusting, mainly for the number of deaths that statistically would have happened anyway in the course of four years. There is also a figure called "total hypothetical demographic loss" that rises to 35 million by including lost wartime births. It grows even larger if the hypothetical loss of post-war children is factored in. For extensive statistics and analysis on wartime losses, see *Naselenie Rossii v XX veke*, t.2, (Moscow: ROSSPEN: 2001), Chapters 2–8.

5 *Sovetskaia povsednevnost' i massovoe soznanie, 1939–1945* (Moscow: ROSSPEN, 2003), 296. Even in 1950, only 38.9 per cent of the Soviet Union was urban. *Naselenie Rossii*, t.2, 198.

6 The military analogy is not only literary for at the time Khrushchev wore the uniform of a lieutenant general. On this, see William Taubman, *Khrushchev: The Man and His Era* (New York: Norton, 2003), 151.

7 *Sbornik zakonov SSSR, 1938–1956* (Moscow, 1956), 383.

8 For a comprehensive discussion of urbanization, industrialization, collectivization, famine, political repression and deportation as causes of falling birth-rate in the 1930s, see *Naselenie Rossii v XX veke: Istoricheskie ocherki, t.1, 1900–1939* (Moscow: ROSSPEN, 2000).

9 David Hoffmann, "Mothers in the Motherland: Stalinist Pronatalism in its Pan-European Context," *Journal of Social History* (Fall 2000): 35–38.

10 I use "child support" rather than alimony to translate the Russian word *alimenty*. For details of the 1936 Law, see *Sobranie zakonov i rasporiazhenii SSSR* 34 (1936): 509–516. For a detailed discussion of the development of the 1936 Family Law, see Wendy Goldman, *Women, the State, and Revolution: Soviet Family Policy and Social Life, 1917–1936* (Cambridge: Cambridge University Press, 1993). On women's reaction to the 1936 anti-abortion policy, see also Sheila Fitzpatrick, *Everyday Stalinism: Ordinary Life in Extraordinary Times, Soviet Russia in the 1930s* (Oxford: Oxford University Press, 1999), 152–155.

11 On discussion of the 1920 law which legalized abortion due to concerns about female reproductive health, see Goldman, 254–255.

12 *Izvestiia*, January 8, 1955.

13 "O naloge na kholostiakov, odinokikh i bezdetnykh grazhdan SSSR," *Vedomosti verkhovnogo soveta SSSR* 42 (December 20, 1941): 3.

14 RGASPI f. 82, op. 2, d. 538, ll. 41–42 and footnote 4 above.

15 RGASPI f. 82, op. 2, d. 538, l. 42.

16 RGASPI f. 82, op. 2, d. 538, ll. 24–27.

17 RGASPI f. 82, op. 2, d. 387, ll. 20–22. The estimate was for the first year.
18 RGASPI f. 82, op. 2, d. 387, ll. 22–25.
19 RGASPI f. 82, op. 2, d. 387, l. 27.
20 RGASPI f. 82, op. 2, d. 387, ll. 35, 38, 39.
21 RGASPI f. 82, op. 2, d. 387, l. 35.
22 GARF f. r-8009, op. 1, d. 497, l. 164. I am indebted to Charles Hachten for his help in locating these documents.
23 One of the few Western scholars who mentioned Khrushchev's direct involvement in the family law was Peter Juviler. In the 1960s anonymous Soviet legal experts had told him that Khrushchev had initiated the 1944 Law and that the delay in amending the 1944 Family Law in the 1950s and 1960s was caused by Khrushchev, once in power, opposing changes to his "brain-child." Peter H. Juviler, "Family Reforms on the Road to Communism," *Soviet Policy-Making: Studies of Communism in Transition*, ed. Peter H. Juviler and Henry W. Morton (New York: Frederick A. Praeger, 1967): 41, 52. The documents on which this chapter is based corroborate Juviler's insider sources regarding Khrushchev's role in the making of the Law and lend credence to their interpretation of the Law's long survival, despite public and internal criticism.
24 However, this was not an entirely new policy. The 1936 law had also expanded childcare facilities and improved medical care for pregnant women, as discussed above.
25 GARF f. r-8009, op. 1, d. 497, l.172.
26 Whether or not the identified mother or father was a genetic parent did not concern the government.
27 Practically speaking, the state replaced fathers as financial providers for single mother families.
28 GARF f. r-9492, op. 1, d. 492, l. 151.
29 GARF f. r-8009, op. 1, d. 497, l. 165.
30 The suggested "birth bonus" for a third child was 500 roubles. For the fourth child and up it ranged between 1,500 and 5,000 roubles. GARF f. r-8009, op. 1, d. 497, l. 165.
31 The 1936 law stipulated that mothers with seven or more children were eligible for governmental aid. The 1943 March version of the draft *ukaz*, prepared under Molotov, reduced the required number of children to five so that a "mid-size family" (four children) would consider having additional children in order to become a "large family."
32 GARF f. r-8009, op. 1, d. 497, l. 165. The emphasis on regeneration also becomes clear since mothers with three to six children would not receive the new aid without producing additional offspring after 1944.
33 GARF f. r-8009, op. 1, d. 497, l. 174.
34 Taxing one-child families was not only a financial measure aimed at tapping a large, previously untaxed portion of the population, but can also be seen as an incentive for young men to marry, since only children sired in registered marriage would count toward tax-free status. Women could satisfy the law with out-of-wedlock children, too. GARF f. r-8009, op. 1, d. 497, l. 174.
35 GARF f. r-8009, op. 1, d. 497, ll. 175–177. Abortionists already faced prison sentences under the 1936 Law.
36 GARF f. r-8009, op. 1, d. 497, l. 175. The post-war proposal to increase punishment for abortion was partially based on the recognition that the 1936 anti-abortion law had failed to stop abortion.
37 GARF f. r-8009, op. 1, d. 497, l. 169. The medical control commission referred to here is probably the same commission that decided which pregnant women would qualify for clinical abortion.
38 GARF f. r-8009, op. 1, d. 497, l. 166.
39 For an extensive discussion of the "new single mother," see Mie Nakachi, "Replacing The Dead: The Politics of Reproduction in the Post-war Soviet Union, 1944–1955" (PhD Dissertation, University of Chicago, 2007), Chapter 4.

40 See "Absconding Husbands" in Fitzpatrick, 143–147. Mothers also left children and had to pay child support. However, in the Soviet discussion of child support, it is generally assumed that it is the father who leaves the family and is therefore responsible for child support.

41 GARF f. r-8009, op. 1, d. 497, l. 167. Monthly payments of 150–300 roubles, depending on the number of children, were meant to match the amount of child support a single mother would receive from a father who made the average Soviet wage, 600 roubles per month. Khrushchev's goal was to provide sufficient aid to single mothers, but the allocations were cut substantially even before the draft became law and again three years later.

42 Fitzpatrick, 142, refers to this as the "residual fuzziness ... about what constituted a marriage," citing the impressive statistic that in the 1937 census one and a half million more women than men considered themselves married.

43 GARF f. r-8009, op. 1, d. 497, ll. 181–183.

44 Those without children paid 6 per cent of total income, the highest tax bracket. Single-child families paid 2 per cent.

45 Miterev's letter to Molotov was dated 3 May 1944. See GARF f. r-8009, op. 1, d. 496, ll. 45–50 and RGASPI f. 82, op. 2, d. 387, ll. 87–89. A copy of the same letter was addressed to Malenkov, Secretary of the Central Committee. GARF f. r-8009, op. 1, d. 493, ll. 48–49.

46 GARF f. r-8009, op. 1, d. 493, l. 48.

47 Born in 1899, in the post-war period Merkov was the head of the department of medical-sanitary statistics within Narkomzdrav USSR. He was trained as a sanitary statistician in Khar'kov, Ukraine in the pre-war period. In 1941 he was evacuated from Khar'kov to Ufa, the capital city of the Autonomous Republic of Bashkiria. During the evacuation, he studied reproduction in Bashkiria and published "The Reproduction of Population in the Bashkiria ASSR on the Eve of the Great Patriotic War." He never continued a post-war analysis of Bashkir population because he was brought to Moscow in August 1943. In Moscow, he continued his study of the reproduction of population, this time, on an All-Union scale. GARF f. a-603, op. 1, d. 324, l. 2.

48 RGASPI f. 82, op. 2, d. 387, l. 91.

49 GARF f. r-8009, op. 1, d. 493, l. 48.

50 RGASPI f. 82, op. 2, d. 387, l. 96.

51 RGASPI f. 82, op. 2, d. 387, ll. 94–95.

52 RGASPI f. 82, op. 2, d. 387, l. 92.

53 This number did not include those children who were born into a common-law marriage and lived with the father.

54 RGASPI f. 82, op. 2, d. 387, l. 93.

55 See Nakachi, Chapter 4.

56 Vladimir Pechatnov, "The Big Three after World War II," *Cold War International History Project Working Paper* 13 (1995).

57 For more on veterans and their proclivities, see Mark Edele, "A 'Generation of Victors': Soviet Second World War Veterans from Demobilization to Organization, 1941–1956" (PhD dissertation, University of Chicago, 2004), Chapter 3.

58 Kazuko Kawamoto, "The Post-war Process of Creating an All-Union Basic Family Law, 1948–1968" (*Dainiji sekaitaisengono sorenniokeru renpokazokukihonho sei-tei katei*) (PhD Dissertation, Tokyo University, 2005), 38.

59 As with all policy measurement, it is impossible to state with certainty what results might have obtained had the law not been adopted. Complicating evaluation in this case, pre-1944 comparisons are impossible since there was no statistical category for illegitimacy with which to compare. The policy, however, was clearly unsuccessful on two grounds. First, the birth-rate never rose to the levels predicted by the planners. In a planned economy, this is failure. Second, the creation of the "single mother" and

"illegitimate child" categories labelled millions of individuals and exposed them to lifelong prejudice. This is failure on humanitarian grounds.

60 See Nakachi, Chapter 4. There are no earlier data for comparison, as this category did not exist prior to the 1944 Law.

61 The 1968 All-Union Basic Family Code made it possible for paternal information to be entered by unmarried mothers into children's birth certificates and ZAGS documents. Actual last names and patronymics could be used with paternal consent or by court order. Alternatively, mothers could use their own last name, in the masculine gender. In either case, no damning blanks would be immediately visible for all to see.

62 On post-war medical surveillance of botched abortions, see Chris Burton, "Minzdrav, Soviet Doctors, and the Policing of Reproduction in the Late Stalinist Years," *Russian History* 27:2 (Summer 2000).

63 For an overview of abortion statistics, including those presented here for 1956 and 1965, see Alexandre Avdeev, Alain Blum and Irina Troitskaya, "The History of Abortion Statistics in Russia and the USSR from 1900 to 1991," *Population: An English Selection* 7 (1995), 39–66. The 1946 and 1954 numbers come from Nakachi, Chapter 3.

64 See Nakachi, Chapter 6.

65 For example, see Starovskii's 16 February 1945 report on marriage and divorce in *Sovetskaia povsednevnost' i massovoe soznanie, 1939–1945* (Moscow: ROSSPEN, 2003), 287–289, where he claimed that the "sharp decline in registered divorces … followed the Law of July 8, 1944." In fact, examination of the data suggests that this was only true in the villages. Other documents indicate that courts in many areas had been told not to process divorce cases until new guidelines for the application of the 1944 Law were issued. This would also dampen divorce statistics in late 1944 and earliest 1945.

66 A demographic institute had existed at Kiev within the USSR Academy of Sciences since the 1920s and in Leningrad since 1931, but both were closed in the 1930s.

67 These included V. S. Nemchinov, M. V. Ptukha, S. G. Strumilin and B. S. Urlanis.

68 RGAE f. 1562, op. 327, d. 114, ll. 32–38.

69 V. I. Ivkin, *Gosudarstvennaia vlast' SSSR: Istoriko-biograficheskii slovar'* (Moscow: ROSSPEN, 1999), 544.

70 For the best treatment of the 1930s TsSU statisticians, see Alain Blum and Martine Mespoulet, *L'anarchie bureaucratique: pouvoir et statistique sous Stalin* (Paris: Decouverte, 2003).

71 *Naselenie Rossii v XX veke. t. 1* (Moscow: ROSSPEN, 2000), 345; Catherine Merridale, "The 1937 Census and the Limits of Stalinist Rule," *The Historical Journal* 39:1 (1996), 225–240.

72 Blum and Mespoulet, 131, 181. A brief biography of Osinskii, whose original name was V. V. Obolenskii, can be found in Ivkin, 446–447.

73 While most experts are in accord that the 1937 census was an impressively accurate affair, Starovskii considered it off by as much as two million souls, both lowering estimates of his predecessors' competence and raising the possibility that the total population was actually higher than originally reported. Additional examples of questionable statistical manipulations by Starovskii can be found in *Naselenie Rossii, t.1*, 351, 355, 363. Blum points out that Starovskii's numbers were basically accurate and his statistical competence "beyond doubt." But his rhetoric and "self-censorship" could be masterfully misleading. For example, low birth-rates in Ukraine in the early 1930s, the result of the famine, were interpreted as the product of runaway abortion, thus highlighting, in hindsight, the correctness of the 1936 decision to prohibit that practice. Blum and Mespoulet, 154, 157.

74 Blum and Mespoulet, 153–154, 156.

75 *Pravda*, 14 March 1946.

76 A careful study of strictly military losses compiled from declassified military archives was published in 1993. *Grif sekretnosti sniat* (Moscow: Voenizdat, 1993) covers all

of the Soviet Union's wars, but three-quarters of the book is devoted to the Second World War. The total number of deaths for men in uniform at the front is 8.7 million (p. 407). Ellman, "Soviet Deaths," 675, revises this down to 7.8 million, not far from Stalin's number.

77 Nakachi, Introduction. In the context of cold-war rivalries, the Kremlin also did not want to reveal the full scale of wartime damage.

78 V. N. Starovskii, "Proizvoditel'nost obshchestvennogo truda i problemy narodonaseleniia," *Vestnik akademii nauk SSSR* 5 (May 1962), 44, 46.

79 The highest organ of the Communist Party, the Politburo, was renamed Presidium by Stalin in 1952.

80 *Prezidium TsK KPSS, 1954–1964* (Moscow: ROSSPEN, 2003), 766, 841. This statement reflects a common 1930s view that statistics was the handmaiden of accounting and industry. In this spirit, TsUNKhU was placed under Gosplan in December 1931. However, at the 16th Party Congress in 1930, Stalin had stated that "there is no accounting without statistics," so that may have somewhat dampened the ardour of accounting enthusiasts. On these discussions, see Blum and Mespoulet, 211–212.

81 *Prezidium*, 869. Mikoian intended to defend Khrushchev by spreading blame to Starovskii.

82 Wesley Fisher and Leonid Khotin, "Soviet Family Research," *Journal of Marriage and the Family* 39:2 (May 1977), 372.

83 In the conclusion to his 1962 discussion of productivity and population, Starovskii calls for intellectual contributions from economists, statisticians, planners, medical personnel, geographers, urban planners, agronomists, technicians, biologists, physicists, chemists, geologists, oil-specialists, machine-builders, transport-specialists and energy-experts. Interestingly, in an article on population, he never mentions demographers. Starovskii, 52–53.

84 In the mid-1960s, attempts were finally made to re-establish demography as an academic field, both at Moscow State University (1965) and in the Academy of Sciences (1967). Alexandre Avdeev, "Avenir de la demographie en Russie" in Jean-Claude Chasteland and Louis Roussel, eds., *Les contours de la demographie au seuil du XXIe siecle* (Paris: Colloque International, 1995), 369–396.

2 The bitter legacy of the 'Great Patriotic War'

Red Army disabled soldiers under late Stalinism

Beate Fieseler

The victory over Nazi Germany made the Soviet Union the world's second superpower and Stalin found himself at the climax of his national and international prestige. However, it was a bitter triumph for which the country and its inhabitants paid a horrible price. Throughout history no victory has ever been accompanied by so much glory and so much misery. The immediate material damage was immense and left large sways of the country in utter devastation. In addition the Soviet Union suffered dramatic demographical losses. The war left not only 27 million dead, but also millions of widows, orphans and invalids. The country emerged victorious, yet domestically extremely weakened, from the fighting. Especially economically the country was far from ready to assume its status as a world power. In order to safeguard its 'spoils of war' the regime had to acquire solid economic foundations very quickly. At the same time, the social consequences of the war had to be 'overcome'. Otherwise social disintegration was to threaten the regime from within and endanger its ideological and structural make-up. Despite the deprivation and losses suffered by the Soviet people the mood in the immediate post-war period was one of optimism and full of hope for a quick return towards normality. In particular, crippled and injured veterans expected significant improvements to their lives – in exchange for the victory, for which they had fought so bitterly and tenaciously. This belief had already been nurtured by official propaganda during the war years. The press had again and again promised 'comprehensive care' to injured and crippled veterans and thus confirmed the Soviet state's commitment to welfare. This essay analyses to what extent the Soviet welfare rhetoric on injured war veterans was translated into reality. The focus will be on veterans' reintegration into work processes, which, according to Soviet understanding, was the most important aspect of social policy. From the perspective of the regime, the immediate return to work of the whole population was one the most important preconditions for quick economic reconstruction and recovery.

Invalidity – a flexible category

According to official data published by the General Staff of the Armed Forces, 2,576,000 soldiers were discharged by the Red Army as invalids up to May

1945.[1] This is about 7.46 per cent of the entire Soviet army, which in 1945 numbered just under 34.5 million soldiers. Taking into account the brutality of the German warfare against the Soviet Union and merciless battle strategies of the Red Army, this ratio seems to be very small indeed.[2] However, the low proportion of invalids compared with the total number of service men hides the true extent of war-sustained injuries and war-related illnesses. Rather, it was the result of a rigid implementation of regulations from above, which stipulated strict limits on who could be recognised as an invalid.

The legal and ideological foundations of recognition procedures had been formed in the 1930s during Stalin's 'revolution from above'. The new order eliminated the previous health-related understanding of invalidity in favour of a purely production-oriented approach.[3] Invalidity status was no longer granted on the basis of just bodily injuries, but judged according to the claimants' complete or partial inability to work. This concept of invalidity was in agreement with the raison d'être of overall Soviet labour policy during the period of forced industrialisation and resulted in a three-grade recognition system. According to the new definition, the first category of invalidity included all people who were entirely unfit for work and in need of full-time care. The second category referred to those who had lost their capacity to work, but required no permanent medical attention. The third category was conferred upon those who were considered partially fit for work in occupations requiring low qualifications and yielding small incomes or who could work under facilitated conditions.[4]

Before 1932 the task of determining a disabled person's level of injury had been the responsibility of a physician. Now medical staff were required to establish the remaining degree of work capacity. The relationship between illness (*bolezn'*) and fitness for work (*trudosposobnost'*) was thus dissolved.[5] Consequently, the reintegration of disabled people into the working process (*trudoustroistvo*) gained enormous priority in all institutions charged with social welfare. One could claim that it became the essence of Soviet social welfare policy.[6] While the new principles apparently made no significant impact on real life in the 1930s, they achieved mass application during and after the Second World War when millions of ill or wounded demobilised soldiers returned from the battlefield, the majority of whom suffered from injuries to the spine or from damage to or loss of limbs.[7] The so-called 'Medical Expert Commissions' (VTĖK) examined the war-disabled of the first category every six months and the two other categories even every three months.[8] Regular check-ups were not only supposed to document any health improvements, but also individual compensation of the handicap, both of which should lead to a downgrading with regard to the invalidity status. As the participation in productive work was considered not merely an effective remedy, but a main tool of rehabilitation,[9] those invalids who had returned to their former jobs ran the risk of losing their recognition as invalids as well as the state pension that came with this recognition.

State pensions

All soldiers acknowledged as Second World War invalids were entitled to receive a state pension, which placed them in a much better position than the invalids of the First World War or the Civil War, whose compensation had been very small indeed.[10] However, even the Second World War pensions were humble. Their amount depended on a number of criteria: former wages, military rank and category of invalidity. Workers received more than peasants, officers and non-commissioned officers were better off than privates, and those who had been drafted right out of school or those who could not demonstrate their previous income (and this was the majority) received only a small lump-sum. At the same time, the pensions for first-category invalids – people who were completely unable to work – were not enough to live on. Disabled soldiers of the second and third categories received an even smaller amount. Unless they decided to make a living from begging, the great majority of war invalids therefore had to return to work in order to survive. By transforming the guaranteed right of the disabled to work in factories, administrative bodies or cooperatives into an obligation in January 1943, politics also worked in favour of the reintegration of war invalids into the workforce.[11]

Training and retraining

Yet most invalids were left alone to find a job after their release from hospital. Official propaganda praised the network of state assistance with regard to professional reintegration, but in fact the majority of the war-disabled did not receive any kind of retraining. In quantitative terms, the military hospitals, which were administered by the People's Commissariat of Health (Narkomzdrav), offered the largest number of retraining courses. Denying this as their duty,[12] however, they merely instructed a fraction of the wounded soldiers.[13] Moreover, the programmes only covered a small spectrum of vocations. There were courses for cobblers, tailors, accountants and clerks, but virtually no training in other, more sophisticated trades or industrial occupations like turner or fitter. On the whole, contemporary observers agreed that most ill or wounded soldiers left the hospitals as unqualified workers,[14] even if they had completed a retraining course. Although the People's Commissariat of State Control (Narkomgoskon) stated in 1945 that the training in the military hospitals could not be considered serious vocational training,[15] the social welfare administration still counted those who had attended any such course, as 'retrained'. In the view of the director of the organisational-instructional department of the Central Committee, however, the qualification programmes for disabled soldiers were still at the 'embryo level of development' by the end of the war.[16] De facto, only those who, in spite of their handicap, were still considered as 'economically useful', had a chance of being retrained. 'Those who had a leg amputated were taught something, but those who lacked a hand – nothing, look for yourself to get along', one disabled soldier remembered.[17]

In the view of the People's Commissariat of Social Welfare (Narkomsobes), the retraining of third-category invalids was to be organized mainly at factory level by the respective People's Commissariats;[18] many companies, however, did not cooperate. They did not regard the enhancement of the working efficiency of the war-disabled as useful enough to compensate the costs and risks of the programmes, but found it more effective to employ them as guards, gate-keepers or cloakroom attendants.[19] After the end of the war, more training was offered by factories, but it is doubtful whether as many as 100,000 invalids participated in such programmes in 1948 alone, as the deputy minister of Social Welfare, M. I. Derevnin, claimed.[20] According to official data, 173,000 invalids of war participated in some sort of retraining in the RSFSR between 1941 and 1945 and, from 1941 to 1948 the number was as high as 350,500,[21] which indicates a significant increase of qualification programmes after the end of the war.[22] But what do these numbers really prove? Fifty-three per cent of all Russian war invalids registered in 1948 (1,520,000 in total)[23] did not complete a vocational training course.[24] At the same time a far larger portion was no longer capable of practising their former vocations. Accordingly, at least one million people would have needed retraining. However, the social welfare organs never calculated the actual need for such programmes and therefore lacked a target towards which to work. As far as the content of the training courses is concerned, they never went beyond simple tasks. They were organised to be of short duration, low cost and easy implementation. On balance, the retraining of hundreds of thousands of war-disabled who desperately needed it did not follow a sound plan. Instead, the retraining efforts may be considered a prime example of the growing gap between propaganda and social performance.[25]

Professional reintegration

According to legal provisions, the return of disabled persons to active working life was one of the prime tasks of the social authorities. Yet it was precisely the authorities' lack of action that becomes apparent in the reports of Narkomgoskon and the authorised officers of the Party Control Commission (KPK).[26] Indeed social welfare bodies spent little effort on reintegrating the war invalids into working life according to their qualification and with consideration to their physical disabilities, but rather left this task to 'automatism' (*samotek*).[27] The same lack of interest was found by the KPK to prevail in the Communist Party and in Komsomol committees, the soviets, the military commissariats and the trade unions.[28] Nonetheless, the employment rate of war invalids, which had amounted to only 57.3 percent in September 1942,[29] was rising steadily and went up to almost 80 per cent before the end of the war (in January 1945), reaching 91.2 per cent in April 1948.[30] This was believed by KPK to be primarily due to the personal initiative of those affected, who had looked for a job on their own accord.[31] About 50 per cent of veterans were employed in agriculture, the rest by public authorities or in the industry. The biggest problems of reintegration were encountered in industrial settings – a consequence of the lack of industrial participation in occupational retraining and of considerable reservations of enterprises to take on war invalids as employees. The percentage of invalids working in industry for the year 1950 was only 22.4 per cent (plus 5 per cent working at shops which were run by the invalids' cooperative), whereas 23.3 per cent of all war-disabled employed continued to work in public administration.[32]

Peasants would usually return to their native villages and try to find a job on the kolkhozes, sovkhozes or Machine-Tractor-Stations (MTS) where they had worked before the war.[33] In 1948, 58.5 per cent of all war invalids in the RSFSR lived in the countryside, accounting for about 2.5 per cent of the rural population.[34] Among them were a number of men incapable of any type of agricultural work.[35] Consequently, about a quarter of those who had previously worked as kolkhoz farmers (or craftsmen) had to change over to the unproductive sector.[36] The number of skilled jobs was not only low, but required a training which most of the war invalids lacked. Therefore, they often faced problems earning a living, even when they were the only males of working age living in the kolkhoz.[37] Since the kolkhoz management generally considered them to be 'useless',[38] many of them had no choice but to work as watchmen, cleaners, errand boys, postmen, herdsmen, gardeners, shop assistants in the village cooperative, agitators or beggars. Some of the war invalids, however, took leading positions on farms, including that of a kolkhoz chairman. In a contribution on 'Care for the War-Disabled in the Soviet Union' social welfare commissar Sukhov gave great prominence to such advancements: 'You may well meet yesterday's soldier and officer of the Red Army as kolkhoz chairman and brigadier'.[39] But in reality, only 4.5 per cent of all war-disabled kolkhoz members worked as chairmen and only 8.6 per cent as brigadiers in the spring of 1948.[40]

Moreover, invalid war veterans in positions of responsibility did not play the affirmative and dominant role which the government had intended for them. War invalids in the villages frequently delivered rebellious speeches or gave full expression to their 'obstinacy'. Some claimed that their kolkhoz membership had lapsed after more than three years' absence, others became self-employed as carters, earning a not insignificant sum of money. Yet others acquired large tracts of land, bravely defending it against all claims. Taking it away from them would not be easy, a party secretary admitted in his report, because 'the invalids will seize a cudgel and attack all representatives of power, claiming: "I offered my health as sacrifice for this country, I have fought for it" '.[41]

Yet this belligerent mood in the villages did not last for long. When famine broke out after a crop failure in large parts of Russia in July 1946, all hopes and expectations collapsed. In the following years, 'hundreds of thousands' of widows, orphans and war invalids who, after being deprived of their food ration cards (September 1946), were able to survive only through occasional petty theft of consumables,[42] are said to have become enmeshed in the wheels of justice and to have disappeared for years in labour camps of the Gulag.[43] Many people left the kolkhozes during that period. The remaining rural population put up passive resistance and met its obligations with increasing reluctance. The political leaders therefore launched a penal campaign against the so-called 'shirkers' in 1948.[44] The law, proposed by Khrushchev, provided for individuals 'maliciously abstaining from work and leading an anti-social parasitic life', to be moved to 'remote places' for eight years[45] and to do forced labour in industrial plants or on large-scale construction sites of the Ministry of the Interior (MVD). In the summer of 1948 more than 12,000 people were banished from their Russian native villages,[46] among them war invalids and their wives, elderly people and single mothers.[47] Some of the expulsion orders directed against these groups were later suspended as 'irresponsible' or 'incorrect', yet by far not all.[48] With our present knowledge, it is not possible to state exact figures on how many of the expulsion orders were directed towards war invalids and how many of them were revoked. But the examples clearly show that by no means did they belong to the 'protected groups of the population' during such campaigns.

Refusal and dismissal

The invalids who had been employed in industry before the war generally tried to find work in their former enterprises, but despite the general lack of human resources, even long-serving factory workers often failed to return to their former collectives.[49] Severely handicapped persons were rejected almost without fail, since factory managers considered them 'cripples'.[50] They were afraid of being unable to meet rigid planning targets with a labour force of limited working capacity and in a state of health so poor that drop-outs were possible at any time. The employment of war invalids was not credited to the factories, and no production plan was reduced accordingly.[51] Many companies refused to employ war invalids entirely: 'I prefer to put a prisoner in the working-place rather than a war invalid', the manager of Dal'stroi commented on a job application.[52] A director informed the social authorities 'that he did not need any disabled workers'.[53] Thus, during the war, many people knocked at the factory gates in vain, hobbling from factory to factory without finding a job – especially, if they were severely disabled.[54] The People's Commissar of State Security (NKGB), Merkulov, informed CC Secretary Malenkov in May 1945 that his office had been provided with information about 'a considerable number of cases of a wrong attitude towards the employment of war invalids shown by the managers of enterprises, public authorities, and kolkhozes'.[55] Although he did not furnish any concrete data, details must have been sufficiently alarming for him to pass them to the Central Committee.

The climax of this practice of rejection was reached after the end of the war. As is apparent from a report of the social insurance department of the trade unions, enterprises were not short of excuses for rejecting war-disabled applicants.[56] But even those veterans who were eventually successful in finding a job could not be sure that they would be able to keep it.[57] In the course of demobilisation, repatriation and re-evacuation, which involved about 20 million people of working age,[58] there was a steep increase in the dismissal of war invalids.

Many enterprises used the return of hundreds of thousands of fully employable males as an opportunity to 'purge' their staff of the war-disabled, dismissing them with flimsy arguments. In view of an increasing number of such dismissals, the social insurance department of the trade unions called for a resolution against this practice. Since ration cards, (workplace-provided) accommodation and all other benefits were cancelled upon dismissal, and even the disability pension was jeopardised because of 'unwillingness to work', the loss of a job meant a threat to one's livelihood.[59] But since the government did not take action to stop the dismissals, their number rose substantially in the second half of the 1940s. Late in 1947, Narkomsobes observed a regular ousting of war invalids from factories, invalids' cooperatives and even from outwork. They were replaced by non-handicapped workers from the redundant labour market that existed after demobilisation had been terminated.[60] It was particularly during the hungry years of 1946–1947 that numerous companies sent away many of those whom they considered to be 'unproductive mouths to feed'.[61]

Beside rejections and dismissals, instances of professional downgrading of war invalids in the hierarchy of the factory shop-floor were frequent after the end of the war.[62] On 3 February 1946, a war invalid from Leningrad, an engineer by profession, sent a petition to the Council of the People's Commissars (SNK), entitled 'For what?', in which he described his experiences.[63] After he had been discharged from the Red Army as an invalid of the second category, he first worked as deputy sovkhoz manager, subsequently as director of an industrial combine until he was able to return to Leningrad in 1944. Here, he was employed in the sales and trade administration as director of a motor vehicle pool until 13 January 1946 and even received bonus payments for his work. On returning from a business trip, he found that a demobilised person was holding his position and he had been moved to a poorly paid job:

> I am now asking for an explanation for what reason I have been removed from my position. Why have I been dismissed? Is a man in this country really worth nothing so that he can be mocked at for no reason? I ask you, to have this matter investigated – I make this request as a member of the Communist Party, as an officer of the Red Army, as an invalid of the Great Patriotic War, and finally as a plain citizen of the USSR.[64]

The growing rejection of the war invalids as members of the workforce even mobilised the Chief Attorney of the USSR, Safonov, at the end of the 1940s. In his report to the Supreme Soviet, he confirmed the existence of 'numerous facts of illegal rejections and dismissals of war invalids by the heads of factories and public authorities'.[65] He, too, did not mention any concrete figures so that the exact order of magnitude of the problem remains unknown. But it can safely be assumed that this was a mass phenomenon. Many of the disabled had to leave their workbenches together with adolescents, housewives and pensioners, who had been mobilised to work in the factories during wartime.[66] They were not considered fit for implementing the ambitious industrial reconstruction programme.

Quantitative and qualitative aspects of reintegration

The occupational reintegration of the war invalids nevertheless presented itself in statistics as a story of success. According to official statements, more than 90 per cent of all war invalids had returned to regular working conditions by 1948–1949, thus having at least disappeared 'from the streets'. The impressive figures do not disclose, however, that for many of them reintegration was accompanied by a decline in occupational status, even if they had no severe physical disablement and had completed a good education. The war invalids complained to different authorities about instances of downgrading and discrimination, which could be found at all industrial and administrative levels. To what extent they were successful, however, has not been established. In fact, the problem was not 'solved', but rather kept on reappearing in the files up to the 1950s.[67] Contemporary public accounts and reports, too, are full of examples of 'unjustified employment' of war invalids far below their level of qualification. Yet Narkomsobes did not want to give the impression that this was a structural problem that had spread over the whole RSFSR, instead, officials talked about certain 'deficits' or 'difficulties'. Since no reliable figures are available, it is hazardous to suggest an estimate of the real dimensions. In 1945 and 1948, Sukhov only provided scanty information regarding the percentage of unskilled workers among the war invalids. Moreover, one should be sceptical about the accuracy of these numbers when looking at how they were compiled. Whereas he stated that the share of unskilled workers amounted to 20 per cent in May 1945,[68] he informed the CC in early summer 1948 in a secret note that only 7.5 per cent of all war invalids working in industry were employed as engineers or technicians, 62.5 per cent as workers, 19.6 per cent as unskilled workers and 8.8 per cent as auxiliary staff (*mladshii obsluzhivaiushchii personal*, e.g. cleaning personnel, janitors, watchmen, etc.).[69] Hence, there was a noticeable increase in invalids working as unskilled labour in industry after the war, reaching a share of 28.4 per cent at the end of the 1940s.

According to statements by Narkomsobes, 16.1 per cent of the war invalids continued to be employed as unskilled workers in 1950, while no information is available regarding the development of the percentage of auxiliary staff.[70] Among those employed as unqualified workers were former workmen who were no longer up to the pressure of quotas in the factories due to their injuries and had to be displaced to (unproductive) jobs,[71] which were paid by the hour and not by the piece.[72] But even skilled workers, engineers and technicians with professional experience ended up in such positions. They were either no longer fit to pursue their previous occupation,[73] but were not allowed to undergo retraining, or were qualified accountants or mechanics who had been put into jobs on a lower level of the corporate hierarchy.[74] All those who had been called up right away from school to join the Red Army without having completed a vocational training were relegated to menial jobs. Once war invalids were put into jobs as unskilled workers, nobody cared for their further career or for an improvement of their qualification.[75] The social welfare offices filed the case away, putting on it a stamp 'job found'. The factories were often not even informed about the total number of war invalids employed by them, and did not pay them any particular attention.[76] An investigation in the year 1949 for the entire RSFSR showed that the share of unskilled labour among the war-disabled in several regions was between 30 and 50 per cent.[77]

Due to the lack of representative investigations, the question of whether these data or the lower figures stated by Narkomsobes may claim greater validity must be left unanswered. On the whole, Sukhov came to a far more optimistic valuation of the situation than other observers. As the official carrying principal responsibility for reintegration, he tended towards glossing things over or fabricating success stories in order to be able to present positive results. There is every reason to believe that the situation was considerably worse than described by the Minister of Social Welfare. While initially the 'wrong attitude' of individual factory managers had been made accountable for the discrimination of war invalids in industry, later the lack of occupational training or war disablement became the principal obstacle for their assignment to a qualified job.[78] Moreover, in many cases, lack or poor quality of prostheses mitigated against qualified employment. With the Soviet Union not paying any attention to the development of functional and modern artificial limbs until the outbreak of war, the models of the 1940s were hopelessly antiquated. They were heavy, functioned only in a very limited way and did not last very long.[79] Moreover, it was impossible to satisfy the ever-growing demand, despite the fact that the number of relevant manufacturers had risen from 38 in 1941 to 108 in 1948.[80] The number of employees, however, rose only from 3,518 to 7,867 and remained inadequate in the face of enormous need and demand.[81] Even though the supply of artificial limbs did not improve much for many years after the end of the war, the topic disappeared from the agenda in the early 1950s.

Income

Many a war invalid employed as a (qualified) worker at a work-bench will have found it hard to improve his or her modest wage by raising productivity. Since Soviet industry was working at a low level of mechanisation, production required great physical effort. At the same time quotas had been raised regularly since 1947 – for unskilled workers often by as much as 20–40 per cent.[82] From the high-wage sectors (mining industry, metallurgy, metal working, oil), war invalids were excluded.[83] Since they were given neither decent prostheses nor any facilities to assist them in their work, there is no denying the fact that invalids were less efficient than non-disabled workmen across all industrial sectors. One cannot say that their workplaces were provided with individual and 'invalid-adapted' equipment. Rather the procurement of a swivel chair for a worker with a disability was celebrated as a special event.[84] Soviet factories were short of light bulbs, window glass, fans, simple tools, spare parts for routine manufacturing sequences and many other items. Under such circumstances the demand for special equipment meeting the needs of people with handicaps was truly a utopian dream. Instead of easing working conditions, some firms treated the war invalids in a reckless manner, putting them into particularly unproductive and thus low-income workplaces, where work was the most unpleasant and targets were the hardest to meet. Often these workplaces demanded too much of their physical strength, in particular when, after frequent breakdowns, a shock labour effort (*sturmovshchina*) became necessary towards the end of a production period in order to master the specified targets 'by storm'. It was precisely because of such chaotic operating sequences that many factories desisted from employing war invalids in their production.

Nevertheless, Narkomsobes always emphasised in its official statements that the average income of invalids, together with their pension, surpassed pre-war incomes. For 1946, Sukhov stated that the rise amounted to between 22 and 50 per cent[85] whereas industrial monthly wages in war years had grown by an average of 42 per cent.[86] The data given by Sukhov were allegedly based on 'large-scale investigations', yet details were not provided. Hence, it is questionable whether this information can be taken seriously. It can be imagined that he invented this increase in order to propagate the idea that the war invalids had been granted 'a fair material compensation', thus participating in the general nominal increase – if not by their own productivity, then by the government's pension payments. The average (Moscow) income of an invalid was estimated by the Minister of Social Welfare in 1946 (before the bread allowance was granted) at 549 roubles, the average pension at 178 roubles, which would result in a total income of 727 roubles for private soldiers and of 1,148 roubles for officers (711 roubles wage, 437 roubles average pension).[87] In that case, the average income of an invalid would even have exceeded the average wage of 626–650 roubles. Yet Sukhov failed to mention on what data the results of Narkomsobes were based. As far as it is known, no serious investigations regarding the income of war invalids were carried out by any public authority. According to statements for various regions provided by trade unions in 1945 and 1950, at least 20–25 per cent of invalids were earning less than before the war, even including their pensions, another 8–10 per cent had as little money at their disposal as in 1940.[88] In 1950, an internal 'conclusion' drawn by Narkomsobes revealed that the war invalids of the third category achieved an average 86 per cent of their pre-war rouble-for-rouble-income.[89]

Against the background of considerable increases in the nominal wages during the preceding decade (one might talk about roughly a doubling)[90] such findings can only be explained by a massive re-employment of veterans into low-wage positions and a failure on their part to meet their respective norms. These figures stand in sharp contrast to those presented by Sukhov in May 1945, namely that only 4.3 per cent of all war invalids of the second and third categories working in industry would not achieve the set production norms, whereas 12.3 per cent would even double them.[91] In several instances Sukhov claimed that invalids were even more productive than non-disabled workers.[92] One year later, he quoted 'information from large-scale investigations' according to which only 3–5 per cent of all war invalids had not met their production norms.[93] In one of his accounts, the information was more precise: in 52 of the 162 enterprises investigated, the actual productivity of invalids was higher than that of the non-disabled. But in fact, only 15 factories and 40 (from a total of 513) war invalids had been checked. The alleged higher productivity applied to just six enterprises and seven invalids (as a rule of the third category and with a high qualification level).[94] Thus, the data supplied by Sukhov had little correlation to reality, but represented what Stalin had once called 'optimistic statistics'.

Apart from such manipulations (or select measuring), the fact that a war invalid complied with the production plan did not mean that his income was at subsistence level. In particular, low-income earners had to fulfil multiple norms in order to make ends meet.[95] Above all, Sukhov's statements on the productivity and income of war invalids hushed up the fact that the increase in gross wages was accompanied by an extremely disadvantageous development of prices and fiscal charges which lead to massive losses in real earnings.[96] Under such circumstances the fact that the minister mentioned high wages of war invalids on every occasion can only be considered cynical.[97] It is indeed possible that there were a number of higher-income earners among veterans, yet permanent insistence on the success of high wages was nothing but a grotesque misrepresentation of conditions as they were. Well-heeled 'high flyers' were an exception even among non-disabled industrial workers[98] and are most unlikely to have found imitators among the war invalids.

The scantiness and unreliability of Soviet statistics make generalising statements on the real income of war invalids difficult. At any rate, the numerous accounts and reports seem to reflect the situation more accurately than the quantitative indices of Narkomsobes generated in a manipulative manner. To judge from the information collected locally, a large if not the largest portion of the war invalids, alongside young workers and single mothers, belonged to the poorest of the poor in a period that was generally characterised by low wages, endemic poverty and miserable living conditions.[99] The political leadership was well informed about the situation by reports from Narkomgoskon and the KPK, but did not undertake any effort to get a grip on the problem. For instance, the low wages of watchmen, which were fixed centrally, were not raised, although this occupation had become a typical job for war invalids in post-war times.[100] Any rise in wages without an increase in performance or qualification was not possible, in fact not even thinkable, within the framework of the restrictive wage policy during the war and post-war years. The gap between the salary of unskilled workers and the industrial top wages paid for workers who were vital for production had become considerably larger during the war. Therefore, poverty among war invalids employed in industry was much more commonplace than among non-disabled industrial workers. In their accounts, many party committees nevertheless praised the successful reintegration of war invalids, being enthusiastic about their high level of productivity[101] and their excellent wages.[102] Such examples, of course, do not refer to 'most of all' or to a 'majority' of war invalids[103] as usually suggested, but to some isolated cases.[104]

In reality many invalids went to work barefoot and dressed in rags. They could not expect any support from the factories or from social welfare bureaus. Distribution of food, clothing or other goods happened according to arbitrary criteria rather than need. War invalids often went away with empty hands.[105] Even donations or gift parcels from abroad failed to reach them.[106] As late as 1948–1951, annual price reductions gradually contributed to alleviating the worst need.[107] But for low-income earners, such as the majority of war invalids, intermittent rise of real wages only allowed a meagre survival. Consequently, in the post-war years the exhausting daily struggle for goods, heating and accommodation consumed all energy. This may well be the reason why acts of protest hardly occurred.

Even before the end of the war Soviet propaganda emphatically confirmed the state's commitment to providing 'comprehensive welfare'. Such rhetoric was designed to make the German enemy alone responsible for the damage of war, while modern and progressive strategies of rehabilitation and medical care were attributed to the Soviet state and its agencies. As such the narrative about Soviet invalids was an important element in post-war Stalinist discourse. State support/state provision for the invalid war veterans became – at least in contemporary discourse – proof for the modernity of the Soviet state and the Stalinist system.

Although Soviet invalids lacked a formal organisation which could lobby the state to meet its self-proclaimed obligations, official propaganda provided them with a gauge against which to measure the shortcomings of the Soviet welfare system. The promises of welfare and care, so forcefully proclaimed by state propaganda, created and raised a certain level of expectation which, under different circumstances, could easily have resulted in a crisis of legitimacy. A legal claim to state awards and services, which could have been pursued in the courts, was outside the realm of the possible in late Stalinism. The victims of state neglect had therefore to take recourse to traditional practices such as petitions and letters of complaint or make use of extra-legal methods such as bribery and blackmail. How often invalids had success with such practices is impossible to estimate. It is also impossible to quantify to what extent the permanent experience of being let down by the state resulted in alienation from or opposition to the Soviet system. Reason for protest was certainly given in sufficient quantity. Examination of the social reality of veterans, which has been carried out here with reference to the integration of invalids into the labour market, reveals severe short-comings and failures on the part of the relevant Soviet authorities. Despite the significant costs of welfare programmes for victims of war, the general level of care was poor and the mechanisms of aid provision malfunctioning.

The welfare provisions for war invalids in late Stalinism represented a modernity, which was mainly achieved through performance. For the large mass of invalids it remained a virtual, since unobtainable, reality. The performance consisted not only of emphatic propaganda, but also of the singling out of a few invalids who, styled as heroes of both war and labour, were the recipients of exemplary care and support. The transformation of an invalid from a symbol of helplessness to a symbol of heroic achievement served not least to legitimise the regime. While the performance of a state devoted to welfare was upheld throughout the entire period, a decision to give priority to the production of heavy industry was taken early on. The provision for invalids became subordinate to the needs of the reconstruction process. The factories endured extreme pressure to achieve maximum performance. Physically disabled people and a workforce whose health was weakened and whose work was hampered by inefficient prostheses did not fit well into a production system designed for the overachievement of production norms. Moreover, its chaotic and dangerous operations depended perpetually on improvisation. Thus the promise of 'comprehensive care' was negatively affected by the pressure to attain planning targets that weighed heavily on the factories. In order to be able to meet production plans, enterprises defended themselves with some justification against taking over the obligations of social authorities. They had trouble employing hundreds of thousands of war invalids as production workers. It was much easier to use them as unskilled work, the more so as the government labour policy was primarily directed at disciplining the labour force through regularly increasing work norms (without any corresponding increase in wages). De facto war invalids had to compete with non-disabled workers on the shop-floor. It was no longer important that they had become disabled while defending their homeland. Determined to reconstruct industries considered vital in a potential future conflict, the political leadership paid little attention to the social consequences of war. Governmental labour policy was more and more in contradiction to the self-proclaimed assertions of the priority of social welfare. For the majority of war invalids living in late Stalinism Soviet welfare remained nothing but a mere fiction.

Notes

1 G. F. Krivosheev (ed.), *Soviet Casualties and Combat Losses in the Twentieth Century*, London: Greenhill Books, 1997, pp. 91–92.
2 See, for example: G. Temkin, *My Just War. The Memoir of a Jewish Red Army Soldier in World War II*, Novato/California: Presidio Press, 1998, pp. 117, 120.
3 *Trudovoe ustroistvo invalidov. Posobie dlia inspektorov otdelov sotsial'nogo obespecheniia*, Moskva: Izdatel'stvo Ministerstva sotsial'nogo obespechieniia, 1952, pp. 29, 56; C. Burton: *Medical Welfare During Late Stalinism. A Study of Doctors and the Soviet Health System, 1945–1953*, PhD Dissertation, University of Chicago/Ill., 2000, pp. 268–269.
4 *Trudovoe ustroistvo invalidov*, p. 56.
5 Ibid., pp. 52–53.
6 N. M. Obodan (red.), *Vozvrashchenie k trudovoi deiatel'nosti invalidov Otechestvennoi voiny. Sbornik postanovlenii i instrukcii*, Leningrad: Lenizdat, 1943.
7 *Ocherednye zadachi organov sotsial'nogo obespecheniia. Materialy Vserossiiskogo soveshchaniia rukovodiashchikh rabotnikov sotsial'nogo obespecheniia (mai 1945 goda)*, Moskva: Izdatel'stvo Narkomsobesa RSFSR, 1945, p. 25; GARF f. 5451, op. 29, d. 166, ll. 116–117; RGASPI f. 17, op. 117, d. 511, l. 107.
8 GARF f. 5446, op. 46, d. 3378, l. 2.
9 Invalidnost', *Bol'shaia Sovetskaia Entsiklopediia*, vol. 17, Moskva: Gosudarstvennoe nauchnoe izdatel'stvo 'Bol'shaia Sovetskaia Entsiklopediia', 1952, p. 611.

10 Postanovlenie SNK SSSR No. 1269, 16 July 1940, *Sobranie postanovlenii i raspori-azhenii pravitel'stva SSSR*, 1984, No. 19, p. 465; 'O dopolnenii postanovleniia Sovnarkoma SSSR ot 16 iiulia 1940', Ibid., 1940, No. 30, p. 729.
11 Postanovlenie SNK SSSR No. 73 'O merakh po trudovomu ustroistvu invalidov Otechestvennoi voiny', in: Obodan (red.), *Vozvrashchenie*, pp. 20–21.
12 RGASPI f. 603, op. 1, d. 12, l. 30.
13 RGASPI f. 17, op. 122, d. 100, ll. 141–142, 228.
14 RGASPI f. 17, op. 122, d. 100, l. 189.
15 GARF (RSFSR) f. A-339, op. 1, d. 1800, l. 108.
16 RGASPI f. 17, op. 122, d. 71, l. 196.
17 V. Bessonov, *Voina vsegda so mnoi*, Moskva: Sovetskaia Rossiia, 1988, p. 71.
18 GARF (RSFSR) f. A-413, op. 1, d. 248, l. 64.
19 RGASPI f. 17, op. 122, d. 100, ll. 125–126, 228; f. 17, op. 121, d. 425, l. 27.
20 GARF (RSFSR) f. A-413, op. 1, d. 1383, l. 80.
21 Ibid., l. 6.
22 RGASPI f. 17, op. 131, d. 36, l. 15.
23 RGASPI f. 17, op. 131, d. 36, l. 10.
24 Ibid.
25 N. P. Paletskikh, 'Sotsial'naia pomoshch invalidam Velikoi Otechestvennoi voiny na Urale v 1941–1945 gg.', in: *Ocherki istorii gornozavodskogo Urala*, Cheliabinsk: ChelGU, 1996, pp. 94–95.
26 GARF (RSFSR) f. 339, op. 1, d. 433, l. 18; f. A-339, op. 1, d. 434, l. 13; f. A-339, op. 1, d. 828, l. 33; f. 339, op. 1, d. 834, l. 22; f. A-339, op. 1, d. 836, l. 10; f. A-339, op. 1, d. 836, l. 64; f. A-339, op. 1, d. 1800, l. 5; RGASPI f. 17, op. 88, d. 604, l. 3.
27 RGASPI f. 17, op. 122, d. 100, l. 172.
28 Ibid., ll. 3, 31, 128, 146–147, 228.
29 GARF (RSFSR) f. A-413, op. 1, d. 234, l. 85; RGASPI f. 17, op. 122, d. 21, ll. 84ff.
30 GARF (RSFSR) f. A-413, op. 1, d. 460, ll. 2, 4.
31 RGASPI f. 17, op. 122, d. 100, ll. 22, 31, 66, 104, 112, 124, 144, 172, 194, 201, 280.
32 RGASPI f. 603, op. 1, d. 3, l. 165; GARF (RSFSR) f. A-413, op. 1, d. 248, l. 63; RGASPI f. 17, op. 122, d. 71, l. 195; GARF (RSFSR) f. A-339, op. 1, d. 1800, l. 126; RGASPI f. 17, op. 131, d. 36, l. 14; GARF (RSFSR) f. A-413, op. 1, d. 1380, l. 13.
33 GARF (RSFSR) f. A-339, op. 1, d. 429, l. 33.
34 RGASPI f. 17, op. 131, d. 36, l. 10; O. M. Verbitskaia, 'Izmeneniia chislennosti i sostava kolkhoznogo krest'ianstva RSFSR v pervye poslevoennye gody (1946–1950)', *Istoriia SSSR*, 1980, vol. 5, p. 126.
35 E. Winter, *I Saw the Russian People*, Boston: Little, Brown and Company, 1945, p. 93.
36 GARF (RSFSR) f. A-413, op. 1, d. 753, l. 29.
37 RGASPI f. 17, op. 121, d. 425, l. 28.
38 *Ocherednye zadachi*, p. 47; Bessonov, *Voina*, pp. 27, 40, 103.
39 GARF (RSFSR) f. A-413, op. 1, d. 582, l. 3.
40 RGASPI f. 17, op. 131, d. 36, l. 14.
41 Paletskikh, 'Sotsial'naia pomoshch invalidam', p. 104.
42 RGASPI f. 17, op. 121, d. 515, ll. 83–84.
43 V. F. Zima, 'Poslevoennoe obshchestvo: Golod i prestupnost' (1946–1947 gg.)', *Otechestvennaia istoriia*, 1995, vol. 5, p. 58.
44 Y. Gorlizki, 'Rules, Incentives and Soviet Campaign Justice After World War II', *Europe-Asia Studies*, 1999, vol. 51, pp. 1245–1265.
45 'Neizvestnaia initsiativa Khrushcheva', *Otechestvennye arkhivy*, 1993, vol. 2, pp. 31–38.
46 V. F. Zima, *Golod v SSSR 1946–1947 godov: Proischozhdenie i posledstviia*, Moskva: Rossiiskaia Akademiia Nauk, Institut Rossiiskoi Istorii, 1996, p. 188.

47 J. Channon: Stalin and the Peasantry: Reassessing the Postwar Years, 1945–1953, in: Ibid. (ed.): *Politics, Society and Stalinism in the USSR*, Basingstoke: Macmillan Press, 1998, p. 191; Zima, *Golod v SSSR*, p. 186.
48 RGAE f. 9476, op. 1, d. 287, ll. 28–29; 'Universal'noe sredstvo', in: V. P. Popov: *Krest'ianstvo i gosudarstvo (1945–1953)*, Paris: YMCA-Press, 1992, pp. 243–249; V. F. Zima, 'Vtoroe raskulachivanie', *Otechestvennaia istoriia* 1994, vol. 3, p. 113; 'Neizvestnaia initsiativa Khrushcheva', p. 32.
49 RGASPI f. 17, op. 122, d. 100, ll. 93, 150; f. 17, op. 122, d. 101, l. 146; f. 17, op. 122, d. 21, ll. 76–77.
50 GARF f. 5446, op. 44, d. 976, l. 128; GARF (RSFSR) f. A-413, op. 1, d. 359, l. 26.
51 GARF f. 5446, op. 44, d. 976, l. 133.
52 RGASPI f. 17, op. 122, d. 100, l. 60.
53 RGASPI f. 17, op. 212, d. 425, l. 28.
54 GARF (RSFSR) f. A-339, op. 1, d. 1800, l. 105; GARF f. 5451, op. 29, d. 165, l. 19.
55 RGASPI f. 17, op. 121, d. 425, l. 28.
56 GARF f. 5451, op. 29, d. 242, l. 2.
57 RGASPI f. 17, op. 122, d. 21, l. 76.
58 M. Hildermeier, *Geschichte der Sowjetunion. Entstehung und Niedergang des ersten sozialistischen Staates*, München: Beck, 1998, pp. 702–703.
59 GARF f. 5451, op. 29, d. 242, ll. 1–4.
60 GARF (RSFSR) f. A-413, op. 1, d. 904, ll. 19–21; GARF f. 5451, op. 29, d. 397, l. 112.
61 *Sbornik vazhneishikh prikazov i instrukcii po voprosam kartochnoi sistemy i normirovannogo snabzheniia*, Moskva: Izdatel'stvo Narkomtorga, 1944, p. 173.
62 GARF f. 5446, op. 47, d. 2976, ll. 44, 39, 8, 7; f. 5451, op. 29, d. 242, l. 1.
63 GARF f. 5446, op. 48, d. 3243, l. 24.
64 Ibid.
65 GARF f. 7523, op. 65, d. 588, l. 37.
66 B. I. Gvozdev, 'Chislennost' rabochego klassa SSSR v pervye poslevoennye gody (1945–1948 gg.)', *Istoriia SSSR*, 1971, vol. 4, pp. 118–119; V. N. Donchenko, 'Demobilizaciia sovetskoi armii i reshenie problemy kadrov v pervye poslevoennye gody', *Istoriia SSSR*, 1970, vol. 3, pp. 98–99.
67 RGASPI f.17, op. 121, d. 217, ll. 59–60.
68 A. N. Sukhov, 'Ocherednye zadachi organov sotsial'nogo obespecheniia v oblasti trudovogo ustroistva invalidov', in: *Ocherednye zadachi*, p. 18.
69 RGASPI f. 17, op. 131, d. 36, l. 14.
70 GARF (RSFSR) f. A-413, op. 1, d. 1380, l. 18.
71 GARF f. 5451, op. 29, d. 242, l. 212.
72 GARF f. 5451, op. 29, d. 128, l. 4.
73 Ibid., l. 26.
74 RGASPI f. 17, op. 122, d. 100, ll. 173–174; GARF f. 8131, op. 22, d. 221; GARF f. 5451, op. 29, d. 242, ll. 159–161.
75 GARF f. 8131, op. 22, d. 103, ll. 138, 195; RGASPI f. 17, op. 122, d. 100, ll. 113, 150.
76 Ibid., ll. 71, 150.
77 GARF f. 5451, op. 29, d. 397, ll. 80–81.
78 RGASPI f. 17, op. 122, d. 71, l. 195.
79 GARF (RSFSR) f. A-339, op. 1, d. 837, l. 4; GARF f. 5446, op. 48, d. 3246, ll. 3–1.
80 RGASPI f. 603, op. 1, d. 4, l. 4; GARF f. 5446, op. 50, d. 4235, l. 76.
81 GARF f. 5446, op. 50, d. 4235, l. 76.
82 D. Filtzer, *Soviet Workers and Late Stalinism. Labour and the Restoration of the Stalinist System After World War II*, Cambridge: Cambridge University Press, 2002, p. 237.
83 Ibid., p. 235.

84 GARF f. 5451, op. 29, d. 130, ll. 57–58.
85 GARF (RSFSR) f. A-413, op. 1, d. 753, l. 7.
86 Hildermeier, *Geschichte der Sowjetunion*, pp. 649–650.
87 GARF (RSFSR) f. A-413, op. 1, d. 753, l. 8.
88 GARF f. 5451, op. 29, d. 165, l. 19; f. 5451, op. 29, d. 397, l. 112.
89 GARF (RSFSR) f. A-413, op. 1, d. 1380, l. 80.
90 J. G. Chapman, 'Real Wages in the Soviet Union, 1928–1952', *The Review of Economics and Statistics* 1954, vol. 26, p. 146.
91 Sukhov, 'Ocherednye zadachi', p. 8. See also: GARF (RSFSR) f. A-413, op. 1, d. 1380, l. 80.
92 Sukhov, 'Ocherednye zadachi', p. 8.
93 GARF (RSFSR) f. A-413, op. 1, d. 753, l. 7.
94 GARF (RSFSR) f. A-339, op. 1, d. 1800, l. 106.
95 Paletskikh, 'Sotsial'naia pomoshch invalidam', p. 93.
96 Hildermeier, *Geschichte der Sowjetunion*, pp. 649–650, 706–707.
97 Sukhov, 'Ocherednye zadachi', pp. 19–20.
98 Hildermeier, *Geschichte der Sowjetunion*, pp. 653–654.
99 RGASPI f. 17, op. 122, d. 156, ll. 1–11, 158–169; Filtzer, *Soviet Workers*, pp. 232–241.
100 A. Pyzhikov and A. Danilov, *Rozhdenie sverkhderzhavy, 1945–1953 gody*, Moskva: Olma-Press, 2002, p. 132.
101 RGASPI f. 17, op. 122, d. 100, l. 270; f. 17, op. 88, d. 470, l. 77; f. 603, op. 1, d. 6, ll. 125–126, 136; f. 603, op. 1, d. 9, l. 90; GARF f. 5446, op. 44, d. 976, l. 133.
102 RGASPI f. 17, op. 88, d. 470, l. 19.
103 RGASPI f. 17, op. 88, d. 338, l. 10; f. 17, op. 122, d. 145, l. 87; f. 603, op. 1, d. 4, l. 167; GARF f. 5451, op. 29, d. 130, ll. 31, 158–166, 170, 173, 175.
104 RGASPI f. 17, op. 122, d. 100, ll. 270–271; f. 17, op. 88, d. 470, l. 90; f. 603, op. 1, d. 6, l. 136; GARF f. 5446, op. 44, d. 976, l. 133.
105 RGASPI f. 17, op. 121, d. 425, l. 30; f. 17, op. 122, d. 44, ll. 9–12; f. 17, op. 122, d. 71, ll. 61–66; f. 17, op. 122, d. 79, ll. 3–5, 186–189; f. 17, op. 122, d. 100, ll. 7, 60, 66, 91, 97, 104, 114, 124, 127, 145–147, 149, 151, 196–197, 226, 228, 283; f. 17, op. 122, d. 107, ll. 96–104; f. 17, op. 122, d. 145, l. 91; f. 17, op. 122, d. 146, ll. 188–193; f. 17, op. 88, d. 470, ll. 20, 82; f. 17, op. 131, d. 182, ll. 1–5.
106 GARF f. 9415, op. 5, d. 95, ll. 59–61; RGASPI f. 17, op. 122, d. 107, ll. 96–104; GARF (RSFSR) f. A-413, op. 1, d. 651, l. 26.
107 Hildermeier, *Geschichte der Sowjetunion*, pp. 50–651.

3   Subversive tales?

War rumours in the Soviet Union 1945–1947[1]

Timothy Johnston

In late May 1945, a Crimean collective farmer read an article in the newspaper 'Red Crimea' about the exiled Polish Government in London. By the next morning Dimitrovka kolkhoz was in uproar. The *kolkhoznik* had come to the conclusion, on the basis of the article, that Britain was at war with the Soviet Union. Rumours about the conflict spread rapidly throughout the collective farm community, before passing to the nearby village of Kishlav. In the ensuing panic, *kolkhozniki* refused to go to work, convinced that a new and bloody conflict had broken out. Only once *oblast'* agitator Oshepkova had explained, in detail, the relationship between Britain and the USSR were the villagers convinced that the war was a figment of their imaginations. Order was subsequently restored.[2]

This essay assesses the prevalence and significance of war rumours between 1945 and 1947 in the USSR. War rumours and even war panics flourished during this period, in direct contrast to the official image of international affairs presented in the Soviet press. I will place these grassroots stories of invasion in their 'functional' context, before assessing them from 'subaltern' and 'thick descriptive' perspectives. I employ the term 'functional' in a Durkheimian sense, enquiring what role these rumours played in post-war Soviet society.[3] I will argue that stories of invasion functioned as 'news', and that their transmission functioned to reinforce social networks in the early post-war period. A 'subaltern' reading of these rumours, drawing on the work of Viola, and others, emphasises their employment as a language of protest for groups opposed to the Soviet regime.[4] I will argue that whilst a 'subaltern' interpretation provides a valuable starting point, it does not offer a full account of the proliferation and content of post-war war rumours. The 'success' and survival of war rumours within the oral sub-culture relied, at least in part, on their plausibility as genuine news, rather than their power to subvert. A 'thick descriptive' analysis seeks to describe the 'webs of meaning' that made war rumours credible as news in this period.[5] I suggest that the fixation of these rumours on an invasion by the Anglo-Americans reflected the widespread conviction that they had betrayed the USSR during the war, and were ready to do so again.

When discussing the content and nature of post-war war rumours I will draw on as wide a variety of sources as possible. This approach aims to mitigate the weaknesses of the individual source groups. It relies on a 'triangulation' effect, whereby the evidence for these rumours in various state, private and interview-generated sources is brought together to provide a composite picture of the rumours under discussion. What this approach lacks in terms of style, it gains by avoiding drawing excessive inferences from individual cases.[6] My major source groupings are: the case files of the State Prosecution Organ of the Soviet Union;[7] letters sent by Soviet citizens to the political leadership in Moscow;[8] 'opinion reports' or *svodki*;[9] information reports such as agitators' accounts of their lecture tours, or records of party gatherings;[10] the material of the Harvard Interview Project (HIP) on the Soviet Social System;[11] and interviews taken by the author between November 2003 and September 2004 in the former USSR.[12] Each of these categories of source attach different significance to the presence of war rumours in post-war Soviet society. 'Rumour' itself, was a rhetorically loaded term within the language of the Bolshevik regime. Spreading rumours was a criminal activity, and persistent offenders faced the threat of prosecution.[13] In practice, however, the government only punished a tiny percentage of those individuals who transferred information orally to their friends, colleagues and family. For the purposes of this chapter, I will be using the term 'rumour' to describe orally transmitted information that provides description or analysis of contemporary events. This definition of rumour is not dissimilar to what is usually considered to be 'news'. Rumour, in the Soviet context at least, is best approached neither as a medium for disinformation, nor as criminal activity, but rather as unofficial 'oral news'.

By 1945 Soviet propagandists were accustomed to presenting the outside world as a threat to the USSR. The external enemy, which had loomed large during the war scares of 1923 and 1927, became an omnipresent feature of Soviet rhetoric in the 1930s.[14] The Second World War heralded the further development of this bipolar worldview. The image of a barbarian German army, bent on destroying the people and culture of the USSR, dominated the films, plays and newspaper articles of the years 1941–1945.[15] The period from mid-1945 to mid-1947 was unusual, therefore, because the Soviet Union had no self-declared enemies. Official news reports were awash with self-confidence, trumpeting the moral authority that the USSR had accrued in the war, and the debt of gratitude that the rest of the world owed to them.[16] In November 1945 Molotov eulogised the Soviet people who had 'played a part of decisive import-ance. Peace has been won for the peoples of the entire world', before alluding to the 'further strengthening of co-operation of the Great Democratic Powers'.[17] Official reports about the First General Assembly of the United Nations and the Nuremberg Trials (December–January 1945–1946) were equally positive, echoing the tone of continued allied co-operation.[18]

The language of the Soviet media shifted by degrees throughout 1946 and early 1947, focusing its criticism on Anglo-American policy in Germany. Yet the thrust of the Soviet press remained the same: no international force threat-ened the Soviet state. Collaboration for progress was the dominant motif. Thus Stalin maintained in 1946: 'I do not believe in a real danger of a "new war".'[19]

In 1946–1947 50 per cent of all stories relating to international affairs concerned peaceful relations with other countries. Over the same period only 7 per cent of international news focused on conflict with the Western Powers.[20] Official Soviet press reports actually downplayed the growing tension between the Allies and presented a more positive image of international affairs than existed in reality.[21] It was only with the rejection of Marshall Aid in mid-1947 that the Soviet media began to emphasise the conflict between the Western Powers and the USSR. In comparison with the anti-German denunciation that preceded it, and the anti-Western vitriol of the Cold War, the message of the Soviet press in the early post-war months was remarkably moderate, lacking a clear image of the enemy, or a developed discussion of the threat of war.

Rumours, hoarding and panics

This message of international progress and collaboration failed, however, to convince significant sections of the Soviet population.[22] Rumours of a new war against the former Allies broke out repeatedly across the USSR in the months following the Nazi capitulation. The fragility of the Grand Alliance was a source of speculation even before peace had been declared in Europe. In January 1945, V. Z.,[23] a resident of Odessa, observed amongst his friends that, 'The Allies will strongly dictate their conditions to the USSR ... There will not be a victory of the Soviet Union in this war'.[24] A Komsomol instructor, following a May 1945 lecture campaign in Ukraine and Belarus, lamented the complete failure to 'explain the question about the relationship between the Soviet Union and the Allies'.[25] A former artillery officer revealed his perceptions of the state of the Alliance in answering his own question: 'Why did the Allies drop the bomb on Hiroshima? ... They wanted not to defeat the Japanese but to show us their strength. We understood it that way'.[26] A Soviet officer who had deserted in 1945 to the Americans complained bitterly that he: 'fled from the SU with an aim to take part in the freedom movement. I wanted to liberate my mother'. He had crossed the lines under the assumption that the Western Powers were on the brink of an attack against the Soviet Union.[27]

Rumours concerning the imminent outbreak of a new war were extremely widespread in the months following May 1945.[28] In September 1946 the government initiated a 'Campaign to Economise on Bread', essentially a rise in the price of cheaper rationed goods and a lowering of the prices of more expensive commercial foodstuffs.[29] These shifts in food prices were interpreted by much of the population as a pre-emptive initiative to conserve food before a new war began.[30] Comrade Karpushin, of the Kolomenskii factory in Moscow, allegedly stated at a workers' meeting: 'I have heard that war is already going on in China and Greece, where America and England are interfering. If not today then tomorrow they will invade the Soviet Union'.[31] A mechanic of 'Forward' Artel, in the city of Tarangog, went even further, allegedly, claiming that: 'On the Soviet Turkish border a war is going on. From there they are sending many wounded. They have begun the evacuation of the cattle from the Caucasus ...

This is the cause of the rise in prices for foodstuffs'.[32] The post-war months were experienced by many Soviet citizens as a period of insecurity. Agitators repeatedly listed 'Will there be a war?' as one of the most popular questions asked at meetings.[33] A former Soviet officer who served in Poland reminisced that he had been relieved when the Soviet Union built their own nuclear bomb: 'They [Britain and America] were looking for a chance to hit us at that time [1946]'.[34]

Rumours of war impacted on the behaviour of the Soviet population in the early post-war period. Hoarding, of either food or money, was very common, as individuals stockpiled supplies in preparation for a future conflict.[35] In October 1946 Fomin, a metalworker, stated amongst his colleagues that: 'The raising of prices on food in all likelihood is a result of the forthcoming war ... now we need to create reserves in order not to be caught out like in 1941'.[36] In October 1947 the Librarian of the Moscow Central Scientific Research Institute of Communications complained at a party meeting: 'You overhear in the shops, on the street that they only have to say: the cabbage has run out – it means that a war has arrived; they are not giving out sugar in the shops – there will be a war. Some are thinking of stockpiling in case of a war'.[37] A group of soldiers in East Germany in 1947 set aside a supply of petrol in preparation for flight once the war began.[38] Churchill's 'Iron Curtain' speech of March 1946 prompted a wave of withdrawals from the local banks in the Crimea.[39] In the five days following the speech, investors demanded 613,000 roubles, from the Savings Bank of Yalta, a fivefold increase in comparison with the period which preceded it. The banks of Saks *raion* (Crimea *oblast'*) ran out of money altogether and had to turn away investors who wanted to extract their savings.[40] There is some evidence that those who opposed the regime were also gathering supplies, for an uprising that would follow the invasion. Secret Police sources alleged that a nationalist group in Estonia was agitating amongst the population in 1945 to: 'gather together weapons necessary for the moment of activisation ... after the collision of the English and Americans with the Soviet Union'.[41]

Other aspects of the behaviour of Soviet citizens also revealed an implicit assumption that peace would be short lived. Senior figures within the Vlasovite movement later expressed their dismay that the Anglo-Americans had dashed their hopes of subsequent war against the Soviet Union. One of them commented: 'There were all sorts of rumours about the Western Allies, and people figured that eventually the Anglo-Saxons and the Soviets would quarrel'.[42] Other Soviet citizens due for repatriation resisted being sent home under the expectation that they would soon return as conquerors alongside an Anglo-American army.[43] Agitators in Yalta province had to convince certain farmers that they should plant wheat, instead of potatoes. The farmers had concluded that potatoes would provide a more durable crop in wartime.[44] Agitator A. I. Nikorev refused to participate in the 1946 elections because: 'When the government has changed I don't want them to say that I was a *komsomolka* and an activist'.[45] Soviet citizens demonstrated by their behaviour, in the early post-war months, that they were planning for a forthcoming war. War rumours were one aspect of this climate of expectation; enduring peace was considered less likely that a fresh outbreak of conflict.

Stories about an imminent invasion led, on some occasions, to full-scale war panics. The reaction to Churchill's speech at Fulton Missouri on 5 March 1946 was exceptional in this regard. The Soviet government's response in *Pravda*: 'Churchill is Rattling His Sabre', on the 11 March[46] precipitated a wave of rumour and hoarding that rapidly degenerated, in some rural communities, into panics and refusal to work. In an attempt to restore order Stalin gave an interview to *Pravda* on the 14 March,[47] which became the basis for thousands of rapidly organised meetings across the USSR on the 15 and 16 March.[48] The records of those meetings demonstrate just how serious the situation had become. On Kuibyshev kolkhoz, Kirov *raion*: 'Amongst the villagers they are gathering their possessions, harnessing their cows and evacuating for Tambov *oblast*''.[49] The Sevastopol *raion* party administration reported that: '... workers in secondary agriculture ... wanted to give up work and flee to the city convinced that the war had already started'. *Kolkhoznitsa* Safonova publicly abused Agitator Bondarenko declaring his words to be 'pure agitation ... you should not hide things from us, the war has already started, we don't want to remain in work'.[50] Only two of the reports use the term 'panic' itself.[51] Whilst they may have avoided the term, the reports from around the Crimea indicate that panic was exactly what had taken place. Over the next few days ten *raions* or *gorkoms* provided 11 lists of questions asked by the population at agitational meetings. The documents reveal a striking uniformity of concern. The population wanted to know whether the speech was a declaration of war, and whether President Truman had supported Churchill's words.[52] A significant number of Soviet citizens lived in expectation of an immanent conflict in the early post-war period. That they were susceptible, not only to war rumours but also war panics, indicates the degree to which the forthcoming invasion had penetrated the collective psyche of the Soviet population.

Rumour as news: a functional context

War rumours made up only a small percentage of the information that was transferred by word of mouth between Soviet citizens in this period. A significant volume of stories, jokes, anecdotes, news and gossip circulated orally in the USSR. The Soviet population relied heavily on these word-of-mouth sources as a means of obtaining information about events of local, national and international significance. The responses of Soviet émigrés to the HIP highlight the importance that the population of the USSR attached to this informal news network.[53] When asked about sources of information in a general manner: 85 per cent of the respondents referred to Soviet newspapers, 47 per cent to 'word of mouth' and 47 per cent to radio.[54] When asked, in the next question, which sources were most important to them: 36 per cent said newspapers, 28 per cent said 'word of mouth' and only 10 per cent radio.[55] Of 271 interviewees who answered which source they considered most reliable: 61 per cent cited oral information, and only 13 per cent newspapers.[56] Word-of-mouth communication provided an important constituent element of the picture that Soviet individuals built up of the world around them. War rumours were only one, albeit important, piece of information, within the context of a large volume of other oral news.

Oral information was transmitted along informal networks of close friends and family; 28 per cent of the interviewees cited family and 77 per cent friends, when asked who were their sources of informal oral communication.[57] Respondents' descriptions of the transfer of information within this unofficial news nexus were often very similar:

> People who heard it would tell it to others and they would tell it again to others and it increased in a geometric progression.[58]

> One didn't have to do much to acquire rumour news – if you had friends.[59]

> All these rumours were only told by very good friends, but as everybody has at least a few good friends these rumours spread.[60]

The extent to which oral news had contributed to respondents' information gathering was determined largely by their social origins. Peasants and unskilled workers relied much more heavily than intellectuals or white collar workers on information obtained via unofficial word-of-mouth transmission.[61] Bauer and Gleicher concluded that: 'In general the word-of-mouth network is the newspaper of the peasant'.[62] This may go some way towards explaining the fact that full-scale war panics were a largely rural phenomena. In the absence of other, instantly accessible, media against which to verify war rumours, village communities reacted by packing up their property and fleeing from the imaginary front line. Despite these social differences, respondents to the HIP expressed caution about the credibility of both officially and unofficially obtained information.[63] They attempted to draw on material from both contexts, cross referencing them against each other:

> Also . . . we could check the rumours we heard, through the press.[64]

> Even the members of the party among themselves don't believe everything that they read in the Soviet newspapers . . . Conversations with members of my family or with friends were very important.[65]

> One must find a middle way. One must search for the truth among all these different sources of information.[66]

Unofficial, orally transmitted information supplemented, rather than supplanted, the official state media.

The chaotic state of the Soviet agitation machine between 1945 and 1947, however, would have increased the reliance of the population on rumour news.[67] Reports from these years bemoan the shortage of qualified political agitators. Even the navy struggled to acquire trained report readers.[68] *Raiony* and *oblasti* often maintained theoretical lists of agitators who never, in practice, read any lectures.[69] Local agitation was characterised by storming, associated with elections, followed by months of inactivity.[70] Official agitation failed, in particular, to satisfy the widespread popular hunger for information about the outside world during these months.[71] In August 1946, the head of MOPR (The International Organisation for the Promotion of the Revolution), wrote to the Agitprop Department, of the Central Committee, begging for more materials for his local agitators to work from.[72] Furthermore, the Soviet press suffered from a lack of credibility amongst its audience.[73] Unofficial war rumours, rather than the state media machine, had forewarned the population of the Nazi invasion before 1941.[74] It was in this context of limited, and to some extent discredited, information that rumours were able to flourish. In the absence of a reliable official press, Soviet citizens were forced to rely on information acquired from private sources, bringing them into contact with war rumours in the early post-war years.

It is possible that the respondents to HIP exaggerated the extent to which orally transmitted information circulated in the USSR. The respondents were an unrepresentative, anti-regime sample of the population of the Soviet Union. Some, though not all, had actively chosen to leave their land of birth. However, the project's authors found that those respondents who were more critical of the Soviet regime were less likely to state that they had received information by word of mouth.[75] Respondents who were well disposed towards the government spoke of actively seeking out oral news as a means of staying well informed, and effectively fulfilling their responsibilities to the Soviet state. Anti-regime respondents placed greater emphasis on the risks attached to passing on oral news. They often claimed that it led to inevitable arrest and imprisonment in the USSR.[76] Bauer and Inkeles concluded, therefore, that their unrepresentative, anti-Soviet, sample had understated the ubiquity of oral communication in the USSR.[77]

As well as functioning to pass on news, the transfer of information within inter-personal networks served as a means of reinforcing social cohesion in the post-war period. Soviet society was a society in trauma after the Second World War. Reconstruction was slow and difficult. Living conditions in the cities were unsanitary and working hours long.[78] Life in the rural communities, which were afflicted by chronic shortages of machinery and manpower, was even harder. Numerous observers reported witnessing women, harnessed as horses, ploughing the fields in 1945 and 1946.[79] The struggle of reconstruction was compounded by a subsistence crisis during the famine of 1946–1947. The trauma of the post-war years generated a variety of social problems. Arkhangel'sk, for example, suffered a series of particularly brutal murders by gangs who controlled the streets at night.[80] In this context of disorder, the transfer of news by word of mouth would have served to reinforce social solidarity. Soviet citizens constituted and affirmed their intimate social worlds by passing on information to those they trusted.[81] War rumours were most common in those areas that had experienced the social stress of German occupation, and so had the greatest need to reaffirm their shared communal identities. These communities also found other ways in this period to consolidate their social unity. 'Collaborators' returning from Germany suffered summary justice at the hands of local mobs, whilst girls who had slept with German soldiers were excluded from work brigades.[82] The transfer of unofficial news within word-of-mouth networks served to reconstruct the damaged social worlds of Soviet individuals, at the same time as they laboured to rebuild the damaged infrastructure of the Soviet state.

Rumour as dissent: a subaltern interpretation

An appreciation of the difficulty of post-war living informs the context within which rumours flourished. It does not, however, account for the prominence of specifically war rumours in this period. Any attempt to go beyond a purely phenomenological description of war rumours requires an assessment of the meanings that those who passed these rumours on might have attached to them. Causal connections cannot be established between descriptions of social conditions and accounts of collective behaviour.[83] Social anxiety may have prompted dramatic rumours of social inversion in the post-war period. Why, however, did they manifest themselves in the form of war rumours? Why, in particular, were they rumours about an Anglo-American invasion of the USSR?

Rumours have largely been interpreted, in a Soviet context, as a vehicle for the articulation of anti-government sentiment. Viola, Fitzpatrick, Davies and others have drawn attention to the employment of rumour as a language of dissent in the 1930s.[84] This 'subaltern' interpretation provides important insights into the meanings Soviet citizens attached to post-war war rumours. An invasion of the USSR held out the promise of a social and political transformation of Soviet society. As Nina Velikova of Crimea *raion* allegedly stated in 1946: 'It is necessary for there to be a war ... Do you understand that if there is a war there will be an exchange of power?'[85] War rumours encapsulated the hopes of those opposed to the Soviet regime. They were the very antithesis of government propaganda, which emphasised the authority and stability of the USSR within the international community. As such they became the linguistic weapon of choice for anti-Soviet activists in the post-war period.

This subversive function of war rumours made them a vital element of the vocabulary of dissent employed by nationalist, religious and anti-kolkhoz groups, opposed to the regime. Slegushkina, a nationalist *kolkhoznitsa* of Starobel'skii *raion* allegedly spread the rumour amongst her colleagues that: 'The government is saying that Ukraine did not want to fight with Germany ... and now the government prefers the Russians. A war is inevitable, indeed without it it will be impossible to live'.[86] In July 1947 Teslenko, an anti-kolkhoz repatriate, claimed that an American officer had told her: 'When the talks with the Soviet Union were agreed about help during the war, then it was stated that at the end of the war the kolkhozes would be abolished, and the communist party too ... and if the Soviets break out of this agreement then the Soviet Union would be wiped off the planet'.[87] Meanwhile, in 1947, the Uniate believers of Bulkhovtsy village gathered systematically to pray: 'in order to spoil the elections and so that the Anglo-Americans would arrive quickly'.[88] It seems possible that the apocalyptic language of religious protest, identified by Viola in the 1930s, had been supplanted by a more earthly day of reckoning for the Soviet government.[89] War rumours functioned as a language of dissent for the disaffected in the post-war period. The USSR's former allies were not threatening to invade on account of the Soviet government's national, agricultural or religious policies in 1946. For those who opposed the Soviet government, however, war rumours offered the promise of liberation. They were a powerful linguistic shorthand for the dream of social transformation.

This 'subaltern' analysis of war rumours provides an explanation of the particularly potent meaning they enjoyed amongst those who hoped for the fall of the Soviet regime. However, war rumours were extremely widespread in the post-war period. They were a feature of both urban and rural communities.[90] They were also current amongst peasant, worker and intellectual groups.[91] A 'subaltern' reading of rumours seems credible in the context of the Soviet village in the 1930s, but less compelling between 1945 and 1947, as the regime basked in the glory of its greatest triumph. War rumours also impacted on the behaviour of individuals who clearly identified themselves with the government. Rudskii, the Vice-Director of Rovenskii Oil Production base despaired in August 1947 that: 'It is clear to all that there will soon be a war ... The population do not support us ... The end is inevitably coming to us, we are destroyed'.[92] Army Colonel Chulkov wrote to the *vozhd'*, in 1947, pointing out to him the importance of anti-aircraft defence in a nuclear context. Whilst his encouragement to increase Soviet artillery defences made no explicit mention of the Anglo-Americans, it was clearly their bombers that loomed in his concerns.[93] War rumours were too widespread and taken too seriously, in the post-war period, to have been purely an expression of anti-regime sentiment.

Rumouring was, of course, illegal in the Soviet Union. But to conclude that the origin and transmission of war rumours was exclusively associated with individuals of anti-government inclination would be to assess them solely within the categories of the official documents themselves. Official opinion reports (*svodki*) routinely laid the blame for war rumours at the feet of repatriates, sect members, nationalist bands and foreigners.[94] Ascribing rumour to these 'suspect' identities provided a vehicle for describing the rumours circulating in the community, whilst at the same time attaching them to groups who were expected to harbour dissent, within the logic of the regime. War rumours were more common in the 'Wild West' of the Soviet borderlands, where nationalist partisans continued to battle the NKVD until the late 1940s. Their scope and credibility, however, extended beyond the language of subversion.

A purely 'subaltern' interpretation also runs the risk of extracting these rumours from the informational context within which they functioned. Soviet citizens seem not to have drawn a distinction, unlike the Secret Police, between subversive rumours and regular oral information. In the written questionnaires associated with the HIP, the respondents were asked how frequently they heard 'rumours' and how frequently they engaged in 'discussion with friends'. The results were almost identical.[95] Despite the efforts of the regime to demonise rumour spreading, Soviet citizens did not differentiate between 'news' acquired via informal discussions and rumours. There seems to be no logical reason for assuming that war rumours were regarded differently. War rumours flourished and were transmitted within the oral news network as information. They survived, within the Soviet era archives, because the state considered them to be socially corrosive comments.[96] Yet it seems unlikely that the women of Saks *raion*, Crimea *oblast'*, who began mourning the fact that they would never see their sons and husbands again in February 1946 were engaging in anti-party discourse.[97] They were simply convinced of the credibility of the rumour that the USSR was, once again, either at or on the brink of war.

Rumour as 'collective problem solving'[98]: a thick descriptive interpretation

Thick descriptive interpretation attempts to describe the imaginative universe out of which social behaviour grows.[99] Rumours are, 'a collective form of problem solving'.[100] They are the product of a group of individuals' attempt to make sense of the world around them. As corporately generated phenomena, rumours must fit within the plausibility structure of the community within which they circulate. Shibutani talks of 'natural selection' as a decisive factor in the spread of rumour. Rumours must be credible to those who transmit them, in order to be passed on.[101] This is the interpretative approach employed by Lefebvre in his classic work on panic in Revolutionary France. The 'aristocratic plot' and the 'brigands' were important because they were perceived to be real by the French population, rather than because of the actual threat they posed.[102] They loomed large within the collective imagination of the population, becoming the focus of first rumour and then panic. Post-war war rumours did not 'succeed' purely on account of their capacity to subvert. They were transmitted as news, and as such, must have made sense to those who passed them on. A full account of the prominence of war rumours in post-war Soviet society leads us, cautiously, into an exploration of how Soviet citizens: 'constructed the world [and] invested it with meaning'.[103]

War rumours were not unique to the months after the end of the Second World War. Whilst they were more prominent in these months, they were a feature of the entire Stalin period, often breaking out in association with increased official rhetoric about the threat of war.[104] The content of these rumours, over time, reveals something of how Soviet citizens understood the relationship between the USSR and the outside world. Davies refers to an outbreak of war rumours, sparked by the expulsion of the Zinovievites from Leningrad in 1935.[105] This domestic political event was interpreted – within the oral news network – as a potential catalyst for invasion. Many war rumours in the post-war period also reflected this implicit assumption that internal Soviet political events might precipitate an assault against the USSR. In early 1946, the rumour circulated in western Belarus that, by striking through the ballot cards, voters would cause the Anglo-Americans to apply pressure for the restoration of the old borders of Poland.[106] An entirely domestic act of protest was assumed to have reverberations on an international stage. The claim that the Allies were threatening invasion if the regime did not abolish the kolkhozes relied on this same assumption: governments beyond the USSR were deeply concerned about Soviet domestic issues. In a similar vein, respondents to the HIP stated that Soviet elections and self-criticism were undertaken purely for the consumption of an external audience.[107] At least some Soviet citizens operated under the assumption that their domestic lives were of profound interest to the world outside.

The presumption that other states were deeply concerned about events within the Soviet Union grew out of a way of thinking about international relations that emphasised the centrality of the USSR within the world community.[108] The Soviet Union was, indeed, an extremely important state after the Second World War. Yet the popular emphasis on Soviet centrality seems to have been neither purely a reflection of national pride, nor a sober assessment of geopolitical realities. The Soviet government encouraged its population to consider themselves as builders of a new and unique Socialist civilisation.[109] That the USSR had 'saved' European civilisation during the Second World War led to the further embellishment of this language of Soviet exceptionalism. War rumours in the post-war period reflected this assumption of Soviet centrality. That other governments were interested in Soviet domestic events, or that their behaviour was conditioned by their relations with the USSR seems to have been a conviction of both those who supported and those who opposed the regime.[110] An attack against the Soviet Union in the post-war period was considered feasible by many Soviet citizens because they regarded their state as a vitally important entity within the international system. British agricultural workers or miners, in this period, did not anticipate an external invasion to defend their personal interests, or wages. Much of the Soviet population, however, seem to have considered their lives to be part of an international drama, in which their state was playing a leading role. It was this way of thinking about international relations that lent credibility to the idea of an external invasion, and enabled war rumours to survive and proliferate within the oral news network.

War rumours also flourished within the context of the Soviet experience of the wartime Grand Alliance as a betrayal. Rumours that the former Allies might invade the USSR after 1945 were credible because of the popular perception that they had been inconstant and unfaithful during the war itself. At an August 1942 gathering of agitators in the city of Arkhangel'sk various agitators complained that they were unable to answer the population's incessant questions about the Second Front. When discussing the popular response to the German advance that summer the chair (Mitin) asked:

> Are they asking why we are retreating?
> Tarasova: They are asking questions about the Second Front; why is there no Second Front? I explained to them that in the near future there must be a Second Front.[111]

The Second Front, or more precisely the Allied failure to open it, became an object of fixation for the Soviet population in the months before June 1944. Soviet citizens assumed that an allied Second Front, in mainland Europe, was a precondition of a Soviet victory. It also had symbolic significance as an expression of Allied sincerity and solidarity with the Soviet war effort. The Second Front in Europe would be the consummation of the Grand Alliance marriage. Disappointment quickly led to cynicism. At the gathering of party activists in Lenin *raion*, Moscow, this written question was submitted: 'When, and in which month will the Second Front be opened? Maybe on the 31st of December 1942 with 15 soldiers and a beaten up tank?' (May 1942).[112] A Senior Sergeant wrote in his diary in late 1943, 'Now the Allies will not wriggle out of their need to open the Second Front ... They are breaking their one promise. It is not Allied behaviour'.[113] Official rhetoric during the war, whilst being somewhat cool towards the Allies, did not openly endorse this view. The 'betrayal' of the Grand Alliance was much more developed within the popular *mentalité* than it was in the official media by 1945. The idea that the former Allies might invade was credible in the post-war period because much of the population did not feel they had been authentic Allies during the war itself.

The predominant image of the Allies during the war seems to have been that they were manipulating the Soviet government, rather than contributing to the war effort. The dissolution of the Comintern in May 1943, coming at the height of frustration about the Second Front, was largely interpreted as a concession (*ustupka*) under the influence (*davlenie*) of the Anglo-Americans.[114] As a report from Gorkii *oblast'* stated: 'A number of questions have been given by the workers ... the majority of which come down to one thing "Isn't the dissolution of the Comintern connected with the demands of our Allies?"'[115] Various Soviet citizens, particularly intellectual groups, greeted the 'liberalisation' of the war years as hopeful signs of progress.[116] Whether they perceived these steps to be for or against the interests of the USSR, however, Soviet citizens considered it credible that the Allies were extracting concessions from the Soviet government, in return for the support they were offering the USSR. The Grand Alliance was understood by the majority of Soviet citizens to be a fragile relationship, characterised by broken promises and manipulation. It was within this conceptual framework that post-war invasion rumours found their origin, and were perpetuated. Rumours of war survived because they were credible. They resonated with the collective understanding of how the world worked, amongst both those who supported and opposed the Soviet regime.

By the end of the Second World War the Soviet regime had lost the capacity to control the manner in which many of its subjects perceived international affairs. Between 1945 and 1947 the state propaganda machine mirrored the chaos that marked the lives of many Soviet citizens. It was also suffering from a shortage of credibility. In this context, war rumours functioned as far more than subversive tales. They functioned as news for many Soviet citizens, to whom the idea of an Anglo-American invasion was entirely credible. They also functioned to reaffirm social unity in the post-war period. They reconstituted relational networks, and assisted in the process of post-war social healing. War rumours were both a product of, and a palliative, to social turmoil in this early post-war window of disorder.

Notes

1 My thanks to the Arts and Humanities Research Council for providing the funding that made this research possible.
2 Gosudarstvennyi Arkhiv v Avtonomnoi Respublike Krym, henceforth GAARK f. 1, op. 1, d. 2414, ll. 67–129.
3 E. Durkheim, *The Rules of Sociological Method*, Glencoe: Illinois, 1938, p. 95.
4 L. Viola, *Peasant Rebels Under Stalin: Collectivisation and the Culture of Peasant Resistance*, Oxford: OUP, 1996; S. Davies, *Popular Opinion in Stalin's Russia: Terror Propaganda and Dissent, 1934–41*, Cambridge: CUP, 1997.
5 C. Geertz, *The Interpretation of Cultures: Selected Essays*, Guernsey: Fontana 1993.
6 For a refutation of the value of comparing various source groups see: J. Hellbeck, *Kritika* 1.2 2000, pp. 439–440.
7 Henceforth Pr. For a discussion see: M. Edele, *A 'Generation of Victors?' Soviet Second World War Veterans from Demobilisation to Organisation 1941–56*, PhD Dissertation: University of Chicago, 2004, pp. 442–450.
8 Henceforth Let. For a discussion see S. Fitzpatrick, 'Public Letter-Writing in Soviet Russia in the 1930's', *Slavic Review* 55.1, 1996, pp. 78–105.
9 Henceforth Sv. For examples of their use see: Davies, *Popular Opinion*, and S. Fitzpatrick, *Stalin's Peasants: Resistance and Survival in the Russian Village after Collectivisation*, Oxford: OUP, 1994. For a statement of the position against employing *svodki* see: S. Kotkin, *Europe-Asia Studies* 50.4, 1998, pp. 739–742. Both the *svodki* and the State Prosecution files relied on material provided by informants, and fulfilled institutional functions within the Soviet system to assess the popular mood, or to secure a prosecution. As such I consider them to be the most problematic of my source groups. I will draw on these sources, however, as part of a composite picture, whilst challenging some of their assumptions about what constituted anti-regime activity.
10 Henceforth Inf. This refers to party generated sources that did not rely on Secret Police material. They roughly correspond to what Jones terms the 'Private Transcript' of party reports. J. Jones, '"People Without Definite Occupation": The Illegal Economy and "Speculators" in Rostov-on-Don 1943–1948', in D. Raleigh, ed., *Provincial Landscapes: Local Dimensions of Soviet Power 1917–1953*, Pittsburgh: UPP, 2001.
11 Henceforth HIP. A collection of interviews conducted in 1950–1951 with Soviet émigrés in West Germany and the USA. See: A. Inkeles and R. A. Bauer, *The Soviet Citizen: Daily Life in a Totalitarian Society*, Cambridge, Massachusetts: HUP, 1959. Quotations from the HIP are from the notes made by the interviewers at the time. They are not verbatim records of the statements of the interviewees.
12 Henceforth Int. It is recognised that interviews of this kind are particularly susceptible to post-rationalisation.
   Archival references that do not allude to rumour content will not be categorised under this abbreviation system.
13 Those who spread rumours were prosecuted under Article 58-10 of the Soviet Criminal Code, for anti-Soviet agitation.
14 J. Brooks, *Thank You, Comrade Stalin! Soviet Public Culture from Revolution to Cold War*, Princeton: PUP, 2000, pp. 38–42.
15 A. K. Pistotis, 'Images of Hate in the Art of War', in R. Stites, ed., *Culture and Entertainment in Wartime Russia*, Indianapolis: IUP, 1995, pp. 141–156. Brooks,

Thank You, p. 170. A. V. Fateev, *Obraz vraga v sovetskoi propagande, 1945–54*, Moscow: Rossiiskaya Akademiya Nauk 1999, p. 17. P. Kenez, *Cinema and Soviet Society. From the Revolution to the Death of Stalin*, London: Tauris, 2001, p. 178.

16 Brooks, *Thank You*, pp. 206–207.

17 *V. M. Molotov Speaks on the 28th Anniversary of the Great October Socialist Revolution. Celebration Meeting of the Moscow Soviet, November 6, 1945*, London: Soviet News, 1946, pp. 5, 17.

18 Fateev, *Obraz vraga*, pp. 36–39.

19 *J. V. Stalin on Post-War International Relations. Full Text of Interview to Press Correspondents and Exchange of Messages 1946–7*, London: Soviet News, 1947, p. 11.

20 Brooks, *Thank You*, pp. 207–8.

21 Fateev, *Obraz vraga*, pp. 37–39.

22 On the grounds of limitation, this essay considers only European Russia, Ukraine, Belarus and the Baltic States.

23 I will omit the full name of the defendants, when citing State Prosecution files.

24 Pr. Gosudarstvennyi Arkhiv Rossiskoi Federatsii, henceforth GARF f. 8131, op. 37, d. 3177 Ж, l. 1.

25 Inf. Rossiiski Gosudarstvennyi Arkhiv Sotsial'no-Politihceskoi Istorii, henceforth RGASPI M-f 1, op. 32, d. 304, l. 14.

26 Int. Andrei Ivanov, Moscow, May 2004.

27 HIP B9, 470, 14. (B schedule Interview subject 9, respondent 470, page 14. Davis Centre Library, Harvard University.)

28 The period under study extends from May 1945 until summer 1947. There is evidence that war rumours continued into the late 1940s (E. Zubkova, Trans., H. Ragsdale, *Russia After the War: Hopes, Illusions, and Disappointments, 1945–1957*, London: M. E. Sharpe, 1998). What is unusual about the rumours of this period is that they were not a product of official hysteria in the government press, but that they existed in direct contradiction with the official line.

29 For details see: Zubkova, *Russia After the War*, p. 41.

30 Sv. Rumours of war are mentioned in reports from: Moscow, the Crimea, Vologda, Ivanov, Novgorod, Pskov, Rostov, Leningrad, Kiev and Estonia. RGASPI f. 17, op. 88, d. 705, ll. 1–137; op. 125, d. 425, ll. 1–53; op. 122, d. 188, ll. 9–29.

31 Sv. RGASPI f. 17, op. 125, d. 425, l. 4.

32 Sv. RGASPI f. 17, op. 88, d. 705, l. 137.

33 Inf. RGASPI f. 17, op. 88, d. 705, l. 73; op. 122, d. 122, l. 37; GAARK f. 1. op. 1, d 2550, l. 7; Gosudarstvennyi Arkhiv Obshchestvenno-Politicheskikh Dvizhenii I Formirovanii Arkhangel'skoi Oblasti, henceforth GAOPDFAO f. 296, op. 2, d. 302, ll. 138, 149, 169.

34 Int. Ilyan Lvoevich, Moscow May 2004.

35 As Julie Hessler points out, hoarding was a common Soviet response to a subsistence crisis such as the harvest failures of 1946–1947: J. Hessler, *A Social History of Soviet Trade: Trade Policy, Retail Practices, and Consumption, 1917–1953*, Princeton: PUP, 2004, p. 11. What is notable, however, is that the post-war hoarders often described their hoarding as preparation for an imminent invasion.

36 Sv. RGASPI f. 17, op. 125, d. 425, l. 4.

37 Inf. RGASPI f. 17, op. 122, d. 289, l. 47.

38 HIP A. 17, 331, 12. (A-Schedule interview, Volume 17, respondent number 331, page 12.)

39 For a discussion of the official response to the speech see p. 8.

40 Inf. GAARK f. 1, op. 1, d. 2550, ll. 13–14, 40.

41 Sv. RGASPI f. 17, op. 122, d. 94, l. 137.

42 HIP B6, 182, 11. Also B6, 382, 24–26. (The opportunism of the Vlasovites, in desperate circumstances, is recognised. Their bitter subsequent reflections on their

naiveté, however, lend some credibility to the claim that they considered collaboration with the Allies to be a genuine possibility.)

43 Inf. GARF f. PO526, op. 1, d. 90, ll. 55 and 111. (These individuals may have stayed out of fear of the consequences of returning. The idea of a future war must have had some credibility amongst them, however, to serve as their declared pretext for staying.)
44 Inf. GAARK f. 1, op. 1, d. 2550, l, 13.
45 Sv. RGASPI f. 17, op. 122, d. 183, l. 36.
46 The article emphasised Churchill's failure to gather support for a racially motivated war. *Pravda*, 11/3/46, p. 1.
47 Stalin denounced Churchill's speech as: 'A dangerous act designed to sow the seed of division amongst the Allied states and complicate their collaboration'. *Pravda*, 14/3/46, p. 1.
48 See: GAARK, f. 1, op. 1, d. 2550, l. 19.
49 Inf. GAARK f. 1, op. 1, d. 2550, l. 15.
50 Inf. GAARK f. 1, op. 1, d. 2550, l. 5.
51 Inf. For use of 'panic' see: GAARK, f. 1, op. 1, d. 2550, ll. 13 (Yalta) and 44 (Soviet *raikom*). 'Panic' was an negatively loaded word in the USSR implying weak mindedness and potential subversion.
52 Inf. GAARK, f. 1, op. 1, d. 2550, ll. 7, 17, 18, 21, 23, 27, 32, 33, 36, 40, 42. As one report noted: 'At nearly every meeting the question was offered whether the speech of Churchill in Fulton was leading to a new war?', l. 26.
53 The majority of the interviews were conducted with individuals who left the USSR before 1945. There seems to be no reason to assume that oral communication became less important in the post-war period.
54 HIP, *Code Book A*, Harvard University: Unpublished, Davis Centre Library, p. 57. The percentages refer to a total of 329 cases.
55 Ibid. pp. 57–58.
56 Ibid. p. 80.
57 Ibid. p. 60.
58 HIP A. 3, 25, 10.
59 HIP A. 31, 1011, 53.
60 HIP A. 32, 1108, 28.
61 Inkeles and Bauer, *The Soviet Citizen*, pp. 165–167. Intellectuals and city dwellers enjoyed greater access to all sources of information, including oral news. Word-of-mouth media played a smaller role in their overall information gathering, however.
62 R. A. Bauer and D. B. Gleicher, 'Word-of-Mouth Communication' in the Soviet Union', *The Public Opinion Quarterly*, 17.3, 1953, p. 305.
63 See: HIP A. 1, 5, 44 and 12, 153, 46 for suspicion of official media; and HIP A. 32, 1091, 35, and 32, 1108, 28 for suspicion of unofficial media.
64 HIP A. 3, 26, 65.
65 HIP A. 1, 8, 74.
66 HIP A. 6, 64, 31.
67 Shibutani claims that rumour is more likely to function as 'a substitute for news' in extreme circumstances. T. Shibutani, *Improvised News: A Sociological Study of Rumour*, New York: Bobbs-Merrill Company, 1966, p. 62.
68 RGASPI f. 17, op. 125, d. 405, l. 30.
69 For example: RGASPI f. 17, op 125, d. 125, l. 15.
70 GAOPDFAO f. 296, op. 2, d. 398, l. 86.
71 See: GAAO f. 5790, op. 3, d. 30, l. 5. In 1946 the *Oblast'* Lecture Buro read only 16.5 per cent of their lectures on political themes, 29 per cent were on literature and 44 per cent on agriculture or medicine. For evidence of widespread interest in foreign affairs see: Zubkova, *Russia After the War*, p. 87. She also attributes this to war anxiety.

72 RGASPI f. 17, op. 125, d. 403, l. 53.
73 See footnote 49, p. 9.
74 The Nazi-Soviet Pact was widely regarded with cynicism in the USSR. HIP A. 4, 31, 38; A. 30, 643, 26. For material about pre-war war rumours see: S. Smith, 'Heavenly Letters and Tales of the Forest: 'Superstition' against Bolshevism', *Antropologicheskii Forum* 1.2 (2005).
75 Interviewers recorded their assessment of how anti-Soviet the respondents were after each meeting.
76 E.g. HIP A. 17, 333, 21.
77 See: Bauer and Gleicher, 'Word-of-Mouth', pp. 301–306; Inkeles and Bauer, *Soviet Citizen*, pp. 161–171.
78 See: D. Filtzer, *Soviet Workers and Late-Stalinism. Labour and the Restoration of the Stalinist System After World War II*, Cambridge: CUP, 2002.
79 HIP B2, 61, 1; A. 30, 641, 42.
80 RGASPI f. 17, op. 122, d. 118, l. 45. See: Zubkova, *Russia After the War,* pp. 38–39, for the same phenomenon in Saratov.
81 Shibutani also points to the socially cohesive function of rumour. Shibutani, *Improvised News*, pp. 22–23.
82 RGASPI f. 82, op. 2, d. 148, ll. 68–70.
83 See Darnton's critique of the historiography of Mentalité: R. Darnton, *The Great Cat Massacre and Other Episodes in French Cultural History*, London: Penguin, 1984, pp. 258–260.
84 Viola, *Peasant Rebels*; Fitzpatrick, *Stalin's Peasants*; Davies, *Popular Opinion*. Viola, in particular, draws on Scott's work on 'subaltern' discourse: J. Scott, *Domination and the Arts of Resistance: Hidden Transcripts*, London: Yale University Press, 1990.
85 Sv. GAARK f. 1, op. 1, d. 2550, l. 25.
86 Sv. RGASPI f. 17, op. 125, d. 517, l. 36.
87 Sv. RGASPI f. 17, op. 125, d. 517, l. 37. The rumour that the *kolkhozy* would be abolished was rife at the end of the war. See: HIP A. 9, 121, 15, and HIP B7, 30, 9.
88 Sv. RGASPI f. 17, op. 125, d. 507, l. 268.
89 Viola, *Peasant Rebels*, pp. 45–61; Smith, 'Heavenly Letters'.
90 War rumours affected 'every kolkhoz' in Bakhchisarai *raion* in February 1946: Inf. GAARK f. 1, op. 1, d. 2550, l. 19. For urban rumours see: Sv. RGASPI f. 125, op. 425, l. 4.
91 Sv. RGASPI f. 17, op. 88, d. 693, l. 2 (worker); Sv. f. 17, op. 125, d. 425, l. 39 (peasant); Int. Ilyan Lvoevich, May 2004 (intellectual).
92 Sv. RGASPI f. 17, op. 122, d, 289, l. 62.
93 Let. RGASPI f. 588, op. 11, d. 896, l. 118.
94 Foreigners: Sv. GAARK f. 1, op. 1, d. 2550, l. 38. Otherwise see above.
95 Bauer and Gleicher, 'Word-Of-Mouth', p. 301. (In the same group of respondents, 27 per cent said that they 'frequently' engaged in rumouring, and 28 per cent in discussions with friends; 28 per cent and 22 per cent said they never engaged in these activities, respectively.)
96 Fitzpatrick, *Stalin's Peasants*, p. 327; Davies, *Popular Opinion*, pp. 10–14.
97 Inf. GAARK f. 1, op. 1, d. 2550, l. 40.
98 Shibutani, *Improvised News*, p. 17.
99 Geertz, Interpretation of Cultures, pp. 3–30.
100 Shibutani, *Improvised News*, p. 17.
101 Ibid. p. 182: 'Ideas are perpetuated as long as they continue to be selected'.
102 G. Lefebvre, Trans., J. White, *The Great Fear of 1789: Rural Panic in Revolutionary France*, London: NLB, 1973, pp. 75–77, 128–131.
103 Darnton, *Great Cat Massacre*, p. 3.
104 See: Viola, *Peasant Rebels*, pp. 57–58.

105 Davies, *Popular Opinion*, p. 94.
106 Inf. RGASPI f. 17, op. 88, d. 693, l. 21. Fitzpatrick records a similar rumour associ-
     ated with the 1930 census. Fitzpatrick, *Stalin's Peasants*, p. 295.
107 HIP A. 1, 8, 13; A. 2, 18, 30.
108 It may also reflect the assumption that other states would mimic the Soviet policy of
     interfering in domestic life overseas, as the USSR had via the Comintern in the
     1930s.
109 S. Kotkin, *Magnetic Mountain: Stalinism as a Civilisation*, London: University of
     California Press, 1995.
110 The widespread popularity of this idea was a great success of the official propaganda
     machine.
111 Inf. GAOPDRAO f. 834, op. 2, d. 203, l. 40b.
112 Sv. RGASPI f. 17, op. 125, d. 82, l. 14.
113 Vasili Ermolenko, *Voennyi dnevnik starshego serzhanta*, Belgrade: 2000, p. 29.
114 Werth describes it as a 'great ideological concession'. A. Werth, *Russia At War
     1941–1945*, London: Barrie and Rockliff, 1964, p. 674.
115 Sv. RGASPI f. 17, op. 88, d. 594, l. 14.
116 HIP B11, 64, 54. See also: A. Werth, *Russia: The Post-War Years*, p. viii.

Part II

Barracks, queues and private plots

Post-war urban and rural landscapes

4 Standard of living versus quality of life

Struggling with the urban environment in Russia during the early years of post-war reconstruction[1]

Donald Filtzer

By any measure the early post-war years were ones of unremitting hardship for the Soviet population. Analysis of this hardship, however, has to be set within the long-term context of demographic shocks and deprivation which punctuated all of Soviet, and indeed immediate pre-Soviet, history up until the late 1940s: the famine of 1891–1892; the First World War; the Civil War and the famines of 1918–1922; the famine of 1932–1933; the Stalinist Terror of 1936–1938; the Second World War; and the famine of 1946–1947. Barely had the population begun to recover from the trauma of one catastrophic event than it was beset by another.

We tend to think of the pre-war Soviet famines as largely rural calamities, but this was not necessarily so. The great hunger of the Civil War years affected the towns before the countryside, and even the famine of 1932–1933 was preceded by mass hunger among urban workers, with death rates in the towns of the Lower Volga reaching a staggering 475 deaths per 10,000 population in the spring of 1932, several months before famine decimated the countryside of Ukraine.[2] Even though food supplies and real wages recovered after this point, even in 1937, the best Stalinist pre-war year in terms of urban living standards, real wages were still a quarter to a third below what they had been in 1928, the last year of NEP.[3]

During the Second World War starvation became part of everyday life not just in besieged Leningrad, but also in the towns of the Soviet hinterland. In Moscow the crude death rate leapt from 156 per 10,000 population in 1940, to 276 per 10,000 in 1942, before receding to 218 per 10,000 in 1943. In Sverdlovsk, the respective figures were 207 deaths per 10,000 in 1940, 266 in 1942, and 243.4 in 1943.[4] Yet even this masks the true impact of wartime deprivation on the adult population. If, by way of illustration, we recalculate the Sverdlovsk mortality figures to measure only deaths of adults and children older than one year, we see that older age groups, who in normal times could have expected to survive into middle age, had now become extremely vulnerable. Non-infant deaths surged from 131 per 10,000 in 1940, to 215 per 10,000 in

1942, 226 in 1943, and 230 in 1944, a rise of 75 per cent, or 2.5 times the increase in the death rate for the population as a whole.[5]

When the Second World War ended, this mass hunger gave way to a modest recovery in consumption during 1944 and 1945. The breathing space this seemed to give the population was to prove short-lived. In 1946 a severe drought in the Western regions caused a harvest failure.[6] The Stalinist leadership could have responded to the bad harvest by releasing food from its reserves; instead, it chose to maintain these reserves and even exported grain abroad. It imposed dangerously high procurements on the collective farms, leaving many with barely enough seed for the next year's planting. In towns it curtailed consumption, first by raising the prices of rationed foodstuffs, and then by removing some 25 million workers, clerical employees and their dependents from the ranks of those entitled to use the rationing system. A famine resulted, in which somewhere between one million and 1.5 million people perished. Relative to the size of the local population, the death toll was highest in the harvest failure regions of Moldavia and Ukraine, but mortality rose right across the USSR.[7] The demographic recovery, which had started in 1944, was again thrown into reverse. A crude calculation of general mortality suggests that deaths per 10,000 population during 1947 rose by about 40 per cent in Ivanovo, 25 per cent in Kuibyshev and around 7 per cent in Moscow.[8] Infant mortality – always a barometer of food crisis – in the urban areas of the RSFSR went from around 9 per cent of live births in 1946 to over 15 per cent in 1947, a rise of 67 per cent. In cities like Ivanovo, Gor'kii, Sverdlovsk, Yaroslavl' and Zlatoust one out of every five newborn babies died in their first year.[9]

Mortality, however, tells only part of the story. Unlike the famine of 1932–1933, the impact of the post-war food crisis fell heaviest on the towns. Medical reports and household consumption surveys from this period all suggest that, outside of the famine's epicentre, peasant families in the RSFSR were better able to maintain nutritional intake than were the families of workers – a fact which strongly suggests that the higher recorded urban mortality was not a statistical artefact due to alleged rural under-reporting.[10] Average per capita calorie intake in workers' families during the first half of 1947 fell to just over 2,100 calories per day in Moscow and less than 1,800 calories in the towns of Moscow *oblast'*, but exceeded 2,400 calories per day among Moscow *oblast'* peasants. This same nutritional advantage is seen in other regions of the RSFSR for which we have comparative worker and peasant data: Gor'kii and Gor'kii *oblast'*, Sverdlovsk and Sverdlovsk *oblast'*, Molotov and Molotov *oblast'*, Tatariya and Bashkiriya. In fact, during January to June 1947, only workers in Moscow and Sverdlovsk cities and in Kemerovo *oblast'* managed to keep per capita daily consumption above 2,000 calories. On the whole, workers' families in the non-famine regions of the RSFSR were consuming only about two-thirds the number of calories which Soviet nutritional standards said they needed. As we would imagine, the diet was also poor in basic nutrients. Workers and peasants alike derived the overwhelming bulk of calories and proteins from starch, that is, bread and potatoes. Workers relied more heavily on bread; peasant families survived mainly thanks to potatoes. There were few sources of animal proteins, and limited consumption of fruits and vegetables. The one great dietary advantage that peasants had over workers was milk. Milk consumption among workers, most notably outside of the Urals and Western Siberia, was minimal, a fact which greatly contributed to the surge of infant mortality in 1947.[11] By early 1948 famine conditions had eased in most of the Soviet Union, but for the rest of Stalin's lifetime and well into the Khrushchev period the urban population remained badly nourished. People did not starve, and protein intake soon approached internationally recommended limits, but the calorie content and the nutritional balance of the diet remained inadequate.[12]

Food consumption may have been the most immediate and most important manifestation of the post-war crisis, but it was not the only factor affecting the short- and long-term well being of the population. Clothing and footwear were almost as hard to come by as food. The average member of a Moscow worker's family, far and away the best provisioned in the country, was able to purchase just 0.7 pair of leather footwear during all of 1947, and just 3.5 metres of loose cloth.[13] The entire USSR clothing industry in 1946 produced a mere one-quarter of a piece of underwear and less than one pair of socks for each Soviet citizen.[14]

This is already a bleak picture of life in this period, but it is far from complete. We need to know something about the larger environment which shaped how people lived, an environment which was largely independent of how much or how little people earned, how they acquired food, or how many months or years they had to wait to buy a set of underwear, a coat or a pair of shoes. The rest of this article will look at two of the most important determinants of that environment: the state of the urban housing stock and urban sanitation. Studying living conditions in this way – what we might call the 'quality of life', as distinct from the standard of living in the more narrow sense this term is generally understood – helps shed light on a number of aspects of workers' experience after the war. First, we shall see that even if wages had been higher and the shops better supplied, life for the average working class family would have remained a struggle, demanding inordinate expenditures of time and physical and emotional energy on such basic tasks as keeping clean, ensuring access to water for drinking and cooking, and maintaining minimal levels of health in cities which, for the most part, were 'sanitation-free zones'. Second, the state of the urban environment helps us to understand other aspects of the period, such as relatively high rates of tuberculosis – the main cause of death among men under the age of 50 and women younger than 40[15] – or the generally higher incidence of infant mortality in towns compared with the countryside. Finally, this discussion may help future researchers to shed light on some of the public health problems that began to beset the Soviet Union only a few decades later, most notably, the decline in life expectancy observed towards the end of the Brezhnev period.

The discussion draws largely, although not exclusively, on the annual reports of the city and *oblast'* offices of the State Sanitary Inspectorate (*Gosudarstvennaya Sanitarnaya Inspektsiya*, or GSI). These were copious documents which described the state of local housing, sanitation, water and air pollution, conditions in children's homes and schools, the health and working conditions of teenage workers, factory safety, the state of medical facilities, cleanliness (or the lack thereof) in food processing plants, local markets and public catering, and a range of related public and environmental health issues. Because they were filed every year, the reports allow us to trace each region's urban environment longitudinally over time, and to compare the conditions in different localities. I have deliberately focused here on regions of the RSFSR which were not under German occupation during the war and which, therefore, did not suffer major devastation (the one partial exception is Moscow *oblast'*). There are extremely valuable and detailed reports for Leningrad, Ukraine, Stalingrad, Rostov-on-Don and other parts of the USSR which saw heavy fighting, but problems of reconstruction there were manifestly different from those which faced the cities of the Soviet hinterland. This was true physically, as whole cities had to be rebuilt, and damaged services and infrastructure restored; and it was true politically, as local political elites had to be reinstated and the authority of both local institutions and the centre had to be reasserted. In the unoccupied territories the war also had a profound effect, but of a different kind. Massive in-migration of refugees and mobilized workers put tremendous pressure on already inadequate infrastructures, which deteriorated further through wartime neglect and lack of investment. Yet these phenomena accentuated systemic problems that already existed before the German invasion and would have persisted even if the war had not occurred. The war may have somewhat distorted the situation, but on the whole the problems we discover there cannot be attributed to the extraordinary circumstances of wartime destruction. They are essentially Soviet problems, and tell us a great deal about the attitude of the Stalinist elite towards its citizens.

The housing stock

Not surprisingly the post-war years saw a generalized housing crisis throughout the USSR. An already inadequate urban housing stock of some 270 million square metres in 1940, fell to just 200 million square metres during the war – partly because of destruction, and partly because of a failure to replace dilapidated dwellings. By 1950 the total stock had risen to 513 million square metres, around 90 per cent more than in 1940, but because the urban population had grown even faster, average living space per resident had actually fallen by around 8 per cent, from 5.1 square metres per person to 4.67.[16] Conditions were sufficiently bad that the post-war Soviet press made little attempt to hide them, or to deny that the housing crisis was a major cause of illegal labour turnover. Health officials were also concerned, because the housing stock was seen as a springboard and incubator of epidemics such as typhus, dysentery and tuberculosis.

The housing stock in almost all Soviet cities was a mixture of low-lying, largely wooden buildings, and larger structures of wood, stone, brick or composite materials. Types of accommodation were also mixed: private dwellings, mostly of wood and with no utilities save perhaps electricity; blocks of flats, usually just two or three storeys high; and dormitories, and barracks, sometimes built of wood, sometimes from cinder block and sometimes from brick. This is not to mention those who right after the war were still living in dugouts and mud huts. During the forced industrialization of the 1930s much of the new housing was located in workers' settlements beyond the city centres or in newly developed districts of previously rural *oblasti*. In the main, each settlement belonged to a particular enterprise or coal mine, which also took responsibility – or more often failed to take responsibility – for supplying water and sewerage.

No matter what type of housing people lived in, overcrowding was almost unbearable. Even families with their own private houses had relatively little room, but for those in dormitories, barracks and even flats, conditions were extremely cramped. In dormitories and barracks this is not difficult to understand. In flats, too, almost no one had separate accommodation. They lived in communal flats, where a family might (if they were lucky) have their own room, but kitchens, and also toilets and bathrooms in the rare circumstances where these existed, were shared. It is essential to bear this in mind when thinking about the real-life repercussions of the conditions we are about to describe. For it meant there was neither space nor privacy to carry out essential household functions like washing clothes and bedding, or bathing.[17]

Outside of Moscow the majority of the population had neither indoor running water nor sewerage. Central heating was an even rarer luxury. This meant very few people had indoor toilets, but had to use primitive facilities located in court-yards which emptied into cesspits. They drew their water from cold water pumps, also located outdoors, either in courtyards or on the streets; in more remote parts of town they might not have even these, but had to rely on wells. Even in buildings that had indoor plumbing, supplies of water were not reliable. Pressure would often drop and the water would not reach the upper floors; some-times it would be cut off altogether.[18] To keep warm people still used traditional wood-burning stoves.

It is now hard to imagine, but up until the early 1950s even Moscow fit this pattern. In 1947, over half of all dwellings – housing roughly 31 per cent of the population – still had neither running water nor sewerage. Over half the housing stock measured in floor space (and probably a similar percentage of the popu-lation) did not have central heating. Average living space was just 4.4 square metres per person. New housing construction was very slow and fell further behind the growth of the city's population, so that by 1950, outside the central districts, with their larger (and older) buildings, the average Moscow citizen had to put up with just three square metres per person. The one area in which the city did make considerable progress was the installation of gas lines into blocks of flats, an improvement with major public health implications, because it allowed people to wash at home, and not rely on infrequent trips to the bathhouse.[19]

The public health authorities in Moscow considered the state of the city's housing 'extremely grave'.[20] Yet if we move just a few kilometres beyond the city limits into Moscow *oblast'*, the housing situation was dramatically worse. Moscow *oblast'* had been partially occupied by the Germans during the war, and although outright destruction of the housing stock seems to have been minimal, neglect and fuel shortages had caused a serious deterioration in the state of basic utilities: water supply, sewerage and central heating. There were also serious long-term structural problems. As of 1947, half of urban residents lived in private dwellings, none of which had running water (and almost certainly had no sewerage) and had to take their water from street pumps or wells.[21] The sudden post-war influx of new workers from outside the *oblast'* placed further strain on the already strapped housing stock and infrastructure. In coal mining districts, for example, average living space per person fell from 6 square metres in 1930, to 3.7 square metres for workers with families and 2.7 square metres for single workers in 1947. Coal mining was not – as it often was elsewhere – an extreme exception here. Dormitories in the major non-mining industrial centres were equally cramped – hardly any came even within striking distance of meeting the accepted sanitary standard of 4.5 square metres per resident. It was common practice throughout the *oblast'* to house workers in factory kitchens, storerooms, 'and other adapted premises'.[22] Nor had the situation improved much by 1949. The Krasnoe Znamya textile mill in Ramenskoe District was still cramming some 15 to 20 families plus 50 to 60 single workers (that is, between 100 and 120 people) all together into single, barracks-style rooms. In Orekhovo-Zuevo overcrowding reached a point where the State Sanitary Inspectorate tried to limit the number of new workers coming into the town by pressing the militia to deny them residence permits (the local textile factories hired the workers – around a thousand in total – anyway). Things were little better in the chemical plants of Stalinogorsk, Voskresensk, Shchelkovo and Klin, or in the steel mills of Elektrostal', all of which were still billeting workers and their families in 'temporary' barracks put up before the war, some of which had been scheduled for demolition as early as 1940.[23] In all these respects housing conditions in Moscow *oblast'* had far more in common with the industrial *oblasti* in the Urals and Western Siberia, than with Moscow city.

If we move further afield into the Central Industrial Region, we see two quite different pictures. The city of Gor'kii had a long industrial history dating back to Tsarist times. Because it was a centre of heavy industry it had another surge of growth in the 1930s. Its housing stock reflected this pattern of development, and was a mix of densely populated, sub-standard, pre-revolutionary dwellings, new blocks of flats in the districts and workers' settlements around its heavy industrial enterprises (most notably the motor vehicle works), and dormitories and barracks. In 1947 and 1948 less than 30 per cent of the population had either indoor running water or sewerage. As in most Soviet cities and towns in this period, new construction did not necessarily guarantee access to these utilities: only half the new housing blocks built by Gor'kii's industrial enterprises during 1947 had both water supply and sewerage; the other half had no sewerage and residents had to take their water from street pumps and hydrants. One consequence of this policy was that the percentage of the population living in housing with sewerage actually fell, to around 26.5 per cent by the end of 1948, and only regained the one-third mark at the end of 1951.[24]

Gor'kii, however, in many ways was privileged because of its heavy industry. Its nearest neighbour, Ivanovo *oblast'*, was an old industrial region and the centre of the country's textile industry. During the early 1930s, when textile workers saw a sharp deterioration in their working conditions and wages, it had also been the scene of major strikes and protests.[25] At the end of 1946 some 60 per cent of the urban population – that is, those living in towns and workers' settlements – lived in private housing, mainly of wood. Most houses, whether private or state-owned, consisted of just one or two storeys. Living space was very cramped, around 4.7 to 5.2 square metres per person. Of greater concern to the public health authorities was the fact that it was almost impossible to modernize the provision of utilities with this kind of housing stock. Virtually no houses had either indoor running water or sewerage. Only 6 per cent of houses in Ivanovo had running water and just over 4 per cent had sewerage. In Kineshma the corresponding figures were 3.5 per cent and 1.1 per cent. The picture was the same in the other major towns: Shuya, Furmanov, Vichuga and Rodniki. Yet they were not the worst. In the textile town of Kokhma not a single dwelling, public or private, had central heating, running water or sewerage.[26] Although we cannot explore this question in detail here, the general lack of sanitation, safe water supply and proper heating was a major factor in the sharp rise in infant mortality during 1947.[27]

It was in the Urals and Western Siberia, however, that the cumulative impact of Stalinist industrialization, the war and the regime's methods of post-war reconstruction was most evident. The Urals cities of Sverdlovsk, Molotov (now Perm'), Chelyabinsk and their surrounding *oblasti* had all experienced feverish expansion during the 1930s, and were dominated by heavy industry: engineering, iron and steel, chemicals, coal mining, ore mining and non-ferrous metallurgy. In almost every case this brought a rapid growth in population without a commensurate investment in housing, roads, sanitation, schools, hospitals and other infrastructure. The same was true of Kemerovo *oblast'* in Western Siberia.[28] These systems – especially housing – came under further pressure during the war, with the mass influx of workers and families evacuated or mobilized from the Western regions of the USSR. The third wave came after the war, as the cities and *oblast'* industrial towns received large numbers of young workers mobilized (for the most part against their will) through the Labour Reserves system, plus so-called special settlers and special contingents under the control of the Ministry of Internal Affairs (MVD).[29]

The pressures this placed on living conditions were almost intolerable. The city of Chelyabinsk, for example, had seen its population swell by several hundred thousand during the war. People found lodging wherever they could: in the kitchens and bathrooms of flats; in the recreation rooms (the so-called 'Red Corners') and kitchens of factories; in basements of public buildings; in schools; in garages; and even in buildings still under construction. Yet this situation still persisted several years after the war had ended, in large part because factories – including the city's largest, the Kirov tractor works – were taking on large contingents of new recruits for whom they had not built or prepared any accommodation.[30] The picture was very much the same in Kemerovo *oblast'*. Some of those worst affected were the 'special contingents', who, not surprisingly in light of their semi-penal status, received little attention from colliery and factory managers. The new arrivals were billeted in clubs, production shops, enterprise or pit canteens, garages, workshops, vegetable storerooms and even stables and pigsties.[31] In addition to the 'special contingents', during 1947 and 1948, the Kuzbass also received some 65,000 young Labour Reserve trainees, many of whom were assigned to remote, undeveloped areas where there was almost no housing or infrastructure. In some towns the influx of new workers exceeded the population already living there.[32]

This mass migration came into a region where housing was already dreadful and then proceeded to deteriorate further. In Molotov *oblast'* (now Perm' *oblast'*) nearly 60 per cent of the available living space at the end of 1948 was in buildings without water supply, and in some cases not even electricity.[33] In Molotov city, which was a centre of the defence industry and thus more favoured than the surrounding *oblast'* towns, it was still the case that only 15 per cent of the population had sewerage and 35 per cent access to water supply at the end of 1951.[34] The forced development of the region affected the housing stock in a number of ways. First, it led to a great deal of spontaneous, unplanned house building, as enterprises built dormitories and barracks anywhere they could, irrespective of whether or not they could provide safe drinking water, roads, pavements, or even transportation to allow workers to travel to work. In Kospash, a mining town in Molotov *oblast'*, if workers wanted water for drinking or washing they had to use snow or scoop water out of contaminated puddles lying on the ground.[35] Second, even where enterprises tried to accelerate construction and to move workers out of unfit premises, they were unable to do so. The Kuznetsk Iron and Steel works in Stalinsk (now Novokuznetsk), the largest and wealthiest enterprise in Kemerovo *oblast'* and the only one which put up significant amounts of modern housing, found that the amount of new construction erected in 1948 would replace less than 75 per cent of the floor space in buildings it had condemned. It rescinded the condemnation and carried on using these buildings for its workers.[36] Third, the region also had to continue to depend on private housing. Much of this, as elsewhere in the USSR, consisted of old wooden peasant houses in an advanced state of dilapidation. Much of it, however, was new, erected as part of an official government policy to encourage private house building as a way to compensate for its own unwillingness to devote adequate resources to residential construction, and as a means to stem illegal labour turnover. The number of people who could be housed in this way was tiny compared with actual need, yet in most Kuzbass towns during 1947 (with the exception of Stalinsk and Prokop'evsk) – and despite the obvious scale of the housing crisis – almost all new housing came from this sector. Like the older private housing stock, the new dwellings usually had no sewerage or running water.[37]

Finally, the Urals and Western Siberia made heavier reliance on dormitories and barracks than the older Russian industrial centres. Conditions in dormitories were almost universally dismal right across the USSR: overcrowding, poor sanitation, lack of furniture and bedding, and shortages of heating fuel were endemic. Even newly erected dormitories which rectified many of these defects still lacked essential utilities and ancillary premises, such as storage rooms, kitchens or rooms where workers could dry their damp work clothes.[38] Yet we need to put this into perspective. In Moscow around 10 per cent of the population were living in dormitories and barracks in 1946, falling to less than 8 per cent by the end of 1949.[39] In Gor'kii in 1948, dormitories and barracks housed only around 4 per cent of the population.[40] In the Urals and Kuzbass, however, dormitories and barracks were mass housing. In Molotov *oblast'* (also in 1948), 18 per cent of coal miners, 21 per cent of workers in iron and steel and non-ferrous metallurgy, and over half of construction workers lived in barracks; roughly another 10 per cent of the *oblast'* urban population lived in dormitories. The distinction between barracks and dormitories is significant, because no matter how bad conditions in dormitories may have been, in barracks they were far worse, since by definition they had no running water, sewerage, central heating, and in some cases not even electric lighting. Yet industrial ministries were still building them.[41]

Urban sanitation

The absence of proper sewerage systems had obvious significance for a whole range of public health problems. One, of course, was the spread of disease: the incidence of typhoid and dysentery in the city of Moscow in 1946 was roughly two to three times higher among those living in buildings without sewerage than for those who lived in buildings which had it.[42] In Kuibyshev, which had extremely primitive systems, and where whole districts of the city had raw sewage running through them in open gullies, around 2.4 per cent of the population suffered some form of serious reportable gastro-enteric infection in 1946, rising to 2.9 per cent in 1947.[43]

Most major cities had small, municipally run sewerage systems in their older, central districts. These consisted of a limited network of pipes and conduits which usually, but not always, led to some form of treatment works. The majority of these core systems dated back to the 1930s or even earlier, but all expansion, and indeed even basic maintenance, came to a halt during the war. Thus when the war ended it was not just a question of picking up where they had left off, but of making often extensive repairs to degraded pipework and mothballed treatment plants, not to mention acquiring essential chemicals, including chlorine.[44]

The situation was further complicated by the fact that the municipal systems were complemented by myriad local systems. Thus many large industrial enterprises had their own systems: some of these were hooked into the municipal network; others treated their wastes to greater or lesser degrees of efficiency and then discharged them into nearby rivers, lakes or ponds; the remainder discharged their wastes untreated directly into a watercourse. Many hospitals also had small, local systems, which collected their wastes and either shunted them into the main urban system, carted them away or allowed them to be absorbed into the ground, sometimes with disastrous consequences.[45]

From this general portrait it should already be obvious that the securing and protection of safe water supplies confronted a number of daunting problems. First, it proved extremely difficult to coordinate and integrate the disparate local sewerage systems into a single, comprehensive and efficient system that would protect an entire city or town. This problem was universal, but it was especially stark in the Urals, most notably in the industrial and mining towns of Molotov *oblast'*. According to the report of the local State Sanitary Inspectorate (GSI), each mine and factory would install just enough sewage pipe to cover its own needs – or sometimes just one building – and then hook this into the small and already-overtaxed town systems. There was no coordination and no general plan: as more and more 'subscribers' spontaneously linked themselves into the system the pipes blocked up and faeces poured out onto the streets.[46]

Second, as towns and cities grew the gap between the demands placed on sewerage systems and their capacity to handle wastes increased, rather than diminished. Towns tended to extend their water supply systems faster than they modernized their sewerage systems and extended them out from the city centre; and they extended the network of sewer pipe faster than they could upgrade or construct treatment plants. This could cause even central districts to be flooded with raw sewage, as happened in Chelyabinsk whenever there were heavy rains or during periods of peak water use.[47] It also impacted on the quality of new housing. One reason why so many new housing blocks had no sewerage was that towns and factories could only expand by locating workers' settlements in outer districts, where there were no sewerage systems and no prospects of creating them. It was perhaps obvious that this would happen in a city like Kuiby-shev, which had sprung up as major industrial centre virtually overnight during the war, but it was no less true of Moscow, whose outlying districts where equally under-served.[48] This also explains why the system in a city of the size and strategic importance of Molotov served the same share of the population, around 15 per cent, in 1951 as it did in 1945 – extension of the system did no more than keep pace with the growth of population.[49]

Third, even where local soviets or industrial enterprises tried to expand their systems, they found themselves thwarted by Moscow, which refused to provide the required funds. Gor'kii city, Moscow *oblast'*, Sverdlovsk *oblast'*, Magnito-gorsk, Shuya and Ivanovo in Ivanovo *oblast'*, and Shcherbakov in Yaroslavl' *oblast'* all had to shelve projects to develop sewerage systems and treatment works because either the central government or the parent ministries of the sponsoring industrial enterprises refused to fund them.[50] Decent urban sanitation was simply not very high on Stalinism's priorities.

Finally, the under-capacity or, as in Gor'kii and Chelyabinsk, the total absence of any treatment plant,[51] caused massive pollution of waterways, as untreated or only partially treated sewage and industrial toxins were dumped into local rivers, reservoirs and ponds. A single factory could contaminate the

water supply for an entire town; the discharges from a single town would contaminate the water supply of the communities downstream. Nothing shows this
more graphically than the fate of Moscow *oblast'*. In 1945 the rivers serving the
oblast' – primarily the Moscow and the Klyaz'ma – were *each day* polluted by
around 350,000 cubic metres of domestic and industrial wastes from the towns
and factories of the *oblast'* itself, plus an additional 500,000 cubic metres of
untreated wastes from Moscow city. These figures remained virtually unchanged
up to the start of 1950. Upstream from Moscow the situation was not so critical,
because the authorities had taken care to construct new sewerage systems and
treatment plants in order to protect the capital's water supply. It was downstream from Moscow that the *oblast'* felt the full brunt of the contamination. As
the Moscow River left Moscow, carrying its generous gift of half a million cubic
metres of untreated sewage from the people of Moscow to the people of
Moscow *oblast'*, it passed a succession of *oblast'* industrial towns (Podol'sk,
Balashikha, Voskresensk and Kolomna, among others), each of which made a
further contribution to the cause. By the time it emptied into the Oka River the
pollution was so great that the Oka remained badly polluted for a further 40 kilometres downstream.[52]

The weak development of sewerage systems impacted on urban hygiene in
another, perhaps less obvious way, namely waste removal. In the West we nor-
mally think of waste removal in terms of garbage, which of itself can pose major
health hazards. Where a large part of the housing stock had no sewerage, however,
human waste went into cesspits, and cesspits had to be emptied and cleaned. Thus
waste removal for Soviet cities was a major undertaking. In fact, it was more difficult after the war than it had been before, because all cities, including Moscow,
now had far fewer resources – horses, horse-drawn carts and cisterns for hauling
away liquid wastes, and trucks – than they had had in 1941.[53] Yet even these
reduced fleets operated far below full capacity. The communal services depart-
ments of Moscow *oblast'* in 1946 received just 10 per cent of the petrol they
needed to run their trucks, while the *oblast'* as a whole (including rural areas) was
given a grand total of six tyres.[54] In Ivanovo in both 1946 and 1947 just over half
the horses were actually fit for work, I presume because the others were either
lame or had no feed.[55] Almost everywhere vehicles were diverted to other jobs,
such as snow removal, hauling fuel, helping with the harvest, and (incredibly,
given the public health implications) even delivering bread.[56]

The end result was that cities were able to remove only a fraction – often a
very small fraction – of all the rubbish and human wastes which the population
generated over the course of a year. Even Moscow could only dispose of around
three-quarters of its rubbish and slightly more than 40 per cent of its human
wastes during 1946 and 1947; the city made considerable progress after that, but
the state of sanitation remained far from perfect.[57] Outside of Moscow the situation was far more difficult. In Gor'kii the amount removed fell in both absolute
terms and as a percentage of need between 1946 and 1948: from roughly 82 per
cent in 1946, to 69 per cent in 1947, and only 40 per cent in 1948, and had still
only reached around 75 per cent by the end of 1951.[58] In Ivanovo the health

authorities estimated that the city cleared around 25 per cent of its solid and human waste in 1946, and just under 60 per cent in 1950.[59] In Kuibyshev in 1947, only about half the rubbish and wastes were hauled away, almost all of it by private citizens using whatever makeshift means they could come up with.[60] The industrial towns of many of the *oblasti* could not match even these meagre results. In Gor'kii *oblast'* and the mining towns of Molotov *oblast'* there was virtually no removal at all.[61]

As dire a picture as these figures suggest, they actually understate the problem. Daily reality was that during most of the year almost nothing was done at all. Almost every GSI report, whether from Moscow or from the remote industrial towns of the Urals, tells exactly the same story. Little rubbish was collected and cesspits were rarely cleaned. In the winter the cesspits tended to freeze, and in most towns there was almost no attempt to maintain basic sanitation. As horrible as this situation may have been, it was compounded by the fact that courtyards did not have enough rubbish skips, so they were filled to overflowing; many were broken and garbage spilled out onto the ground. Cesspits were not hermetic and so leaked. In the worst cases people were afraid to use the toilets, especially at night, and so simply relieved themselves on the ground, but of course this only made matters worse.[62] The winter freeze tended to mask the worst effects, but when the spring came the situation became intolerable, as all the frozen faeces, urine and rotting garbage began to thaw out. Thus every single Soviet city, without exception, relied on a mass spring cleaning campaign, during which they mobilized every available vehicle, every spare draught animal, and tens of thousands of local residents to work weekends for a solid month (and sometimes longer) cleaning the cities and towns. Rubbish and human waste would be carted to local dumps, sent to collective or state farms to be used as fertilizer, or burned. Towns used the campaigns to try to repair skips, toilets and cesspits, and to install new ones. The spring cleaning campaigns were more or less successful, but once they were over the problems would reappear. Streets, courtyards and cesspits again became littered and fouled, people carted whatever they could to unofficial, makeshift dumps, and so the campaigns would be repeated in the autumn, although not always as effectively.[63]

This puts into context the public campaigns, in the factory newspapers, for example, to teach people the basics of personal hygiene. Lessons on the need to wash hands before preparing food and after using the toilet, and even to use the toilet in a 'civilized' manner,[64] all implied that it was people's lack of 'culture' which led them to foul the streets with their own excrement. No doubt countless people had long ago given up caring about where they relieved themselves, but even the most slothful attitude was conditioned by a much more basic fact: proper 'civilized' facilities simply did not exist.

The impact on daily life

It is one thing to describe the state of the urban environment, and quite another to try to imagine what this meant for the daily lives of those who lived in it.

Those of us even remotely familiar with the former Soviet Union remember it as a shortage society where people invested countless hours and energy just to acquire food, not to mention elementary consumer goods. With the emergence of Soviet sociology in the 1960s, we also began more fully to appreciate the burden that the absence of basic household appliances, most notably washing machines, placed on women, who continued to perform the lion's share of domestic labour.[65] But for Westerners, and even for many younger Russians who no longer have elderly relatives in the countryside, the way that housing and the lack of sanitation shaped everyday life right after the war is almost impossible to grasp. The overwhelming majority of urban residents did not have indoor running water. They had to fetch water from pumps or standpipes in the streets or in building courtyards. Supplies were not always regular and the pumps often froze up during the winter. If people lived in multi-storey buildings they had to carry heavy buckets of water up the stairs. Washing clothes, sheets, dishes or one's body was labourious and time-consuming, especially in a crowded communal flat with no bathroom. All of this was made infinitely worse by the absence of sewerage. As few people had indoor toilets and the outdoor toilets and cesspits were in a wretched state, during the winter, at night and for small children the year-round, people had to use buckets (children had chamber pots) and then empty them the next day – but because there was no hot water, or even cold running water, cleaning out the buckets was no straightforward task. In short, just maintaining basic levels of cleanliness and hygiene for yourself and your family was drudgery.

People's travails might have been eased had there been an even remotely comprehensive network of public services, but there was not. Most large towns had at least a limited number of laundries run by the city soviet or industrial enterprises (some hospitals also had their own laundries, but these were strictly for their own use). Conditions in the laundries were far from ideal. Few were well mechanized, so most, if not all, the work had to be done by hand; ventilation was poor; and maintenance of plant and premises had been neglected during the war.[66] What most affected the population, however, were not the terrible working conditions of the laundry workers, but the fact that the laundries had such limited capacity that they could take in little or no washing from the general public. Instead, they could only serve institutional customers, mainly dormitories, children's homes, schools and hospitals. Only in Moscow did they accept appreciable amounts of private laundry, but this met only about 8 per cent of actual need.[67] Virtually everywhere else people were left to their own devices.

The other battle, of course, was to maintain some degree of personal cleanliness. Here people relied heavily on the public bathhouse, some of which were owned by the municipality, others by industrial enterprises. Like laundries and indeed the rest of the urban infrastructure, the state of bathhouses had badly deteriorated during the war. Boilers needed major overhaul or total replacement. Pipework had to be renewed and ventilation systems restored. Walls and ceilings had to be replastered.[68] This gave rise to two sets of disparities. In most localities the capacity of the baths would not have met the needs of the population, even if

Table 4.1 Average number of washes (*pomyvki*) per resident provided by municipal and enterprise bathhouses per year and per month

Town and year	Washes per year	Washes per month
Moscow, 1948	15	1.25
Moscow *oblast'* industrial towns, 1947	12–15	1.0–1.25
Ivanovo, 1946	19	1.6
Gor'kii, 1947	40	3.3
Sverdlovsk, 1947	12	1.0
Chelyabinsk, 1946	17	1.4
Kuibyshev, 1947	14	1.2
Polovinka (Molotov *oblast'*), 1947	36	3.0
Stalinsk (Kemerovo *oblast'*), 1947	7	0.6

Source: GARF-RSFSR, f. A-482, op. 47: d. 7669, l. 105 (Moscow City); d. 6347, l. 94 (Moscow *oblast'*); d. 4925, l. 202 (Ivanovo); d. 6358, l. 8 (Sverdlovsk); d. 4960, l. 49 (Chelyabinsk); d. 6345, l. 349 (Polovinka). GARF-RSFSR, f. A-482, op. 52s, d. 224, l. 89 (Kuibyshev). GARF, f. 9226, op. 1: d. 895, l. 101 (Gor'kii); d. 932, l. 24 (Stalinsk).

the bathhouses had all been working properly. The bathhouses, however, did not work at full capacity: closures for repair, boiler breakdowns, irregular fuel deliveries, soap shortages and even disruptions to the supply of mains water all reduced the population's access. Each bathhouse in Chelyabinsk, for example, averaged an entire month out of action during 1946. Even if they had been working flat out they would have provided people with the chance to bathe only 20 times year, but the real figure was 17, or just 1.4 'washes' (*pomyvki*) a month.[69] This picture was fairly typical, as Table 4.1 shows. It gives the actual number of washes and baths that each resident of the cities listed had per year and per month during the early post-war years.

There is both more and less to these figures than meets the eye. They present a roughly accurate picture of the services provided by public and factory bathhouses, but in theory people had access to other facilities not included in these data, most notably factory showers and baths or showers that people took in their own homes. The situation just after the war was such that neither of these greatly affected the figures. Even in Moscow only around 9 per cent of residential buildings in 1947 had bathrooms – the health authorities estimated that between them they added up to only around 80,000 bathrooms out of an officially registered population of over four million. Yet even half of these were unusable because the water or gas supplies to the buildings were out of order.[70] The same was true of factory showers. Right after the war many of these factory installations were themselves out of service. In Gor'kii, public health officials estimated that even if full use of private baths and factory showers were included in the figures, people would still have no more than three baths a month. Only by 1951 did the restoration and expansion of showers at work make a major difference.[71] The other thing to keep in mind is the relative needs of the populations in different places. Residents in Polovinka in Molotov *oblast'* in theory enjoyed twice as many baths a month as people living in Moscow.

Polovinka, however, was a mining town, and the idea that coal miners could take a bath only once in ten days was alarming, to say the least. As awful as this sounds, there were other mining districts in the Urals and Kemerovo *oblast'* which had no bathhouses at all.[72]

It is obvious that, with such limited facilities, people made do as best they could. The point here is not to claim that people went around incredibly dirty, but that to maintain even basic levels of cleanliness for yourself and your family involved an enormous effort. We should keep in mind that to wash or bathe at home was hampered not just by the lack of hot water and terrible overcrowding. During the whole of the early post-war period the country suffered from a soap shortage of truly crisis proportions, which reached its nadir during the typhus epidemic which broke out in 1947. In many places you could only find soap at the bathhouse – but, as already noted, even bathhouses were not guaranteed supplies, and sometimes had to limit their operations because they had no soap.[73] By the time the soap shortage began to ease after 1948 or 1949, people faced a new obstacle: public baths were put on a commercial basis and had raised their prices, and at least two reports claim this caused people to stop using them.[74]

Conclusion

What broader conclusions can we draw from the above discussion? The first, and perhaps most obvious, concerns issues of public health. Overcrowded housing, unsafe supplies of drinking water, poor and sometimes non-existent urban sanitation, and the shortage of soap made the population highly vulnerable to a whole host of diseases, including tuberculosis, typhus, typhoid, dysentery and upper respiratory infections. The latter two were the main causes of infant mortality, while serious skin infections, another indicator of poor hygiene, were a leading cause of time lost from production due to occupational illness.[75] We should be clear, however, that sanitation and hygiene were only part of a much larger pattern of health risks. A comprehensive analysis of these would have to include such factors as prolonged periods of poor nutrition, food safety, the breadth and quality of medical care (including lack of hygiene in hospitals and clinics), working conditions and work hazards, sanitation and hygiene in schools, and the ever-growing volumes of industrial air and water pollution.[76]

The second conclusion concerns the political relationship between the Stalinist regime and its citizens. The central argument of this article has been that we need to broaden our concept of the standard of living and see the urban environment as a whole (that is, not just housing) as a vital component of popular consumption. From the account we have given it might appear that the lack of urban infrastructure was essentially an inheritance from Tsarist times and the USSR's belated industrialization. After all, workers had endured miserable housing and urban conditions long before the 1930s, and much of what limited infrastructure they had available was initiated after the Bolshevik Revolution. This, however, is not quite the whole story. Stalinist industrialization was financed by ruthlessly curtailing consumption in both town and countryside. The most visible manifestations of this were the collapse of real wages, the famine of 1932–1933, and the dire shortage of even the most basic consumer goods, from shoes to kerosene. Yet these were not the only elements of consumption which came under pressure: if little of the surplus product created by Soviet workers came back to them in food and clothing, the same was true of housing, safe drinking water and urban sanitation. The decision to erect factories and coal mines at breakneck speed, to bring millions of peasants into the towns to build and staff them, but not to provide these workers with accommodation, infrastructure and public services was a political decision, a reflection of the regime's essential priorities and its view of its citizens as expendable objects, valuable only for their capacity to generate the surplus from which the Soviet elite drew its privileges.

Even before the Second World War this policy, reinforced by terror, had brought the population to an almost permanent state of exhaustion. The war and the post-war hardships only deepened this trend. Literally every aspect of daily life was a battle for survival: finding food, staying warm, avoiding major illness, getting a decent night's sleep, washing clothes, going to work and back, and just trying to stay reasonably clean dominated all aspects of existence outside the workplace, which was itself an extremely inhospitable and often hostile environment. A population living under such stress proved politically useful to the Stalinist regime. Even without the intimidation caused by an omnipresent secret police, the reality was that people had little time or energy to reflect on the causes of their predicaments, much less to rebel. They responded instead with political passivity and demoralization, while seeking individual ways to circumvent the system.

This was the legacy which Stalin bequeathed to his political heirs. Khrushchev, Malenkov and Beria were all perceptive enough to see the need to raise consumption, rehouse the population, and even to allow a limited degree of political reform. We know that under Khrushchev the standard of living increased appreciably. Extreme poverty was reduced. Diet improved. Tens of millions of Soviet citizens moved into better accommodation. Life expectancy rose, and came closer than it ever had done – or would ever do in the future – to matching average life expectancy in the West. Yet the curious thing about this is that it did not end popular demoralization or provide durable political support for the system. The combination of political relaxation and the new-found emphasis on the 'consumer society' led to an important change in people's perceptions, as Soviet society shifted from what we might call absolute to relative deprivation.

One example of this is housing. Between 1955 and 1965, a year after Khrushchev was ousted from power, some 84 million Soviet citizens – roughly a third of the Soviet population – had moved into newly built accommodation. The annual rate of urban housing construction virtually tripled between 1950 and 1965; moreover, the largest increases were in precisely the types of cities we have discussed here: Gor'kii, Molotov (by then again known as Perm'), Stalinsk (Novokuznetsk), Kuibyshev and Sverdlovsk.[77] For all the improvement this brought, it did not come close to solving the housing crisis. People continued to be crowded into communal flats, and it was still a long while before people outside of Moscow and Leningrad could expect reasonable access to running water and sewerage.[78] The new housing even became famous for its shoddy construction, and quickly acquired the epithet, 'Khrushchev slums' (*khrushchoby*, a pun on the Russian word for slums, *trushchoby*) – despite the fact that for those who moved into them they almost certainly represented an amelioration of their housing situation.

An even starker illustration was food. For all the failures of Khrushchev's various agricultural policies, the fact remained that during his rule the production of grain, meat, milk and vegetables all rose by around 50 per cent.[79] Yet this merely kept pace with the increase in the urban population, so that the structure of the urban diet showed a far slower rate of change. We have already noted that well into the 1950s the average working class family was still consuming fewer calories than dietary norms called for and still drawing most nutrition from starch. It was only towards the end of the 1950s that we begin to see a modest shift towards consumption of animal proteins and fruit and vegetables, although we should keep in mind that these increases were measured against a very low starting point and were regionally unbalanced.[80] Discontent grew, culminating in workers' risings in Novocherkassk and other Soviet cities during the summer of 1962, which the leadership suppressed by force of arms.[81] From that moment onwards any political rapprochement between the Soviet regime and its working class became impossible. Under Khrushchev's successors the country continued to take on more and more of the external trappings of a modern industrial society, while proving less and less able to meet its people's aspirations or to reverse their political demoralization.[82] The Stalinist system – for this is what it remained, even under Gorbachev – was inherently incapable of housing, feeding or clothing its people. Having exhausted its last possibilities, it simply disintegrated from within.

Notes

1 Research for this article was financed by a project grant from the Arts and Humanities Research Council in the United Kingdom. I want to thank Natasha Kurashova for her valuable comments and suggestions on earlier drafts of the paper, and Chris Burton for ongoing discussions and advice about the work of the State Sanitary Inspectorate.

2 R. W. Davies, *Crisis and Progress in the Soviet Economy, 1931–1933* (Basingstoke, 1996), p. 187; R. W. Davies and Steven G. Wheatcroft, *The Years of Hunger: Soviet Agriculture, 1931–1933* (Basingstoke: Palgrave, 2004), pp. 402–407.

3 Janet Chapman, *Real Wages in Soviet Russia Since 1928* (Cambridge, Mass.: Harvard University Press, 1963), pp. 144–145.

4 GARF-RSFSR, f. A-482, op. 47, d. 3443, l. 7 (Sverdlovsk), and op. 47, d. 4941, l. 11–11ob. (Moscow).

5 GARF-RSFSR, f. A-482, op. 47, d. 3443, l. 7. We can express this trend another way. In 1940 in Sverdlovsk a newborn baby was 30 times more likely to die in that year than an adult or small child over one year of age. In 1944 the gap between the two death rates had fallen to just 6:1.

6   The grain harvest in 1946 was 39.6 million tons, compared with 95.5 million tons in 1940 and 47.3 million in 1945. Eugene Zaleski, *Stalinist Planning for Economic Growth, 1933–1952* (London: Macmillan, 1980), pp. 582–583, 618–619.

7   The classic study of the famine itself remains V. F. Zima, *Golod v SSSR 1946–1947 godov: proiskhozhdenie i posledstviya* (Moscow: Institute of Russian History, 1996). Estimates of famine deaths are in Michael Ellman, 'The 1947 Soviet Famine and the Entitlement Approach to Famines', *Cambridge Journal of Economics*, vol. 24, no. 5 (September 2000), pp. 603–630. The food policies of the regime and the attenuated recovery are detailed in Donald Filtzer, *Soviet Workers and Late Stalinism: Labour and the Restoration of the Stalinist System After World War II* (Cambridge; Cambridge University Press, 2002), chs. 2 and 3.

8   Calculated from mortality figures in RGAE, f. 1562, op. 329, d. 2230, l. 4 (1946), and d. 2648, l. 212 (1947), and population estimates in GARF-RSFSR, f. A-482, op. 47, d. 4925, l. 23 (Ivanovo), f. A-482, op. 52s, d. 224, ll. 50–1 (Kuibyshev), and f. A-482, op. 52s, d. 224, l. 172 (Moscow). In absolute terms, deaths rates per 10,000 population went from 119 to 176 in Ivanovo, 125 to 157 in Kuibyshev, and 125 to 134 in Moscow.

9   RGAE, f. 1562, op. 329, d. 2229, ll. 1, 4–11, d. 2230, ll. 3–12, d. 2648, ll. 196–8, 204–13, 242.

10 For a discussion of this issue see Ellman, pp. 614–615.

11 The detailed calculations are in Donald Filtzer, 'The 1947 Food Crisis and Its Aftermath: Worker and Peasant Consumption in the non-Famine Regions of the RSFSR', PERSA Working Paper No. 43 (April 2005), at http://www.warwick.ac.uk/go/persa, which also gives full references to the statistical sources. Soviet nutritional standards were higher than Western dietary norms, both then and now. If we adapt Western norms to take account of the heavy physical labour that all but young children performed, the intense cold of the climate and the inadequate heating of factories and homes, the shortfall in daily calorie intake is around 25 per cent, adjusted for the age and gender composition of the average family.

12 In 1955 per capita calorie consumption in workers' families in the RSFSR was still only about 88 per cent of the recommended daily level, and still some 200 calories less than in peasant families. Bread and potatoes remained the major source of nourishment. GARF-RSFSR, f. A-482, op. 30, d. 7221, l. 13. I am grateful to Andrei Markevich for bringing this report to my attention.

13 GARF-RSFSR, f. A-374, op. 3, d. 2231, ll. 2–2ob., 3–3ob.

14 Calculated from production series and population estimates in Zaleski, pp. 614–633.

15 GARF-RSFSR, f. A-482, op. 52s, d. 245, l. 56.

16 Zaleski, pp. 592–593, 608, 630–631, 633.

17 It was still the case in 1970 that roughly two-thirds of workers' families lived in shared accommodation. Murray Yanowitch, *Social and Economic Inequality in the Soviet Union* (White Plains: M. E. Sharpe, 1977), p. 42.

18 This situation was universal. See, for example, GARF-RSFSR, f. A-482, op. 47, d. 4941, l. 117 (Moscow, 1946); op. 47, d. 4925, l. 163 (Ivanovo *oblast'*, 1946); and op. 49, d. 3243, l. 8 (Kuibyshev, 1951).

19 GARF-RSFSR, f. A-482, op. 47, d. 4941, ll. 143, 144, 144ob., 150ob., d. 6351, ll. 106ob.–107ob., and d. 7669, l. 137; and op. 49, d. 111, ll. 64–5. The installation of gas was not an unmixed blessing. In 1950 there were 352 cases of domestic gas poisoning in Moscow, and in 1951 the public health inspectors turned down nearly a third of applications to install gas appliances on safety grounds. GARF-RSFSR, f. A-482, op. 49, d. 3249, ll. 50–1.

20 GARF-RSFSR, f. A-482, op. 47, d. 6351, l. 108.

21 GARF-RSFSR, f. A-482, op. 47, d. 6347, l. 106; RGAE, f. 1562, op. 329, d. 4591, l. 32.

22 GARF-RSFSR, f. A-482, op. 47, d. 6347, ll. 106, 109–10, 116.

23 GARF-RSFSR, f. A-482, op. 49, d. 103, ll. 68–74.

24 GARF, f. 9226, op. 1, d. 798, ll. 26, 26 ob., 28ob.; d. 895, ll. 94–5, 108ob.–110. GARF-RSFSR, f. A-482, op. 49, d. 3240, l. 36.

25 Jeffrey J. Rossman, *Worker Resistance under Stalin: Class and Revolution on the Shop Floor* (Cambridge, Mass.: Harvard University Press, 2005).

26 GARF-RSFSR, f. A-482, op. 47, d. 4925, ll. 218, 221, 227–8.

27 GARF-RSFSR, f. A-482, op. 52s, d. 221, ll. 77–8.

28 On the planless character of pre-war expansion see GARF, f. 9226, op. 1, d. 899, ll. 56–60 (Molotov *oblast'*), and d. 932, ll. 1–4 (Kemerovo *oblast'*). Just how underdeveloped these regions were can be gauged from the fact that as late as 1948 Molotov *oblast'* had virtually no paved roads, and not a single one of its 11 industrial towns had an adequate water supply.

29 The 'special contingents' were part of a larger group of semi-penal and conscripted (although notionally free) workers whom I have likened to indentured labourers. The post-war reconstruction of the Stalinist economy relied disproportionately on this category of toilers, who probably provided the equivalent of the entire net growth in the number of industrial and construction workers between 1945 and 1952. See Filtzer, *Soviet Workers and Late Stalinism*, ch. 1.

30 GARF-RSFSR, f. A-482, op. 47, d. 4960, l. 9, and d. 6363, ll. 12–13. At the end of 1947 new workers at the Kirov factory were living in the factory bathhouse, disused railway coaches, and 'semi-destroyed' barracks and factory offices which had been quickly renovated to accommodate them.

31 GARF, f. 9226, op. 1, d. 932, ll. 8, 14–18.

32 GARF, f. 9507, op. 2, d. 418, l. 17, and d. 420, l. 34. GARF-RSFSR, f. A-482, op. 47, d. 7659, ll. 1–3.

33 GARF, f. 9226, op. 1, d. 899, l. 184.

34 GARF-RSFSR, f. A-482, op. 49, d. 3250, ll. 5, 8, 21.

35 GARF, f. 9226, op. 1, d. 900, ll. 110–13, and d. 932, ll. 1–4.

36 GARF, f. 9226, op. 1, d. 932, ll. 5–11. GARF-RSFSR, f. A-482, op. 47, d. 7659, ll. 17–18.

37 GARF, f. 9226, op. 1, d. 932, l. 10; GARF-RSFSR, f. A-482, op. 49, d. 1628, l. 77; Filtzer, *Soviet Workers and Late Stalinism*, p. 96.

38 For a general account of dormitory conditions, see Filtzer, *Soviet Workers and Late Stalinism*, pp. 93–95, 138–139.

39 GARF-RSFSR, f. A-482, op. 47, d. 4941, l. 150ob., and op. 49, d. 111, l. 66ob.

40 GARF, f. 9226, op. 1, d. 798, l. 46, and d. 895, l. 111.

41 The 1948 figures are from GARF, f. 9226, op. 1, d. 899, ll. 178, 184. The 1947 GSI report lists 98,000 people living in dormitories (GARF-RSFSR, f. A-482, op. 47, d. 6345, l. 330). According to RGAE, f. 1562, op. 329, d. 3152, l. 48, the urban population of Molotov *oblast'* in December 1947 was around 990,000, but this was a very imprecise estimate.

42 GARF-RSFSR, f. A-482, op. 47, d. 4941, ll. 22ob., 25.

43 GARF-RSFSR, f. A-482, op. 52s, d. 224, ll. 49, 55.

44 For accounts of the impact of the war, see the examples of Moscow *oblast'* in GARF, f. 9226, op. 1, d. 691, ll. 124–6, and the city of Sverdlovsk, in GARF-RSFSR, f. A-482, op. 47, d. 3443, ll. 67–8. For problems with chemicals during the early post-war years see GARF, f. 9226, op. 1, d. 932, l. 32 (Kemerovo *oblast'*), and GARF-RSFSR, f. A-482, op. 47, d. 3443, l. 63 (Sverdlovsk). Yet chemicals were still in short supply during the early 1950s – see GARF-RSFSR, f. A-482, op. 49, d. 3240, l. 9 (Gor'kii), and d. 3261, l. 14 (Chelyabinsk).

45 In Sverdlovsk, for example, seepage from the cesspits at several of the city's infectious diseases hospitals, including its sanatorium for children with tuberculosis, was polluting the ground water from which they drew their water supply. GARF-RSFSR, f. A-482, op. 47, d. 3443, ll. 88–9.

46 GARF, f. 9226, op. 1, d. 899, l. 285.

47 GARF-RSFSR, f. A-482, op. 49, d. 3261, l. 15. Notably, this report is from 1951.

48 For Kuibyshev see GARF-RSFSR, f. A-482, op. 52s, d. 224, ll. 84–5; for Moscow, see GARF-RSFSR, f. A-482, op. 49, d. 3249, l. 28.

49 GARF-RSFSR, f. A-482, op. 49, d. 3250, l. 21.

50 GARF, f. 9226, op. 1, d. 798, l. 34ob. (Gor'kii); op. 1, d. 693, l. 62 (Sverdlovsk *oblast'*). GARF-RSFSR, f. A-482, op. 47, d. 4937, ll. 36–7 (Moscow *oblast'*); op. 49, d. 1628, l. 68 (Magnitogorsk); op. 47, d. 4925, l. 183, and op. 49, d. 1610, l. 11 (Ivanovo *oblast'*); op. 47, d. 7685, l. 94 (Yaroslavl' *oblast'*).

51 In 1952 both these cities were still awaiting construction of their very first treatment plants. GARF, f. 9226, op. 1, d 798, l. 34–34ob., and GARF-RSFSR, f. A-482, op. 49, d. 3240, ll. 26–7 (Gor'kii). GARF-RSFSR, f. A-482, op. 47, d. 4960, ll. 39–43, and op. 49, d. 3261, ll. 15–17 (Chelyabinsk).

52 GARF, f. 9226, op. 1, d. 691, ll. 130, 139–40. GARF-RSFSR, f. A-482, op. 47, d. 4937, ll. 34–5, and op. 49, d. 103, ll. 26–7.

53 The number of horses belonging to the cleaning trusts of the main towns in Moscow *oblast'* fell from 270 in 1941 to just 80 in 1945; the number of trucks dropped from 49 to 19. GARF, f. 9226, op. 1, d. 691, l. 154. In Moscow city the number of horses went from 201 in 1941 to 95 in 1946; the total number of waste removal vehicles of all kinds dropped from 2,006 to 1,000 in the same period. GARF-RSFSR, f. A-482, op 47, d. 4941, ll. 124ob., 126.

54 GARF-RSFSR, f. A-482, op. 47, d. 4937, l. 45. By 1950 the *oblast'*'s fuel needs were still being met by only 20 per cent. GARF-RSFSR, f. A-482, op. 49, d. 103, l. 41.

55 GARF-RSFSR, f. A-482, op. 47, d. 4925, l. 192.

56 GARF-RSFSR, f. A-482, op. 47, d. 4941, l. 126; d. 4960, l. 41; d. 6335, l. 78. GARF, f. 9226, op. 1, d. 693, l. 75; d. 798, ll. 34ob, 35; d. 895, ll. 95, 95ob.; d. 932, ll. 50–1.

57 GARF-RSFSR, f. A-482, op. 47, d. 4941, l. 124; op. 47, d. 6351, l. 88; and op. 49, d. 111, l. 50ob. According to the last of these, the city was officially removing over 90 per cent of rubbish and human waste by 1948 and 1949, yet the city's residential dwellings were still dirty. The GSI explained the paradox by the fact that official calculations of the volume of waste being generated – and hence the percentage being collected – were based on figures for the legally registered population; they took no account of those living illegally in Moscow, but who nonetheless contributed to the accumulation of waste.

58 GARF-RSFSR, f. A-482, op. 47, d. 4923, l. 37, and op. 49, d. 3240, ll. 35–6; GARF, f. 9226, op. 1, d. 798, ll. 34ob.–35, and d. 895, l. 95–95ob.

59 GARF-RSFSR, f. A-482, op. 47, d. 4925, l. 92, and op. 49, d. 1610, l. 15.

60 GARF-RSFSR, f. A-482, op. 52s, d. 224, ll. 86–7.

61 GARF-RSFSR, f. A-482, op. 47, d. 6335, ll. 77–8, and d. 7656, ll. 68–9 (Gor'kii *oblast'*); GARF, f. 9226, op. 1, d. 900, ll. 116–17 (Molotov *oblast'*).

62 GARF, f. 9226, op. 1, d. 932, ll. 48–9. GARF-RSFSR, f. A-482, op. 49, d. 3261, l. 19.

63 References to the annual campaigns in every locality are too numerous to list. Some of the more detailed descriptions appear in GARF, f. 9226, op. 1, d. 693, ll. 73–5 (Sverdlovsk *oblast'*, 1945); GARF-RSFSR, f. 482, op. 52s, d. 224, ll. 62, 88 (Kuibyshev, 1947); GARF-RSFSR, f. A-482, op. 47, d. 6347, ll. 79–81 (Moscow *oblast'*, 1947).

64 *Golos Dzerzhintsa* (Dzerzhinskii Cotton Textile factory, Ivanovo), 7 March 1946, 24 June 1948; *Stalinets* (Kuibyshev Motor Vehicle and Tractor Parts factory), 2 August 1950.

65 This has long been well document in the translated literature, not to mention Western studies. Soviet women performed roughly three times as much domestic labour as Soviet men, a figure which remained remarkably constant from the 1930s to the 1980s (*EKO*, no. 8, 1988, pp. 144–145 – Tat'yana Boldyreva). In the mid-1980s the

Soviet population spent more time queueing than the total number of hours worked in Soviet industry (*Ekonomicheskie nauki*, no. 1, 1990, p. 34 – A. Chuikin).

66 Only four of the 41 laundries in Molotov city in 1951 were fully mechanized, and not a single one met even the most elementary 'sanitary' requirements. GARF-RSFSR, f. A-482, op. 49, d. 3250, l. 24.

67 GARF-RSFSR, f. A-482, op. 47, d. 4941, l. 136. Only one of the 62 laundries in Sverdlovsk did work for ordinary residents; the laundries in Moscow *oblast'* also began to accept private washing in 1947, but the amounts were insignificant. GARF-RSFSR, f. A-482, op. 47, d. 6358, l. 9 (Sverdlovsk); and d. 6347, l. 98 (Moscow *oblast'*).

68 This is the list of repair work needed to Moscow's baths during 1947, but these were among the best in the country. GARF-RSFSR, f. A-482, op. 47, d. 6351, l. 101.

69 GARF-RSFSR, f. A-482, op. 47, d. 4960, l. 48. The term *pomyvka* covered anything from a full bath or shower, to just sponging oneself down at a sink.

70 GARF-RSFSR, f. A-482, op. 47, d. 6351, l. 107.

71 GARF-RSFSR, f. A-482, op. 49, d. 3240, l. 39. In Ivanovo, by contrast, even as late as 1950 factory shower rooms could provide no more than 80,000 showers a years for a population of probably over 300,000 people. GARF-RSFSR, f. A-482, op. 49, d. 1610, ll. 22–3.

72 GARF, f. 9226, op. 1, d. 900, ll. 113–16, 124 (Molotov *oblast'*); and d. 932, ll. 24–5 (Kemerovo).

73 On the soap shortage and typhus epidemic, see Zima, pp. 171–178.

74 The reports are from Moscow *oblast'*, GARF-RSFSR, f. A-482, op. 49, d. 103, l. 55; and Ivanovo, GARF-RSFSR, f. A-482, op. 49, d. 1610, l. 23. In Ivanovo attendance started to go back up once they lowered the charges.

75 Skin infections accounted for nearly one out of every seven days lost to occupational illness in Molotov *oblast'* in both 1947 and 1948. In some industries in the *oblast'* they were more costly than industrial accidents. GARF, f. 9226, op. 1, d. 929, ll. 11, 17, 20, 22, 29.

76 On the problem of water supply, see Donald Filtzer, 'Environmental Health in the Regions During Late Stalinism: The Example of Water Supply', PERSA Working Paper No. 45, available at http://www.warwick.ac.uk/go/persa.

77 *Narodnoe khozyaistvo SSSR za 60 let* (Moscow, 1977), pp. 492–501.

78 A survey of women industrial workers in Sverdlovsk in 1961, for example, found that over a quarter were still living without any amenities at all, even running water. M. A. Korobitsyna, *Zhenskii trud v sisteme obshchestvennogo truda pri sotsializme.* Candidate dissertation (Sverdlovsk, 1966), pp. 115–117.

79 Good discussions of Khrushchev's agricultural policies are available in a number of sources. One of the best and most accessible remains Zhores A. Medvedev, *Soviet Agriculture* (New York: W. W. Norton, 1987), ch. 6. The figures on food production are from p. 199.

80 *Byulleten' nauchnoi informatsii: Trud i zarabotnaya plata*, no. 12, 1960, pp. 31–33. In 1959 workers in Gor'kii and Ivanovo still consumed only three-quarters as much meat as workers in Moscow. Although workers in Ivanovo were nearly equal to Moscow workers in consumption of fruit and bread, workers in Gor'kii consumed only 40 per cent as much fruit and 40 per cent more bread.

81 The events in Novocherkassk have long been well documented. For a concise account, see Vladimir A. Kozlov, *Mass Uprisings in the USSR: Protest and Rebellion in the Post-Stalin Years* (Armonk, New York: M. E. Sharpe, 2002), chs. 12–13.

82 I remember watching a popular Russian television programme in 1998, which every week recreated life in a flat from a particular decade of Soviet history. During the episode devoted to the 1970s two former officials from the Ministry of Agriculture argued that their efforts during the Brezhnev years had by no means been a failure. After all, they said, no one had starved, and to this extent their work had been a

success. By all appearances these were two conscientious and decent people who could by no means be dismissed as simple 'nostalgiacrats'. At one level what they said was perfectly true, especially if measured against the famines still within living memory of many people. It equally shows, however, just how little dynamic and vitality the Soviet system possessed as a distinct social formation or 'mode of production'.

5 'Into the grey zone'

Sham peasants and the limits of the kolkhoz order in the post-war Russian village, 1945–1953

Jean Lévesque

The collective farm system created during the collectivization drive of 1929–1930 was consolidated prior to the Second World War. Throughout the 1930s, complementary features designed to improve its workings were progressively introduced, the most important of which being the adoption in 1935 of the Model Statutes of the Collective Farm *artel'*, which served as a pseudo-constitution regulating most aspects of the relationship between the collective farm peasantry and the Soviet state. Furthermore, the Stalinist party-state continued its clampdown on the last remnants of peasant individual agriculture not only through the enactment of increasingly harsh legislation and the implementation of fiscal measures against individual householders, but also by developing methods of sharing kolkhoz income that would force 'idlers' to participate actively in the economic life of the collective farm. While it can be argued that more energy was spent on collecting grain and other foodstuff than on policing Soviet villages, the campaign methods used during the collections eventually left an imprint on other policies directed at the countryside. The campaign-type (*kampaneishchina*) of policy implementation usually brought short-lived effects and made it very difficult to sustain efforts that could settle core problems in a more effective way.

Many such shortcomings developed in the post-collectivization period and worsened during the war. For instance, the rules concerning membership were set in the Model Statutes in 1935, but quickly peasants started manipulating them to their own advantage. Although illegal in theory, seasonal work outside the farm (*otkhodnichestvo*) intensified during the post-war period and created real labour problems on collective farms. Despite administrative harassment and short-lived repressive episodes, the labour-day minimum was barely fulfilled by a majority of collective farmers. In fact, the recorded number of violators never diminished. In some regions, it could reach almost half of all able-bodied farmers.[1] Moreover, the struggle against private plots undertaken in September 1946 hardly produced any results; the gross agricultural output of the plots grew throughout the post-war period, and regional studies highlight the increased time spent on them after the war.[2]

The importance of the private plot for kolkhoz members' survival was nothing short of universal, but the strategies to avoid labour discipline varied from one segment of Soviet rural society to another. Labour policies helped targeting elements considered to be improperly collectivized, but at the same time revealed the discrepancy between what type of behaviour was considered anti-kolkhoz by representatives of state power and what kind of actions were perceived as such by rural communities. What Sheila Fitzpatrick called the margins of kolkhoz society in the 1930s included migrants, individual householders, craftsmen and various outsiders.[3] Individual peasants, who numbered close to five million households in the wake of the collectivization drive, virtually disappeared from the Soviet countryside during the war thanks to unfavourable taxation policy. Migrants remained and their numbers seem to have swelled at the end of the 1940s. The involvement of peasants in collective farming considerably varied and, besides migrants, other segments of the collectivized peasantry loosened off the grip the state had established on them through minimal or non-involvement in collective labour. Thus there were different shades of grey between the few individuals who lived in the midst of the farms but who refused involvement in farm activities, and those who fully participated and whose subsistence depended on the kolkhoz. On one end of the spectrum, a minority of peasants cumulated high labour-day earnings with the help of intensive, specialized work throughout the year. On the other extreme, there were those who pretended to be members of collective farms but who tried to avoid most of the duties associated with membership. These peasants were perceived by state organs as representing a threat to collective agriculture and considered a sham (*mnimyi kolkhoznik*), a false one (*lozhekolkhoznik*) or loosely collectivized elements (*okolokolkhoznyi element*), all of whom were considered variations on the same theme.

Unlike in the pre-war period, when individual householders represented an alternative to the kolkhoz order (albeit a fiscally unsustainable one with little appeal), the post-war sham peasants appeared from within the boundaries of the collectivized peasantry. The new scapegoat for the deficiencies of Soviet agriculture was difficult to eradicate for it constituted a grey zone, constantly forming out of different segments of the collectivized peasantry but falling out of the grasp of the state, as revealed by the implementation of labour discipline policies. Women, youngsters, members of splintered households and migrant workers provided the grey zone with peasants who learned to avoid the increasing pressure of the collective order in the village, a pressure which became unbearable as the Stalinist leadership saw no other remedy to the ills of the kolkhoz economy than strengthening discipline and perfecting its functioning. Ironically, under the guise of a hegemonic collective sector in agriculture, peasant society was disintegrating.

Fading elements: splintering households and reluctant teenagers

Pressure by peasants to receive a larger personal plot, which had already existed before and during the war, continued to be applied after it. The main difference between these periods lies in the fact that during the war, chairmen tended to be fairly liberal in letting peasant households enlarge their plots at the expense of the socialized land fund. Any attempt at reducing the plots to their previous size would have been perceived as an assault on peasant well-being. However, in September 1946 the top party-state leadership sent a clear message to collective farmers that limitations on the size of the plots would be the order of the day. Collectivization was aimed at 'destroying the concept of the peasant *dvor* (household)',[4] but peasants did not see it that way. In fact, the number of households seems to have been the only element of agricultural life displaying continuous rates of growth during and after the war. As each household was entitled to rights over a private plot, peasants advanced ingenious means to be granted extra shares of land. The so-called 'fictional division of households' was clearly one of the weapons of the weak.[5]

In October 1949, G. Safonov, the General Procurator of the Soviet Union, made a report to Malenkov, the Secretary of the Central Committee, on the results of a general investigation carried out by regional procuracy organs into the illegal practice of fictional division of families.[6] Lamenting over its frequent occurrence, Safonov explained that peasant households were divided, or registered as separate kolkhoz households, with the purpose of receiving supplementary private plots and in order to maintain in private possession 'illegal' numbers of heads of cattle. Extended families were the usual culprits. Although Safonov's examples came mostly from regions such as Smolensk (888 cases) and Kursk (1,232 cases), he claimed that the control operation had uncovered the same phenomenon in Voronezh, Chkalov, Poltava, Orel, Velikie Luki, Gomel, Bobruisk, Minsk, Polessa in the Belorussian Republic, in the Mordvinian Autonomous Republic and in Samarkand in Uzbekistan. He gave the example of the female collective farmer Ivanova, from Smolensk, who had received a private plot of 0.25 ha after registering independently from her family while living under the same roof. Apparently, her younger brother had done the same.[7]

This type of fictional division, argued Safonov, was very easy to register and to cover with a semblance of legality. The absence of norms in the existing legislation was a serious hindrance to further control. Articles 73–84 of the Land Code of the Russian Republic had been adopted in 1922 and the instructions of the People's Commissariat for Agriculture and of the People's Commissariat for Justice, which regulated these articles, dealt only with individual householders. They did not reflect, wrote Safonov, the situation created in agriculture by the victory of the kolkhoz order.[8] In this sense, Safonov was correct. In the post-war period, the effective legal definition of a peasant household (*dvor*) was the one inherited from the 1922 Land Code of the RSFSR and of other republics, which was later enshrined in the 1935 Model Statutes and the 1936 Soviet Constitution. Regarding the partition of households, the Land Code stated that 'they were only permitted if independent farming was possible on the partitioned plots. If not, only the family property could be divided while the land was to remain in one household allotment'.[9] During the post-collectivization period, this definition of the household remained practically unchanged. In 1960, the Soviet legal scholar I. V. Pavlov defined the *dvor* as 'primarily a union between family members and without this family component the concept of *dvor* cannot be'.[10]

Apparently, Safonov had written about this legal problem earlier in 1948 but no answer to his request for legal modifications had ever been given. Seemingly embarrassed by the situation, Safonov nevertheless recommended that the problem be solved by judicial means and suggested that People's courts or general assemblies of the kolkhoz membership be granted the right to intervene in such matters. Among suggestions for discouraging the practice, there were some grassroots proposals for a real allotment of plots on a per-eater basis (*na edoka*). Such was the proposal of the rank-and-file *kolkhoznik* Mikhail Zhelud-kov from Penza, who suggested it in his 1947 letter to A. A. Andreev, the chair-man of the Council for Collective Farm Affairs. Zheludkov noted that many young peasants broke the rules with impunity. He considered it to be a common occurrence that many sons broke away from their extended family as soon as possible upon marriage so as to receive a supplementary plot, a garden and a cow.[11] This tended to create small households alongside more traditional or extended ones, all using the advantages of the same size of land. Fictional splin-ters were practically impossible to detect. This was a conclusion Safonov, along with I. V. Benediktov, the Minister of Agriculture, reached in May 1950 when they reported to G. M. Malenkov that resolution drafts to combat fictional splin-ters were pointless unless a new procedure regulating formation and disband-ment of households replaced the current one released in 1927.[12]

Young people splintering away from their family in order to gain private plots was only one expression of the difficulties of integrating young people into the collective farms. Recruitment of youngsters into collective farms became a major problem after the war. In the 1930s, younger *kolkhozniki* 'could and did leave the village in large numbers'[13] and there are no reasons why young people should have stopped after the war to spearhead the exodus. Yet, those aged 12–16 were first subjected to a labour-day minimum in 1942. They certainly played an important role on farms during the war, but their contribution to collective labour clearly dwindled as soon as the conflict ended.[14] Most were reluctant to work on the farms after the war and simply refused to join the farms upon their sixteenth birthday. In September 1949, Andreev requested proposals for modifications to the Model Statutes of Collective Farms from a select group of kolkhoz chairmen. They replied by sending lists of statute clauses which required, in their view, urgent improvements in order to tighten state rule over the collective farm peasantry.[15] Among these proposals, one of the most frequent complaints was that of young villagers refusing to join the kolkhoz upon their sixteenth birthday. In fact, the language of the Statutes was quite vague in that regard. The text stated that young members of collectivized households could join (*mogut vstupat'*) the kolkhoz and that this decision required the approval of the collective farm general membership assembly. Many rural teenagers, certainly upon the advice of their parents, used the legal gap to live a life on the

Table 5.1 Comparison between the number of rural teenagers and the number of peasants joining the kolkhoz, 1942–1953

Year	Teenagers	Joined the kolkhoz (number of households)
1942	4,249,000	
1943	5,701,200	
1944	6,447,400	
1945	6,064,400	Not avail.
1946	5,412,000	Not avail.
1947	5,226,400	842,041
1948	5,193,000	639,332
1949	5,926,100	115,553
1950	6,612,200	693,293
1951		310,400
1952		264,300
1953		340,800 (persons)

Source: RGAE, f. 1562, op. 324, d. 406, ll. 1–14; d. 632, ll. 7–14; d. 884, ll. 1–14; d. 1369, ll. 1–13; d. 1774, ll. 1–20; d. 2568, ll. 1–34; d. 3068, ll. 1–27.

fringes of the kolkhoz system and were also reticent to be drafted in the various labour reserves. Many deserted despite the threat of criminal prosecution.[16]

On the farms, the problem of membership mentioned above can be better assessed by an analysis of the size of the population segment that chose not to join the kolkhoz in the period covered by this study. Table 5.1 is based on incomplete kolkhoz labour statistics, but it is nonetheless possible to reconstruct a fragmentary picture of the reality of young people in the post-war countryside. A simple comparison between the second and third columns shows that the number of households joining the farm yearly does not correspond to the pool of potential candidates, namely young people. Despite the fact that these statistics do not provide data for the number of households joining the collective farms before 1947, it is possible to compare the number of new members with the number of teenagers living and working in the village. It appears that the number of people joining the collective farm, given in numbers of households, was, year over year, lower than the potential number of young peasants reaching sixteen years of age and expected to become full members of kolkhozes. The fact that data concerning new members is given in number of new households creates a supplementary difficulty, but it has been shown that post-war kolkhoz households were fairly small with an average of 3.5 members.[17] Should this average prove accurate, it would mean that the number of new members who joined the kolkhoz reached, in 1947, a maximum of 2.9 million people, a year of famine in both rural and urban areas; the abyss was reached in 1953 with only 340,800 new kolkhoz members, while the teenage population numbered between five and six million during the post-war period. Also, one must keep in mind that some adults, especially the overtaxed individual householders, joined the kolkhoz along with young people from collectivized families. Young people thus had a simple choice: if they did join the kolkhoz, it was to their best advantage to split away from the extended family and set up on their own with a separate private plot and livestock. This would explain the large number of small households in the Soviet countryside. If they did not, however, they still enjoyed the possibility of having access to their parents' plot while engaging in a variety of non-kolkhoz activities. Data thus suggest that a substantial number of rural youth chose to avoid the economic, social and legal disadvantages linked to collective farm membership, but further research is needed to determine what their actual occupations were. Significantly, no post-war rural teenagers could have had any memory of pre-collectivization life, nor had they ever known any other economic system.

The 'loosely-collectivized' elements

With the disappearance of the figure of the kulak prior to the war, the new target of official propaganda in the village was rapidly oriented towards those who collaborated with the German occupier during the war. However, in a majority of regions which had not experienced occupation, the new enemy was represented by those rural elements who profited from wartime circumstances and enriched themselves on the black market and via kolkhoz enterprises. After the war they were ideologically merged with those who simply enlarged their plots during the war and contributed a minimal effort to their collective farms. This can be seen in the September 1946 Central Committee and Council of Ministers joint resolution attacking 'illegal enlargements of private plots at the expense of collective farm lands' which was even considered 'exploitation of fellow farmers by a few elements'.[18] At the same time, the government had tremendous difficulty controlling the attribution of private plots, which tended to be granted to households (*dvory*), notwithstanding the number of able-bodied farmers.[19] Soviet agricultural organs were somehow trapped by the concession made in 1935 of granting households the free use of a plot and had a hard time finding a mechanism that would force *all members* of households to work in the kolkhoz. There is no strong evidence that such efforts ever succeeded during the period covered by this study.[20] When peasants worked in the kolkhoz, they were usually rewarded the same pay by the kolkhoz management, no matter how hard or how long they worked.[21]

As it was shown earlier, the number of peasants not fulfilling the minimum amount of labour-days numbered a few millions every year. It was argued by representatives of the central bureaucracy that both local powers and Prosecutor's organs did not enforce the existing laws with zeal, as the impact of this legislation on peasant labour was not tremendous. The events of 1948 showed that the problem of labour discipline and fake households was obviously too complex to be solved by open repression. Inspectors from the Council for Collective Farm Affairs gathered data on members of collective farms who were not involved in collective farm work in 1948 and 1949. That same organ produced a long report describing the loopholes in the system of membership. The same year the Central Statistical Administration informed Malenkov of important deficiencies in the data collection on the rural population. These reports did not result in any major policy reviews and remained largely dead letters. They indicated, however, a growing official awareness of the threat posed to overall agricultural performance by rural de-population.[22] For the purpose of this study, they provide meaningful examples of the multiple ways grey zones were formed on the margins of kolkhoz society.

In September 1948 the main official responsible for the implementation of the Statutes, N. Diakonov, warned the Chairman of the Council for Collective Farm Affairs, A. A. Andreev, of the existence of a large number of rural elements in rupture with their former kolkhozes (*otorvavshiisia*). He made sure to use the phrase 'loosely collectivized elements' and not 'de-collectivized'. Diakonov pointed out that the overall number of collectivized households decreased by a million between 1940 and 1947. Furthermore, this situation was especially intolerable, since the number of households of '*rabochie i sluzhashchie*' (workers and white-collar workers) living in rural areas had increased to 50 per cent of all households in 'purely agricultural areas' like the regions of Omsk, Stavropol', Penza, Mogilev and the Udmurt autonomous republic. The households worrying Diakonov consisted of 'former' collective farmers breaking away from their collective farms, but keeping their private plots through deceit. They were no longer fulfilling their obligations to the state, to which they had been subject as individual householders.

Using the example of two peasants living in the same kolhoz in Cheliabinsk, Diakonov provided striking data about the advantage of 'breaking away from the collective farm' (Table 5.2).[23] It is clear that Elizarov would have been choked by taxes, if he were to register as an officially registered individual householder. He made ends meet by inhabiting the grey zone of 'peasants breaking away'. Elizarov was simply designated a 'worker living in a rural area'. In fact, there were good reasons to avoid being registered as an individual householder.[24] Soviet policies towards individual householders were effective and led to the almost complete disappearance of *edinolichniki* after the Second World War. V. F. Zima in his book on the 1946–1947 famine suggests that there were close to 200,000 individual households compared with the 232,000 kolkhozes in the Soviet Union. They occupied 9.4 per cent of all arable lands in 1940 against 1.3 in 1950.[25] They had clearly become an endangered species by the time Stalin died.

The reasons behind Elizarov's behaviour are not too difficult to find. As a follower of a well thought-out and well-trodden strategy for coping with the postwar crisis in collective agriculture, Elizarov was by no means an extreme oppositional. His 0.15 ha private plot paled in comparison with other 'former' collective farmers using a more comfortable position and plots of up to one hectare. Indeed, Elizarov is representative of 40–47 per cent of all peasants in some districts in Khabarovsk, 54.8 per cent in some farms in the Udmurt ASSR and close to a third in the entire country.[26] The appearance of 'grey-zone farmers' annoyed Soviet agricultural authorities and constituted a direct blow to

Table 5.2 Obligations of two categories of peasants from Cheliabinsk *oblast'*

	Kolkhoznik Mel'nikov E. P.	'Former' kolkhoznik Elizarov, G. K.
Dimensions of the private plot	0.15 ha	0.15 ha
Number of cows	1	1
Number of sheep	4	–
Poultry	–	10
Money paid in agricultural taxes	470 roubles	350 roubles
Tax in kind: milk	260 litres	260 litres
Tax in kind: potatoes	320 kg	85 kg
Timber cut for the State	20 cubic metres	0
Days spend on road repairs	6	0
Labour-days earned	647	0

Source: RGAE, f. 9476, op. 1, d. 727, 11. 8–9.

the very foundations of the collective farm system. They provided an example for those wishing to avoid many of the disadvantages that kolkhoz member-ship usually implied. Supplementary evidence in the form of schemes of land tenure in the collective farm surveyed by Diakonov's staff clearly shows that 'former' *kolkhozniki* were not in any way expelled from the farms, but rather could go on enjoying the benefits of the same private plots they used as full members of the kolkhoz.[27] Furthermore, repressing or eliminating these ele-ments required resources the top state-party leadership could not or would not spare. The 1948 campaign against idlers was 'legally' intended only against registered *kolkhozniki*.[28] Diakonov may have qualified the appearance of 'former collective farmers' as 'unnatural' (*nezakonomernoe*), but he could only target representatives of local power for the unbearable slack in commit-ting 'such mistakes and perversions of party politics'.[29] He failed, however, to point out all the loopholes which made this strategy possible for peasants and collective farm chairmen.

Another possibility for enlarging the grey zone was created by the poor state of local statistics on collective farm population. It is clear that they were of great economic importance since planning organs used yearly reports from collective farms processed by the Central Statistical Administration in order to assess the amount of work to be extracted from individual collective farms and establish planning targets to be met. There was a strong tendency to under-report cat-egories of the collective farm population for which labour quotas were set up in regional and national plans. The results of a survey conducted in the regions of Poltava, Kiev and Dnepropetrovsk in Ukraine and Vologda, Iaroslavl', Molotov Riazan, Kursk and Rostov in the RSFSR are shown in Table 5.3. It is interesting to note that the unrecorded presence of some peasants on collective farms estab-lished yet another grey zone, since unaccounted peasants were hard to control. 'It is not accidental', the Chairman of the Statistical Administration Starovskii wrote to Malenkov in August 1949, 'that the majority of able-bodied individuals

Table 5.3 Results of a survey conducted by the Central Statistical Administration showing differences in population accounting in nine regions of the Ukraine and Central Russia, 1948

Category of population	According to data from sel'sovet	According to data from yearly reports of collective farms	Difference (nedouchet)	Difference (%)
1 Able-bodied men from 16 to 60 and women from 16 to 55	4,614	3,908	706	15.3
2 Teenagers from 12 to 16 years	790	637	153	19.4
3 Elders and disabled	1,469	1,195	274	18.7

Source: GARF, f. 5446, op. 53, d. 4415, ll. 34–38.

had access to a private plot and half of them work only on this plot while the other half make their living as wage earners'. Aside from family members of the collective farm administration this group was made up of peasants who had been excluded from the kolkhoz 'on paper' without having lost their plots – in most cases brides who had joined their husbands without properly registering their membership.

Finally, another type of 'marginal' element was created by local implementations of rules concerning membership in the collective farms. Although the Model Statutes of 1935 defined membership as voluntary, the absence of internal passports for peasants greatly reduced peasant mobility and hindered their capacity for working where they wished. Peasants consciously violating important clauses of the Statutes could be expelled, but this occurrence was rare and usually did not lead to a situation where peasants could join a new collective farm. In short, mobility was not supposed to be a feature of the collective farm system. Thus when reports started appearing in 1948 about the question of *kolkhozniki*'s free choice of place of work, the Council for Collective Farm Affairs once again targeted regional and district authorities for their neglectful attitude towards peasant mobility.[30] The main reason invoked for departure was better earnings in the new kolkhoz. While peasants who worked in a kolkhoz different from their home kolkhoz remained collective farmers *de facto* and *de jure*, the practice strongly displeased central agricultural authorities who saw in it a main cause for disorganization of production, weakening of work discipline and misuse of labour in collective farms. Moreover, there was only one step between this type of migration and the unofficial status of 'former' *kolkhoznik*, because those peasants could easily break away from their farms and simply work as agricultural day-labourers. First noticed in the Republics of Central Asia, where economic inequality between kolkhozes was high, migration between collective farms took on 'massive proportions'. The phenomenon eventually appeared in other regions like Kiev, Riazan, Molotov, Stavropol', Omsk, Groznyi and Novosibirsk.[31]

Seasonal migrants

The phenomenon of *otkhodnichestvo* (seasonal migration) is certainly one of the central characteristics of the Russian peasantry in the post-emancipation period and it had thus been the object of many remarkable studies.[32] However, we know very little of the post-collectivization period and it is sometimes wrongly assumed that the harsh passport regulations issued in 1932 had somewhat succeeded in reducing the flux of rural migrants to Soviet cities. It is also assumed that the pattern of migration was from the village to the cities, which is certainly not *otkhodnichestvo*, but simply rural exodus.[33] Newly accessible archival material reveals that the decrease in rural population started earlier than 1950,[34] but that there were clearly movements back and forth between villages before that date, after which peasants obviously lost any hope of any improvement of their condition in the farms and so left permanently. Migrant workers undermined the grip the kolkhoz system had on them, since their behaviour was hard to control by chairmen and they were little dependent on kolkhoz income for their subsistence. As such they constituted another segment of the grey zone formed within the kolkhoz – one that was especially difficult to define, let alone control.

Kolkhozniki dealt with economic hardship by combining the income coming from their private plots with payments in labour-days from collective farm work and earnings gained from extra-agricultural activities. Given the extraordinary hardship provoked by the post-war agrarian crisis, the collective farmers' first reaction was to spend more time and energy on their private plots even if it became less profitable in absolute terms. A second reaction was to seek employment outside the collective farm. Work outside the farm and the private plot remained statistically important, as shown by the results of a July 1951 survey of 21,000 peasant families in 1940 and 12,600 families in 1950.[35] Three major conclusions can be drawn. First, the income from the private plot was reduced (54.2 per cent of total income in 1940 and 43 per cent in 1950) by growing taxation,[36] but still represented the most important source of income for *kolkhozniki*. Second, the income coming from the collective sector slightly grew, from 15.1 per cent in 1940 to 19.4 per cent in 1950, but remained one-fifth of the total income for the budgets included in the study. Finally, work outside the farm grew slightly after the Second World War (from 15.5. per cent in 1940 to 19.4 per cent in 1950), but – somewhat surprisingly – represented the same share of the total income as the collective farm sector. Of course, the results of this budget study only apply to a sample of families. Yet this provides an insight into a contradiction enshrined in the reality of the post-collectivization village: the kolkhoz was meant to be the main provider of income for its members but remained only one source among many. It also suggested that an important part of the able-bodied kolkhoz population engaged in one form or other in seasonal work.

Kolkhoz population statistics can provide further insight into this phenomenon. First of all, collective farm work did not employ *all* peasants for the *whole year*. As an average for the Soviet Union, in 1947, 50.1 per cent of all able-bodied *kolkhozniki* did some work on the farms in January, while kolkhozes reached full employment in the peak of the summer rush in August. Three years later, only 39 per cent of all peasants participated in collective farm work in January against 74.6 per cent in July 1950.[37] This meant that in 1950 between 7.8 and 16.6 million peasants were free to engage in work outside the farm at some point in the year as they did not participate in any work in the kolkhoz. Preliminary data established for 1947 and 1949 show 'official *otkhod*' in small numbers. Nonetheless a relatively large contingent of peasants was temporarily absent from their collective farms (when compared with the total able-bodied population) and their low average of labour-day earnings provides supplementary evidence of their minimal involvement in collective agriculture (Tables 5.4 and 5.5).

Table 5.4 Comparison between registered able-bodied *kolkhozniki* and *kolkhozniki* working in sectors other than agriculture in various regions of the USSR, 1947 (in thousands)

Region	A/b men	A/b women	Temporarily absent[a]	Other occupation[b]
RSFSR				
North	136.9	317.7	71.4	62.1
North-West	140.4	266.3	26.8	28.7
Cent. Non-Black Earth	1,649.5	3,430.0	369.2	456.7
Cent. Black Earth	876.9	1,842.0	208.8	181.1
Middle Volga	559.1	1,060.8	163.5	61.2
N. Caucasus:				
Crimea	577.4	1,616.2	202.5	24.5
Urals	564.7	1,087.3	179.0	53.8
West Siberia	508.7	933.8	166.2	25.9
East Siberia	235.2	352.7	64.4	9.5
Far East	46.7	54.7	13.9	1.7
Central Asia	1,291.8	1,648.8	396.1	38.6
Caucasus	636.8	766.4	110.8	51.4
West Ukraine	64.1	90.7	84.8	5.7
East Ukraine	1,883.0	3,939.2	634.4	308.0
Byelorus. SSR	355.0	730.9	67.5	61.2

Source: RGAE, f. 1562, op. 324, d. 406, ll. 1–14; d. 632, ll. 7–14; d. 884, ll. 1–14; d. 1369, ll. 1–13; d. 1774, ll. 1–14; d. 2179, ll. 1–20; d. 2568, ll. 1–34; d. 3068, ll. 1–27.

Notes

a Temporarily absent from the kolkhoz for work in industry or transportation, but who did accomplish some work during the course of the year.

b Living on the kolkhoz and registered as *kolkhozniki*, but working full-time in transportation, state enterprises or industry.

Table 5.5 Comparison between registered able-bodied *kolkhozniki* and *kolkhozniki* working in sectors other than agriculture in various regions of the USSR, 1949 (in thousands)

Region	A/b men	A/b women	Temporarily absent	Other occupation
RSFSR				
North	133.2	298.5	53.3	60.7
North-West	140.7	273.0	21.8	31.8
Cent. Non-Black Earth	1,593.2	3,343.7	404.2	596.3
Cent. Black Earth	903.6	1,891.8	239.6	273.5
Middle Volga	516.2	1,006.4	150.3	87.2
N. Caucasus:				
Crimea	551.0	992.0	107.3	40.4
Urals	540.9	1,043.4	141.8	62.2
West Siberia	504.6	913.5	11.4	24.0
East Siberia	243.1	361.6	44.8	12.8
Far East	47.5	60.5	44.8	3.6
Central Asia	1,387.0	1,717.5	282.9	63.7
Caucasus	633.7	774.5	101.1	97.5
West Ukraine	684.5	968.1	112.0	138.4
East Ukraine	1,875.6	3,886.4	411.5	408.3
Byelorus. SSR	439.0	829.6	93.9	89.6

Source: see Table 5.4.

Passport regulations made *otkhodnichestvo* an illegal activity if enacted without the authorization of the kolkhoz management. The government repeatedly attempted to regulate migrant work. In July 1938 Sovnarkom issued a resolution aimed at regulating recruitment of workers from the kolkhoz through the institution of Orgnabor (Organized Recruitment). This legislation, among other things, fixed the number of representatives per region and district in order to increase central control. The resolution paid specific attention to a list of 32 *oblasti*, mostly in the Ukraine, Belorussia, in the Middle-Volga, Central Black Earth and non-Black-Earth regions where *otkhodnichestvo* was fairly common.[38] Yet legislation restricted the power of chairmen to exclude peasants engaged in this type of labour migration, in order to maintain large contingents of *kolkhozniki* available for the fulfilment of plans. A resolution to this effect was passed in April 1938.[39] This protection, however, played against the very concept of labour discipline since the preservation of the peasants' rights and access to a private plot tended to hinder the effectiveness of any other coercive measure. The following year, a different resolution from Sovnarkom legalized expulsion for non-fulfilment of the labour-day minimum. In 1947, the Central Committee paid particular attention to the regions of Briansk and Gor'kii, where thousands of peasant households were expelled for reasons of non-fulfilment of the prescribed labour minimum (30.6 per cent of all cases) and for temporary departure to industrial sites of at least one member of the family (46.3 per cent).[40] In 1947, a new resolution with the same objectives came to clarify many of the dispositions of the previous measure. Significantly, it ordered kolkhoz chairmen not to interfere with the work of the Orgnabor, and not to oppose the will of *kolkhozniki* who wished to extend the duration of their contracts or of members of their family willing to follow them.[41]

In total, Orgnabor statistics indicate that 773,000 rural workers were recruited in 1946 and 667,000 in 1948. Together with recruitment by enterprises and FZO/RU and Railway Schools, a total number of 2.1 million was reached in 1946 and 1.5 million workers in 1948, much lower than the data obtained from kolkhoz statistics.[42] This means that the 'unofficial' *otkhodnichestvo* was much more important than the official one. For the farms, however, the loss was painfully resented.

Kolkhoz chairmen tended to complain *en masse* about labour problems stemming from migrants. For instance, when the chairman of the Council for Collective Farm Affairs A. A. Andreev sent a circular letter to all collective farm chairmen in September 1949, with a request to provide the organ with proposals for modifications to the Model Statutes of 1935, he received answers that pinpointed the rules of membership as marred by loopholes which facilitated seasonal work outside the farm. In these proposals, the need to increase the labour-day minimum, the need to institute a special clause that would automatically register teenagers as kolkhoz members upon their sixteenth birthday (the minimum age for kolkhoz membership), and the need to strengthen chairmen's discretionary power to exclude *otkhodniki* from the kolkhoz and deprive them of their private plots were common occurrences.

Chairwoman Kuklina, from Kirov *oblast'*, wrote in great detail about the problematic lack of a labour force in her region. She gave concrete examples of collective farms with a sown area of 543 ha tilled by 18 able-bodied peasants, while a neighbouring village could rely on 38 *kolkhozniki* to till 463 hectares of land.[43] This inequality was to be solved, in her view, by forcing all peasants to take a more active part in kolkhoz work through a reduction of the size of their plots.[44] There were also entire households who had refused to join the kolkhoz for more than 20 years and led a 'parasitic way of life', often working as day-labourers in the collective farm or in industry. Solutions to this problem varied. A chairman from Saratov mentioned that the Charter contained a clause granting the kolkhoz management the right to allow peasants to leave the kolkhoz for seasonal labour outside the farm. In his view, this led to abuses by chairmen who tended to favour their relatives at the expense of other peasants. He suggested that only the general membership assembly should grant peasants the right to leave the kolkhoz temporarily.[45] The same opinion was expressed by a chairman from Orel, who suggested that uncontrolled departure from the kolkhoz be banned without the explicit agreement of the general membership gathering. The violators should simply be excluded from the kolkhoz. This was justified by the chairman's statement that the 'illegal departure from the kolkhoz has taken on massive numbers'.[46]

A chairman from Stalino *oblast'* in Ukraine, V. T. Litvinov, described the peculiar problems of his kolkhoz being located near an industrial site:

The peculiarity of our kolkhoz is that it is located near a few important coal mines and stone-pits. The majority of the population of the village of Kurakhovka included in our kolkhoz works in industry. In the kolkhoz, there is no single household not having a member working in the industry. As a rule, the majority of able-bodied peasants is working in this sector and only elderly peasants, housewives and school children in the summer work in the kolkhoz. There are 146 households counting 296 able-bodied peasants, 123 of them being males. But only 113 people work in the kolkhoz, among them 25 men. In essence, this situation leads to the fact that the main source of income for most families is the wages earned in the industry, and the kolkhoz appears to be just a subsidiary source of foodstuff. Those peasants tend to have private plots of 0.4–0.5 ha, privately-owned livestock, and use kolkhoz facilities. As workers, they pay for their livestock neither taxes in kind nor in cash. This considerably weakens labour discipline.[47]

As a remedy, Litvinov recommended raising the labour-day minimum in kolkhozes close to industrial centres to 300 days for men and to 250 for women, applying the same laws against absenteeism that existed in industry and prosecuting industrial managers who hired labourers from collective farms without the chairman's explicit agreement.[48] This required a firmer attitude from kolkhoz management, and more control over the management from the membership assembly.

In general, it has been demonstrated that many peasants kept one foot in the village and another outside of it – whether on another collective farm, on a sovkhoz, in a factory or in any other form of employment. Many kolkhoz chairmen were not satisfied with attempts to tackle this problem, and recommended harsh measures against fellow collective farm chairmen who allowed such practices, and against factory managers who hired rural migrant workers. However, resisting the constraints and the poverty resulting from full-time kolkhoz participation, peasants engaged in a 'self-decollectivization' against which chairmen could do little. As a result seasonal migrations turned into a permanent exodus after 1950.

Conclusion

Two decades after the onset of collectivization the collective farm system had trouble socializing the younger generation of peasants, who were in no position to remember village life before the advent of the kolkhoz. It is surprising that, despite the rigour of the collectivization campaign, the labour-day minimum for kolkhoz members – the backbone of the whole system – was not properly enforced. Expelled members were allowed to stay in kolkhoz territory and make a living in direct contradiction to socialist emulation. Furthermore, it is surprising to discover that the rigid passport regulations set in 1932 did not stop millions of collective farmers from seeking employment outside their village, which set the stage for their more definite exodus after 1950. Yet it was precisely these dynamic migrants and young people who would have been capable of instilling energy into a system that desperately needed it. Instead they shared the fatalistic attitude so well-described by Ol'ga Berggolts, who visited a village in the region of Novgorod and wrote in her diary in 1949:

> Everyone in this village is a victor, and these people are a people of victors. But as some may say, what does it bring? Well, there are postwar difficulties, a Pyrrhic victory (at least for this village) – but are there any perspectives? I am somewhat struck by the situation of oppression and submissiveness of those people, who are almost reconciled with the absence of perspective.[49]

Such observation might help to explain the passivity of post-war rural society. Yet passivity might indeed be a misleading term. Everyone coped in their own way and the variety of means peasants used to survive made the collectivized peasantry something very different from a homogenized mass. Indeed, the diversity of their survival strategies was such that control of their behaviour was beyond the grasp of the late Stalinist state.

Notes

1 *Krest'ianstvo i gosudarstvo (1945–1953). Sbornik dokumentov*, V. P. Popov (ed.), Paris: YMCA Press, 1992, esp. pp. 233–242.
2 Most notably M. N. Denisevich, *Individual'noe khoziaistvo na Urale (1930–1985 gg.)*, Ekaterinburg, 1993 and *Sel'skoe khoziaisvto Urala v pokazateliiakh statistiki (1941–1950)*, V. P. Motrevich (ed.), Ekaterinburg, 1993.
3 Sheila Fitzpatrick, *Stalin's Peasants: Resistance and Survival in the Russian Village after Collectivization*, New York: Oxford University Press, 1994, pp. 152–173.
4 Ibid., p. 112.
5 James C. Scott, *Weapons of the Weak: Everyday Forms of Peasant Resistance*, New Haven: Yale University Press, 1985.
6 RGAE, f. 9476, op. 2, d. 36, ll. 4–7.
7 Ibid., l. 5.
8 Ibid., l. 7.
9 V. P. Danilov, *Rural Russia Under the New Regime*, ed., trans. Orlando Figes, London: Hutchinson, 1988, pp. 247–253, 224, 229–230.
10 I. V. Pavlov, *Kolkhoznoe pravo*, Moscow: Izdatel'stvo iuridicheskoi literatury, 1960, pp. 334–340, esp. 338–339. Furthermore, Pavlov defines the characteristics of the *dvor* as: the right to a private plot, the existence of labour relationships between the household members, and their participation in the collective sector. According to Pavlov, the household could be enlarged in cases of marriage or birth, but the inclusion of outside members (*postoronnye* or *litsa so storony*) – described by the generic term *primachestvo* – has to be agreed upon by all members of the household. Pavlov considers this practice to be a relic of the pre-collectivization period, rarely encountered at the time he wrote his treatise. Only death or long periods of absence from the household (with the exception of military service, study or long-term medical treatment), could be sufficient justification for excluding a person from the household. Pavlov does not look at any other causes such as the problem evoked by Safonov. Pavlov, *Kolkhoznoe pravo*, pp. 338, 340–341.
11 *Krest'ianstvo i gosudarstvo (1945–1953)*, p. 272

118   *Jean Lévesque*

12 RGAE, f. 7486, op. 7, d. 1056, ll. 13–23.
13 Fitzpatrick, *Stalin's Peasants*, pp. 98–101.
14 RGAE, f. 1562, op. 324, d. 406, ll. 1–14; d. 632, ll. 7–14; d. 884, ll. 1–14; d. 1369, ll. 1–13; d. 1774, ll. 1–20; d. 2568, ll. 1–34; d. 3068, ll. 1–27. On the North, see the data in the study by M. A. Beznin, T. M. Dimoni, L. V. Iziumova, *Povinnosti rossiiskogo krest'ianstva v 1930–1960-x godakh*, Vologda, 2001, pp. 117–121.
15 RGAE, f. 9476, op.1, d. 864, 865, 866.
16 Donald Filtzer, *Soviet Workers and Late Stalinism. Labour and the Restoration of the Stalinist System after World War II*, Cambridge: Cambridge University Press, 2002, pp. 122–123, 156. Filtzer later concludes that young workers from the countryside carried within themselves the strong resentment harboured by the peasantry towards Stalin and the Soviet regime.
17 Jean Lévesque, 'Part-Time Peasants: Labour Discipline, Collective Farm Life, and the Fate of Socialist Agriculture in the Soviet Union after the Second World War, 1945–1953', PhD dissertation, University of Toronto, 2003, p. 206.
18 *Istoriia kolkhoznogo prava. Sbornik zakokondatel'nykh materialov SSSRi RSFSR, 1917–1958gg.*, T.II, Moscow: Gospolitizdat, 1958, pp. 291–294.
19 See *Krest'ianstvo i gosudarstvo (1945–1953)*, pp. 275–280.
20 This point is made clearer in Lévesque, 'Part-Time Peasants', pp. 45–118.
21 GARF, f. 5446, op. 51, d. 2264, ll. 11–109; GARF f. r-8300, op. 24, d. 358, ll. 211–216.
22 Compare to *Krest'ianstvo i gosudarstvo (1945–1953)*, pp. 111–114.
23 RGAE, f. 9476, op. 1, d. 727.
24 During the war for instance, taxes, bonds and collections (*sbory*) made up 31.8 per cent of the income of collective farmers in 1943, while individual households would pay 80.4 per cent. M. A. Vyltsan, *Krest'ianstvo Rossii v gody bol'shoi voiny, 1941–1945. Pirrova pobeda*, Moscow: Rossiiskii nauchnyi found, 1995, p. 137.
25 *Sovetskaia derevnia v pervye poslevoennye gody, 1946–1950*, I. M. Volkov (ed.), Moscow: Nauka, 1978, p. 214.
26 RGAE, f. 9476, 1, d. 727, l–2. In their report on fictional splinters of households, Safonov and Benediktov claimed in 1950 that about half of the rural population were not members of collective farms. RGAE, f. 7486, op. 7, d. 1056, l. 17. According to Verbitskaia, 32 per cent of the rural inhabitants of the RSFSR in 1951 were not members of the collective farms. O. M. Verbitskaia, *Naselenie Rossiiskoi dereveni v 1939–1959 gg.*, Moscow: INI RAN, 2002, pp. 62–63. It is very difficult to know, however, how many of them have been *kolkhozniki* at some point.
27 RGAE, f. 9476, op. 1, d. 727, ll. 16–32.
28 Lévesque, 'Part-Time Peasants,' pp. 119–183.
29 RGAE, f. 9476, op. 1, d. 727, ll. 11–13.
30 RGAE, f. 9476, op. 1, d. 730, ll. 15–16.
31 Ibid., ll. 17–18.
32 On the pre-revolutionary period see, among many others, Jeffrey Burds, *Peasant Dreams and Market Politics. Labor Migration and the Russian Village, 1861–1905*, Pittsburgh: University of Pittsburgh Press, 1998; Robert E. Johnson, *Peasant and Proletarian: The Working Class of Moscow in the Late Nineteenth Century*, New Brunswick, 1979; Barbara Alpern Engel, 'The Woman's Side: Male Out-Migration and the Family Economy in Kostroma Province,' *Slavic Review* 45, 1989, pp. 257–271. On the NEP see V. P. Danilov, 'Krest'ianskii otkhod na promysly v 1920-godakh', *Istoricheskie zapiski* 94, 1974; on the collectivization period see Nobuaki Shiokawa, 'The Collectivization of Agriculture and Otkhodnichestvo in the USSR, 1930', *Annals of the Institute of Social Science* 24, 1982–1983, pp. 129–158 and more generally David Hoffmann, *Peasant Metropolis; Social Identities in Moscow, 1929–1941*, Ithaca: Cornell University Press, 1994. For the later 1930s see Gjis Kessler, 'Krest'ianskaia migratsiia v Rossiiskoi Imperii i Sovetskom Soiuze. Otkhod-

nichetsvo i vykhod iz sela', *Sotsial'naia istoriia. Ezhegodnik 1998/199*. Moscow: Rosspen, 1999, pp. 309–330 and Sheila Fitzpatrick, *Stalin's Peasants*, pp. 164–173. On repression towards *otkhodniki*, see Lynne Viola, 'The Second Coming. Class Enemies in the Soviet Countryside, 1927–1935', *Stalinist Terror. New Perspectives*, J. Arch Getty, Roberta T. Manning (eds), New York: Cambridge University Press, 1993, pp. 82–96.

33 See for example, V. F. Zima, *Golod v SSSR 1946–1947 gg.: proiskhozhdenie i posledstviia*, Moscow, 1996. Zima claims that close to 10 million peasants fled the countryside in the wake of the 1946–1947 famine.
34 Mark Edele, 'A "Generation of Victors?" Soviet Second World War Veterans from Demobilization to Organization, 1941–1956', PhD Dissertation: University of Chicago, 2004, pp. 39–107.
35 *Krest'ianstvo i gosudarstvo*, p. 215.
36 The post-war tax burden on a single household has dramatically increased, as demonstrated recently by V. P. Popov, going from an average 112 roubles in 1940 to 523 roubles in 1951. Popov, *Ekonomicheskaia politika sovetskogo gosudarstva, 1946–1953 gg.*, Moscow, 2000, p. 201. In spite of this, the private plot economy demonstrated considerable stamina. See Lévesque, 'Part-Time Peasants', pp. 184–254.
37 RGAE, f. 1562, op. 324, d. 406, ll. 1–14; d. 632, ll. 7–14; d. 884, ll. 1–14; d. 1369, ll. 1–13; d. 1774, ll. 1–14; d. 2179, ll. 1–20; d. 2568, ll. 1–34; d. 3068, ll. 1–27.
38 'Ob uporiadochenii dela nabora rabochei sily iz kolkhozov', Postanovlenie SNK SSSR ot 21 iuliia 1938g., *Istoriia kolkhoznogo prava. Sbornik zakonodatel'nykh materialov SSSR I RSFSR*, Tom. II, Moscow: GosIurLit, 1958, pp. 46–48.
39 RGASPI, f. 17, op. 122, d. 313, ll. 59–62.
40 Ibid.
41 'O poriadke provedeniia organizovannogo nabora rabochikh', Postanovlenie Soveta Ministrov SSSR ot 21 maia 1947g., *Direktivy KPSS I Sovetskogo pravitel'stva po khoziaistvennym voprosam*, T. III., Moscow: Gospolitizdat, 1957, pp. 205–209.
42 GARF, f. 5446 r, op. 53, d. 4415, pp. 51–52.
43 Ibid., l. 19.
44 Ibid.
45 Ibid., ll. 164–165.
46 RGAE, f. 9476, op. 1, d. 865, l. 10.
47 RGAE, f. 9476, op. 1, d. 864, ll. 268–269.
48 Ibid., l. 266.
49 Ol'ga Berg'golts, 'Iz dnevnikov (mai, oktiabr' 1949)', *Znamia*, no. 3 (1991), p. 16.

Part III

The corrupted state
War and the rise of the second economy

6   A "campaign spasm"

Graft and the limits of the "campaign" against bribery after the Great Patriotic War

James Heinzen

On 3 May 1946, P. I. Minin sent a desperate letter to Josef Stalin. In it Minin described an "epidemic" of bribery and its debilitating effects.[1] Minin, a communist working in the Political Administration of the Baku Military *okrug*, sent the letter to the *Osobyi sektor* of the Central Committee, certainly aware that through this special channel the letter was more likely to come to the attention of Stalin himself. In a tone of anger and frustration, Minin's letter, now preserved in the archive of the USSR Ministry of Justice, vividly portrays the breakdown of both formal state mechanisms and social norms during and after the Great Patriotic War. Ultimately, this letter resulted in the launching of a post-war "campaign" against bribery. In five brief but powerful paragraphs, Minin alleged that bribery was running rampant in all walks of Soviet life.

Minin did not describe the bartering, shadow-market exchanges and other types of informal "wheeling and dealing" common among managers in industrial firms scrambling to obtain scarce materials and fulfil ambitious plans.[2] Rather he referred to a tidal wave of everyday graft endured by the majority of the population desperate to make ends meet, but forced to placate state functionaries demanding payments for regular services. Minin wrote that, because of the difficulties experienced during the war, "bribery has now become an extremely widespread phenomenon." Since 1943–1944, he notes, there had been an upswing in this kind of crime among representatives of the state, too many of whom had become greatly emboldened. Bribes "are given and accepted by people of the most varied professions and in the most varied forms." Mail carriers take a "reward" for delivering mail and telegrams, and if one refuses to give the bribe, one's correspondence will be lost or delivered only in the distant future. Fitters demand payments to connect individuals to the network of gas and water pipes. Railroad employees take bribes to allow passengers to ride or claim packages. Teachers and professors take bribes to admit students into institutes or to pass exams. In sum, bribery "has become a serious evil and danger, against which we must lead a decisive battle."

To be sure, Minin's was not the first letter with such allegations that had been sent to the party leadership. Since 1943, the central prosecutor's office and the Ministry of Justice had been reporting an increase in convictions for the offering and acceptance of bribes. It is not wholly clear why *this* letter in particular inspired a campaign against graft. Most likely, it caught the attention of Stalin, without whose approval such a campaign could not have proceeded. Certainly, the letter's graphic, anguished description of a cascade of bribery at all levels of Soviet society grabbed the interest of someone at the highest reaches of the party. (It would be Zhdanov who demanded that the legal agencies respond to Minin's charges.) Doubtless, the author's position as a high-ranking party member carried weight. His position in the military district of Baku may have been an important factor. Perhaps this letter was merely the last straw, another piece in a flood of correspondence from the localities. Importantly, the letter was signed ("with communist greetings"), and therefore not one of the innumerable anonymous denunciations received every month. Moreover, the fact that Minin made allegations without mentioning names, describing instead a general phenomenon, leads one to believe that he did not simply hold a grudge against certain individuals – a common reason provoking accusations of "corruption."[3]

An analysis of the internal discussions surrounding the "struggle against bribery" that ensued in 1946 provides insight into a number of aspects of late Stalinist state and society, including the growth of a variety of informal, but technically illegal, practices and the reasons for their existence; attitudes toward pervasive graft; and an apparent hesitation by state officials to press forward with measures to control, if not eliminate, bribery. This study also allows insight into how a post-war campaign was launched, and how institutional interests shaped the parameters of the campaign.[4] Ultimately, the campaign against bribery was quite disappointing. The campaign's weaknesses – and the reasons for them – highlight significant features of the late-Stalinist state and its interactions with a society struggling to recover from the catastrophe of war. The opening of state and party archives now allows an exploration of the campaign's origins, the way that the campaign was discussed, formulated and received by various state actors, and its rather haphazard implementation. Such sources were, of course, unavailable previously, forcing observers to depend primarily on the Stalinist press.[5] This research examines bribery particularly as viewed through the legal agencies such as the *prokuratura* and courts. These agencies were most deeply involved in the investigation and prosecution of bribery cases. Yet this crime also affected the legal organs in a disproportionate way, since they could dispense an extremely valuable commodity in the late Stalin period – the mitigation of legal penalties, including freedom from prison. Individual prosecutors, judges and other personnel in the legal agencies could charge handsomely for this service. The courts were therefore at the centre of the anticorruption campaigns – in an uncomfortable, dual position as both a key institution in the implementation of the campaign and as one of its main targets.

Background: bribery in the Second World War

A critical, largely unrecognized turning point in the blossoming of the kind of corruption that was a hallmark of the Brezhnev era was the Great Patriotic War and its aftermath. The years between 1943 and 1947 also represent a pivotal period in the state's efforts to control corruption. Certainly, such criminality existed before 1941. The introduction of the five-year plans acted as a major impetus for the growth of the black market and informal mechanisms inside the socialized economy. It was the period of war and reconstruction, however, that created conditions to cement and accelerate existing patterns.[6] Wartime conditions created an atmosphere conducive to official crime. During the war, the various permutations of official graft, theft of state property by officials and corruption became more deeply entrenched, especially after 1943. The midpoint of the war seems to have been a crucial moment for many forms of crime. In the words of one Ministry of Justice report, the "general moral uplift in the country" was one major reason for the decline in crime at the beginning of the war. By 1943, however, arrests for many types of crime, including crimes by officials, had leapt significantly.[7] By the end of the war, law enforcement agencies issued new measures to combat bribery, together with renewed measures against networks of theft and speculation. Extraordinary shortages, dislocation and famine put officials in the position to benefit at their posts. Money was tight, and much of the population lived on the brink of poverty.[8]

To take the case of the legal professions, a number of factors were in play. As the Minister of Justice N. M. Rychkov pointed out in a letter to Stalin, during the war a huge turnover among judges took place due to death, injury and relocation of personnel. The majority of officials in place by the end of the war were new, often quite young, usually without juridical training or experience, or, for that matter, a well-developed professional ethic.[9] In many places, "it was necessary to build the entire judicial apparatus all over again, from scratch." Such was the case in Ukraine, Belorussia, Lithuania, Latvia, Estonia, Moldavia, Karelo-Finlandia and in the occupied parts of the RSFSR. At the same time, the pay of judges and prosecutors dropped during and after the war, making them more amenable to consider accepting, or soliciting, bribes.[10] According to OBKhSS records, for example, the police in Krasnodar *krai* in 1946 uncovered the case of a judge who accepted 15,000 roubles in exchange for the acquittal of a warehouse employee who stole six bags of flour.[11] When attempting to explain the corruption of their personnel, the *prokuratura* and judicial authorities focused on these issues – pay and professional ethics. One of the most common types of illegal payoffs emerged from the USSR's extraordinary post-war housing problems. Bribes were often offered to gain access to housing (or not to be removed from it).[12] This matter was especially pressing for demobilized soldiers or for people who lived in the wrecked areas undergoing reconstruction. People returning from evacuation wanted their apartments back.[13] At the same time, rural officials found fertile ground for graft. Peasants paid bribes to employees of the Ministry of Food Collection (*Ministerstvo zagotovok*) to escape obligatory deliveries of food to the state. Employees sold fictitious receipts that declared that peasants had met their obligations.[14] The police responsible for using informants to expose graft confirmed that bribery appeared to be widespread in state and cooperative supply and trade organizations. Bribes were given to silence inspectors who uncovered cases of theft by bookkeepers and other responsible officials.[15] Also widespread was the practice of doctors taking bribes to release people from military service or jobs.[16]

A category of graft that affected judges, prosecutors and other law enforcement personnel in particular was the offering of payments to free from prison persons arrested for theft or violating labour laws, or for illegal trading or speculation.[17] In one case, a certain Olga V. Sprimon between June 1943 and the end of 1944 arranged for 11 people to get out of jail. As a secretary in the Military Tribunal in Moscow, she had access to case files and the stamp of the Military Tribunal, and was thus able to send secret correspondence. She regularly sent to NKVD camps and colonies false copies of decisions of the Military Collegium of the Supreme Court and the military tribunal of the Moscow Military District about the need to release certain people from serving the rest of their sentences. For this service, she received bribes from the relatives of the prisoners and from the convicts themselves. The payments were sometimes monetary, but also took the form of food, products and valuables. Over the course of 18 months, she received over 200,000 roubles worth of payments. In exchange, she freed 11 people, including five who had received ten-year sentences under the 7 August 1932 law on theft of public property.

Many people who offered such bribes were acting on behalf of defendants who had been given very long, mandatory sentences for theft of state property. The regime had created a class of potential bribe-givers among the families of imprisoned thieves. This provided ample opportunity for average people to become involved in bribery, as they attempted to buy the freedom of a family member. Many officials in the law enforcement organs, of course, could not resist the temptation to accept these offers. Provisions, housing and freedom from imprisonment: these diverse commodities were scarce, in extremely high demand, and worth paying for. People with the ability to distribute these commodities could earn sizeable sums. Not surprisingly, these forms of graft continued into the post-war period.

Minin's complaints

It was against this backdrop of wartime and post-war crises that Minin wrote his impassioned letter about the prevalence of bribery. Minin's observations, largely confirmed by law enforcement officials, provide insight into the bribery phenomenon after the war. Perhaps most disturbing to Minin was the degree to which people in positions of authority either tolerated or actively participated in bribery: "Unfortunately, certain responsible workers have no qualms about bribery, because most often it takes the form of a gift, offered either in kind or as money." Minin's concern that officials believe that a "gift" is something less than a bribe, and certainly nothing to concern oneself with, is critically important. The lack of conscience is a very dangerous sign, indicating that officials have become inured to the stigma that should, in Minin's mind, accompany graft. In descriptions of graft, legal agencies also often described bribery as akin to an "epidemic," a disease that spread like a contagion; corruption begat corruption. Yet it was a disease against which many officials felt no need to develop a resistance. Taken together – the epidemic nature of bribery in combination with the high level of tolerance for it among responsible officials and society at large – created the possibility for a very dangerous situation. Party legal experts and ideologists – echoing Lenin – often stated that bribery was an "especially disgraceful crime," or "one of the most disgusting of crimes."[18] Yet Minin notes that the social controls that should be in place to stop it were absent. Potential criminals have no sense of shame, and eyewitnesses feel no moral outrage about such crimes (and therefore do not report them). One can see this troubling "conspiracy of silence," for example, in the arrests of groups of people in courts, warehouses, housing offices and other workplaces, where criminal collectives worked together and protected each other. With a bit of digging by the *prokuratura* and the police, interlocking criminal relationships linking other participants were uncovered, exposing the "epidemic" – but only if law enforcement decided stubbornly to pursue the matter. If just one person in the workplace had informed to the authorities instead of participating, the gang would have been exposed.

This wave of bribery had grave ramifications both for the regime and for the state's relationship with the population. Minin wrote: Bribes "corrupt both the giver and the receiver, corrupt the work of government institutions and enterprises, become a serious obstacle in our construction, [and] provoke legitimate dissatisfaction and indignation among the labouring masses." Having noted the demoralizing effect of the need to give bribes, Minin argues that graft goes further, creating a rift between the population and the state. In his opinion, the perception among the population that bribery typically goes unpunished created hostility toward the state and served to de-legitimize the organs of government. Only a serious, all-Union campaign against bribery, including strict new laws, can begin "to uproot bribery and everything connected with it." All levers of action and *obshchestvennost'*, including the *kontrol'* and punitive organs, must be used in a crackdown. Minin believed that the cornerstone of the battle should be the promulgation of either a *Postanovlenie* of the Council of Ministers or a harsh new *Ukaz* of the Supreme Soviet (either of which would have been public and had the force of law) aimed specifically at eradicating bribery.

Although crime statistics were not published in the Stalinist USSR, of course, reports available in Ministry of Justice archives suggest that there were surprisingly few convictions for bribery in the late-Stalin period. The archival material discussed below further indicates that a large majority of bribery went unreported and unpunished, and that the legal agencies were well aware of this fact.[19] The largest number of convictions for bribery in any year between 1937 and 1956 was about 5,600 (in 1947).[20] In 1946 and 1947 within the borders of the USSR, for example, approximately the same number of people were convicted of premeditated murder as were convicted of bribery.[21] According to Ministry of Justice figures, in 1946 citizens' courts convicted 4,695 people of premeditated murder in the Soviet Union; about 3,900 were convicted of giving or receiving bribes. The figures for 1947 are similar: 5,657 people were convicted of bribery, and 6,215 were convicted of murder. If we keep in mind that 1947 marks the high point of convictions for bribery in the period, such figures suggest the huge gap between the actual level of bribery and the limited way that phenomenon was reflected in Soviet courts.

To be sure, law enforcement faced a number of problems in prosecuting bribery. There were usually no missing stock or products, no falsified account books or empty cashier's drawers. Unlike murder, there was no body or missing person as evidence that a crime had been committed. A bribe was a private agreement, typically between two people, that left no obvious trace. The transaction was likely to go unreported by either side. In nearly every case, the authorities discovered the existence of a bribe only when one party became dissatisfied with the "deal" that had been struck, feeling that they had been "deceived," that "their money vanished in vain." As one *prokuratura* document marvelled, "Without fearing threat of punishment for offering a bribe, they [bribe givers] informed various organizations and demanded the return of their money."[22] In other cases one of the actors panicked or experienced a sudden twinge of conscience. Therefore, unless one of the parties turned in the other (or if an official immediately reported someone who approached them with a bribe), instances of bribery remained hidden.[23] Sources agree that the proportion of bribes that were somehow discovered *and* reported to the authorities *and* prosecuted reflected only the tip of the iceberg.

Ministry of Justice reaction to Minin letter

On 15 May 1946, Zhdanov forwarded copies of Minin's letter to two of the relevant legal agencies: the USSR Ministry of Justice and the USSR Supreme Court. Zhdanov, in his capacity as Central Committee Secretary, requested that the head of each agency respond to Minin's letter, including "their opinion and how they evaluate the situation concerning bribery."[24] Zhdanov's request made it clear that in this case the initiative for examining the issue of bribery came from the highest reaches of the party, not from the legal agencies themselves. The Minister of Justice for the USSR, N. Rychkov, wrote to Zhdanov with his reaction to Minin's observations in a letter dated 23 May 1946.[25] Although "the author of the letter, perhaps, generalizes the facts of bribery too much," Rychkov acknowledged that "lately, especially during the war, it [bribery] has acquired a widespread character." He further conceded, "It is also doubtlessly the case that the struggle against it has been undertaken extremely hesitantly."[26]

In a separate letter to Central Committee member N. S. Patolichev, Rychkov, echoing Minin's complaints, stated that "in many organizations, primarily connected with serving the population, and with supply (railways, housing organs, apartment administration, bases for supplying food and manufactured goods, etc.), bribery has become an almost ordinary, everyday phenomenon."[27] Such language describing bribery as "ordinary" is startling indeed. Rychkov goes further, candidly implicating the legal system itself in accepting bribes: "Even the organs of the court, the *prokuratura*, and the police are frequently infected with bribery." The metaphor of corruption as a contagious disease arises again, in this case infecting the very agencies in charge of eradicating it. Rychkov continues, arguing that to fight bribery it is also necessary "to overcome party members' conciliatory attitude to this disgraceful phenomenon."[28] According to Rychkov, Party, Komsomol and trade unions had failed to do educational work to change this attitude. Many party members, made aware of cases of bribery, stay silent rather than reporting to the investigative organs.[29] Infrequently, there are even instances when leading party workers stand up for bribe-takers. Officials in responsible positions cover up for other party members, because, for example, they need the rare skills of the guilty cadres. "We have to finish off this tolerant attitude towards bribery and the passivity of party organizations and party members in the struggle against bribery." To combat this tolerance among party members, the Minister of Justice called on the Central Committee to issue a decree on the struggle with bribery, a draft of which he appended.

An explanation commonly offered by the legal agencies for pervasive bribery during and after the war was that persons convicted of the crime were not punished severely, if they were punished at all. For several years, Rychkov had acknowledged that some judges had done a poor job waging the struggle against bribery, failing to imprison a significant proportion of those convicted. In 1944–1945, the Ministry of Justice accused local judges of "liberalism" for their tendency to assign sentences that were too "soft," encouraging bribe-takers and -givers alike. Rychkov implied that some corrupt jurists were accepting bribes in exchange for light sentences. The Commissariat of Justice, Rychkov notes, had responded to this laxity by sending instructions to judicial organs, demanding the intensification of the struggle with bribery, including an order on the inadmissibility of weak sentences.[30] Inconsistency in punishment was a feature of the post-war campaign against bribery. Evidence indicates that, during the war, those who *offered* bribes were punished by courts significantly less strictly than those who *accepted* bribes. Ministry of Justice statistics show that during the war average punishments for bribe-givers became significantly lighter.[31] In 1941, 71.7 per cent of persons convicted of giving bribes were sentenced to prison time. By 1945, that figure had dropped to less than half, or 48.6 per cent. Of those convicted of taking bribes or acting as an intermediary under articles 117 and 119 in 1945, about two-thirds (64.8 per cent) were given jail time; the figure for 1944 was approximately the same.[32] According to at least one source, bribe-givers and intermediaries were often not charged even if prosecuting authorities knew their identities.[33]

The Law Code itself enshrined this difference in treatment of those who took and those who gave bribes, prescribing less severe penalties to the giver. Yet it is possible to speculate that judges were sympathetic to the plight of common people forced to give bribes during the chaos of wartime and reconstruction. Judges may have believed that many instances of bribe-giving simply either did not rise to the level of a crime or could be excused, considering the circumstances. In late 1945 and 1946, the satirical magazine *Krokodil* published cartoons that served to reinforce the impression that the party wanted to challenge common assumptions about the acceptability of giving bribes. They seem to reflect – while simultaneously attempting to contradict – a belief among the population that those forced to offer bribes would not be (or at least *should* not be) punished. The thrust of the cartoons is to instruct readers that not only the accepters of bribes, but also those who offered them, would be arrested and imprisoned. One cartoon, entitled "Irregular [literally, Incorrect] Verb" takes the form of a grammar lesson, making fun of one man's idea that "I give [a bribe], He takes [the bribe], He sits [in prison]." The *official* who accepted the bribe will have to do time, not he, the cartoon's hapless bribe-giver tells us. This use of the verbs is "incorrect," since they will *both* "sit." Two more *Krokodil* cartoons that appeared around the same time repeat the theme of the bribe-giver who believes himself innocent. (*Krokodil* 10.11.1945, p. 7. See also *Krokodil* 30.11.1945 and 20.8.1946.)

Those who solicited bribes also often got off without punishment. The actions of the courts sometimes gave birth to feelings of impunity among corrupt officials. Officials felt that they would not be apprehended, and if they were caught, that they would not be punished. Minin mentioned the case of a group of Baku doctors tried in March 1946, charged with accepting bribes to exempt men from military service. One of their bookkeepers was convicted of taking 1.5 million roubles. Minin alleged in his letter that the citizens of Baku, made cynical by widespread official corruption, predicted that nothing would happen to this bookkeeper: "Well, they'll sentence him to be shot," they said, "but then they'll change the execution to ten years in prison, and then with the help of money and friendships he'll be free in two or three years." In fact, the court did sentence the bookkeeper to death, but then reduced the punishment to ten years, just as the "citizens of Baku" had foreseen. While acknowledging lenient sentencing by judges, Minister of Justice Rychkov also took every opportunity to highlight the shortcomings of prosecutors and police in the battle, deflecting blame from the judiciary. In his response to Minin's letter, Rychkov pointed the finger at the *prokuratura* and the police, arguing "the quantity of exposed facts about bribery is insignificant . . . *The organs of the prokuratura and police wage the struggle against these crimes very weakly*" [emphasis in original]. The police fail to follow up on information supplied by informants and make arrests.[34] The *prokuratura* gathered evidence haphazardly. While Rychkov admitted the difficulty of exposing guilty parties in cases of bribery, he insisted that finding criminals was possible with careful investigations (by the *prokuratura*), and especially with good undercover work (by the police). The efforts of prosecutor General Konstantin P. Gorshenin since the beginning of 1945 were judged to have been ineffective. ". . . There is no noticeable strengthening in the struggle against bribery" and "this creates an atmosphere of impunity."[35] In a bit of institutional defensiveness, the Minister in charge of the courts thus laid the bulk of the responsibility for the epidemic of bribery at the feet of the police and *prokuratura* for their failure to find guilty parties and to prosecute them.

The *prokuratura* and the campaign

The central prosecutor's office reaction to accusations that bribery was widespread and insufficiently prosecuted can also be characterized as a combination of candor and political defensiveness. A *prokuratura* Commission appointed to investigate the causes of bribery and to design measures to combat it laid out several factors contributing to the persistence of bribery in the post-war USSR.[36] Although this report was produced in 1947–1948 as a part of an ongoing self-examination by the *prokuratura*, its conclusions would certainly seem to apply to the entire post-war period. Moreover, although the Commission focused on eradicating bribery inside the *prokuratura* itself, its observations apply to state employees in many branches of government and economy. In a report to Gorshenin, the Commission candidly pointed out that prosecutors often did a half-hearted job of investigating themselves; not surprisingly, this resulted in an atmosphere that tolerated malpractice. Hinting at "protectionism" among prosecutors at the top of the hierarchy, the commission insisted that the *Sledstvennyi otdel* of the USSR *prokuratura* should supervise investigations into bribery, regardless of "the high position of the individuals suspected or accused in this matter."[37] A related problem was the slack ethical training of *prokuratura* personnel. Much like the judiciary, the *prokuratura* had undergone a great deal of turnover during the war. New staff (and not only the new) too often failed to understand the right and wrong of bribes and "gifts." Such misperceptions were to be countered, in part, by the publication of articles in the *prokuratura*'s journal, *Sotsialisticheskaia zakonnost'*, dedicated to reinforcing the appropriate ethics of personnel. Articles were to emphasize such simple things as the obligation to excuse oneself from cases that involved friends and family. Case studies of crimes by court and *prokuratura* employees should be described, heightening the sensitivity and vigilance of prosecutors. It is worth noting that, in fact, very few articles concerning bribery actually appeared in *Sotsialisticheskaia zakonnost'*. After a minor flurry in 1947, articles about bribery essentially disappeared from the pages of this journal until after Stalin's death.

A further point made by the *prokuratura* Commission was arguably the most critical. The commission noted that the state could increase penalties, strengthen procedures and shame wrongdoers in the press and in trials. Yet as long as the Soviet population (including officialdom) felt itself in short supply of money, food and access to justice (and the formal channels could not supply these commodities), officials would continue to solicit illegal payments and the population would continue to pay. Low pay contributed to employees succumbing to temptation. *Prokuratura* employees were in an "extremely difficult material situation," and this circumstance demanded a "fundamental review of the material and everyday conditions of their work." Prosecutors and investigators needed higher pay and better pensions. Yet for prosecutors, it was not only money that was in short supply – prestige was also lacking. Prokuratura employees, the Commission argued, were underpaid not only in an absolute sense, but also relative to other professionals. The Commission's report noted that in pre-revolutionary Russia, legal and judicial officials were very well compensated in comparison with other functionaries; in the last years of tsarist Russia, prosecutors received even more than the Guards Officers. In Great Britain, France and the USA, prosecutors were also paid at a level higher than other professionals. Especially in light of the special place of the legal organs in the USSR – "its high ideological and political level and its service to the interests of the people and the state" – the Commission argued that changes must be made to the "intolerable" material conditions of *prokuratura* personnel. As a consequence of poverty, public prosecutors often "ended up in direct material dependence on local traders and cooperative organizations, as a result of which the soil was created for the intermingling of *prokuratura* employees with the employees of these organizations . . ." Especially in the provinces, such relationships were based on "mutual concessions and favours." In these murky circumstances, "entrepreneurs," operating in the shadows of the official economy, and law enforcement officials started doing each other favours, swapping some food or scarce manufactures (and the occasional cash payment), for some assistance with the law. This arrangement was, of course, the classic definition of "*blat*," informal relationships based on reciprocal social favours. What is perhaps most surprising is that, with very rare exceptions such as this, *blat* is never referred to in the documents that emerge from the anticorruption campaign. It is almost as though *blat* was considered by all to be a normal component of social relationships, so utterly a part of the fabric of everyday life that it was not even mentioned.

The *prokuratura* thus attempted to use the campaign against bribery – and the embarrassing description of instances of graft among its own employees – as an opportunity to lobby for greater pay and heightened status. Judges and prosecutors essentially argued that they were compelled to break the law. Elements of the Commission's candid observations – that prosecutors were overly tolerant of graft especially in their own ranks, that some lacked a keen sense of ethics, and that they were underpaid and tempted to steal – were certainly correct throughout the late Stalin period. Such observations, however, did not become part of the language of the anti-bribery effort. Instead, the language that framed the campaign became watered down, a victim of self-censorship. Agencies scrambled to protect their institutional interests or to defend the prestige of the party and state. A growing divergence between the Commission's frank comments and the official language eventually used in the campaign is visible in the posturing of the *prokuratura* during the process of drafting of the campaign.

Prokuratura editing

On 6 July 1946, Rychkov's office forwarded for review drafts of several documents addressing a proposed campaign against bribery to the office of Procurator General Gorshenin. The Central Committee requested that each of these documents, including the draft of a letter addressed to the Central Committee and a draft resolution of the Central Committee,[38] was to be composed jointly by the *prokuratura*, Ministry of Justice and MVD.[39] The *prokuratura* response to these documents prepared by the Ministry of Justice took the form of editing. An examination of some of these edits enables us to reconstruct the *prokuratura*'s objections to certain language. A comparison of the originals and the edited versions indicate some of the ways the leaders of these institutions did – and did not – want the sensitive questions of bribery to be discussed. Rychkov's office drew up a draft Central Committee *postanovlenie*, entitled "O bor'be so vziatochnichestvom." This version represented a major shift in emphasis from earlier correspondence between the Ministry of Justice and the Central Committee. The draft begins: "Recently, numerous reports and signals from below about the growth of bribery have poured into the Central Committee. Exploiting the difficulties of wartime and the post-war period, criminal and morally unstable elements in the state apparatus have moved onto the path of receiving and extorting bribes."

The Ministry of Justice thus shifted away from the notion that bribery had become an "almost ordinary, everyday phenomenon"[40] (as written in an earlier letter to Patolichev in the Central Committee) to an assertion that only "morally unstable and criminal elements" took bribes.[41] The *prokuratura* revised the language in the formative documents of the anti-bribery campaign even further. For instance, Rychkov's original letter to the Central Committee stated that bribery "corrupts the state apparatus, paralyzes its normal work and creates the soil for every possible kind of illegality. . . ." *Prokuratura* editing dropped this language, instead stating much more narrowly that "certain employees in the state and economic apparatuses" were corrupt.[42] The rather explosive allegation that bribery had "paralyzed" the government was also excised. In another place, the *prokuratura* editors removed the phrase: "The quantity of persons charged with receiving and giving bribes is insignificant."[43] This original statement would have been an affront to *prokuratura* success in charging people with crimes, while also highlighting that the great majority of bribery went undiscovered and, therefore, unpunished. The editing thus removed a key point. Editing also served to reduce the scope of the phenomenon by changing language, alleging that bribery had reached "epidemic" proportions, to the notion that bribery is "alien by nature to the Soviet government." Placing the accent not on party members or even "responsible employees," the edited version implicated lower-level employees as the main culprits.[44] Final versions acknowledged, but did not emphasize, that party members were involved in graft, mostly through their passive tolerance of it, rather than their active participation. Perhaps most strikingly, the final version of the *prikaz* eliminates any reference to corruption inside the legal agencies themselves. Although both the Ministry of Justice and the *prokuratura* acknowledged that bribery had "infected" their own personnel, and such language was contained in the Ministry of Justice's letter to the Central Committee, no mention of this was made in the *prikaz*. Rather, the *prikaz* noted that bribery was widespread, "especially on transport, in trade, supply, and sales organizations."

The impulse to downplay the extent of bribery is also revealed in the discussion over whether to issue a public decree (an *ukaz*) intensifying the battle against bribery. Such a decree would have been issued outside the regular law code, but would have been highly publicized and held the force of law. An *ukaz* demanding stronger enforcement and tougher penalties would have made graft the subject of a very public campaign, published in national and local newspapers. In all likelihood, it was the Ministry of Internal Affairs under S. N. Kruglov that pushed most strongly for a special *ukaz* on bribery. In the 1930s, the police organs had argued in favour of issuing extra-legal decrees, often succeeding against the opposition of the legal agencies. Such a public *ukaz* would pressure prosecutors to demand longer sentences for bribery and force judges to assign them. A draft decree of the Presidium of the Supreme Soviet of the USSR, dated 4 July 1946, was discussed by the Ministry of Justice and the Chair of the Supreme Court. The draft decree provided for a significant increase of criminal responsibility for the receipt and rendering of bribes. The *ukaz* would have provided for a five-year minimum prison term with confiscation of property for the acceptance of a bribe by any *dolzhnostnoe litso* (person holding office).[45] In particularly aggravating circumstances, the receipt of bribes could be punished with execution. The giving of bribes or acting as an intermediary would result in a term of no fewer than three years plus the confiscation of property. This represented a significant increase in minimum penalties, which rose from a six-month minimum prison term (and two-year maximum), to a three- or five-year minimum term.[46] The legal agencies, however, opposed the creation of a special *ukaz* increasing penalties against bribery. The Ministry of Justice argued that it was not necessary to issue a new *ukaz*.[47] The existing Criminal Code was strong enough, Rychkov argued, for it stipulates that a person who accepts a bribe will be punished with two years in prison. If an official of the state were involved or if extortion took place, the sentence could be increased to ten years with confiscation of property.[48]

The response of the Supreme Court

The Supreme Court's 17 May 1946 response to Minin's letter offered a different perspective on the question of a public decree on fighting bribery.[49] Addressed to Zhdanov, the memorandum was written by I. Goliakov, the Chair of the Supreme Court. While agreeing in large part with the *prokuratura* and Ministry of Justice, the Supreme Court added a number of significant new objections. Like the heads of the *prokuratura* and the Justice Ministry, Goliakov argued that the current criminal statutes on bribery were sufficient. The police, *prokuratura* and courts should therefore enforce present law "correctly, reasonably and with consequences." Moving beyond the response of the Minister of Justice, Goliakov also offered a defensive denial of Minin's accusations. The wartime problems had inspired among most people "a great upsurge of patriotism, directed toward overcoming these obstacles and defeating the enemy." At the same time, unfortunately, it had "corrupted the unstable and mercenary elements". Minin had overstated the problem, painting the situation "with overly gloomy colours, considering bribery in our time to be an extremely widespread phenomenon that practically takes the shape of a natural disaster." Despite the author's good intentions, Goliakov found Minin's assertion that "everyone takes bribes" to be exaggerated and unconvincing.

Beyond this assertion that Minin had distorted the real situation, Goliakov introduced a new dimension to the objections to a public *ukaz*. Minin's allegations, he writes, amount to "slander directed against Soviet society." An *ukaz* (or a published *postanovlenie* of the Council of Ministers) would be embarrassing – and even dangerous – to the regime. "The publication and wide promulgation of such a law could create a false, distorted picture, both in our country and especially abroad, about the moral character of Soviet society, and could be used by hostile elements with anti-Soviet aims." Goliakov thus worried about the negative reactions both of Soviet citizens and of "hostile elements" abroad. Capitalist and fascist states, after all, were tainted with bribery. Bribery was officially a nearly vanished relic of capitalism. A public *ukaz* decrying the stubborn persistence of bribery would amount to an admission that 30 years after the revolution, socialism had not penetrated as deeply into the consciousnesses of the Soviet people, even of leading cadres, as the regime claimed. Goliakov thus pressed for secrecy. Stalin (and others in the Central Committee) may have agreed that to discuss a wave of bribery openly would be embarrassing abroad and potentially dangerous at home.

Ultimately, such reasoning won the day. Neither a public decree nor a public resolution of the Council of Ministers was issued. Rather, an *internal* and *secret prikaz* was issued jointly by the Ministry of Justice, the *prokuratura* and the Ministry of Internal Affairs (accompanied by an *internal* and *secret postanovlenie* of the Central Committee.)[50] The press "campaign" was also muted. A key theme in discussions of the anti-bribery efforts was that the press must "mobilize society" to fight bribery. In particular, legal agencies called for the necessity of placing articles about bribery trials in newspapers. Yet, despite the calls for a major press campaign,[51] in the end the results were rather minimal. According to a Ministry of Justice complaint, only 23 articles about bribery trials appeared in newspapers throughout the country during the second half of 1946. A review of the press indicates that after an initial upsurge in the number of articles in central newspapers in the second half of 1946, the quantity quickly dropped off.

Campaign and reality

The anti-bribery campaign did have certain results in the short term.[52] More severe punishments were applied to people convicted of bribery after the *prikaz* was issued than before. In addition, the proportion of people sentenced to prison for accepting bribes rose to 88 per cent in 1947, from 74 per cent in 1946. For offering bribes, that proportion rose to 75 per cent in 1947, from 67 per cent in 1946.[53] Nevertheless, nearly a quarter of people convicted of rendering a bribe still received no jail time. These results were not always so evident, however. A 14 February 1947 letter signed by the deputy of the Central Committee's *Upravlenie kadrov* Nikitin alleged that neither Minister of Justice Rychkov nor

Prosecutor General Gorshenin had taken the *prikaz* seriously enough, as they had failed to design and implement necessary measures.[54] Worst of all, judges and prosecutors were still taking bribes, and they were not seriously disciplined by the agencies for which they worked.[55] Moreover, the letter declares, "signals" from below with information about instances of bribery in court and prosecution organs and in numerous other organizations and enterprises are arriving "in large quantities." Indeed, official data indicate that relatively small numbers of people were arrested for bribery in the late Stalin years. Ministry of Justice statistics show a rise in the total number of convictions for bribery in 1946 and 1947, followed by a gradual decline beginning in 1948. In the long run, the campaign did not result in increased convictions. According to *prokuratura* figures, the number of people charged Union-wide with accepting bribes dropped each year, from 3,291 people in 1948, to 2,499 in 1949, to 1,903 in 1950, to 1,298 in 1951. The 1951 total thus amounted to far less than half the 1948 number. The number of people charged with offering a bribe declined in similar fashion: from 3,080 in 1948, to 2,716 in 1949, to 2,003 in 1950, to 1,863 in 1951.[56] This trend followed the general pattern of Soviet campaigns – an initial upsurge in arrests and convictions, followed by a decline back to, or below, the initial levels.

Within two years of its launch, the campaign had clearly deflated. Eighty-two people were charged with giving or receiving bribes in the city of Moscow in the first half of 1950. Yet prosecutors in seven of the city's regions brought absolutely no charges. In the second half of the year, eight of Moscow's regions similarly sent no cases of bribery to court.[57] In the entire year 1951, *prokuratura* investigators in all of Moscow *oblast'* sent to the courts 31 cases of bribe-taking involving 50 people.[58] A total of only 35 people were charged with offering bribes in Moscow *oblast'* in 1951.[59] Legal agencies were well aware that these official crime statistics did not reflect the full extent of the phenomenon. The *prokuratura* tended to produce the most thoughtful reports, moving beyond piles of self-congratulation and formulaic self-criticism to more serious analysis. Their reports indicated that bribery continued to be a pervasive but still rarely prosecuted phenomenon. For its part, the *prokuratura* did not claim that this decline in convictions illustrated the effectiveness of the anti-bribery measures, or that it reflected some inevitable decline in criminality as Soviet society progressed toward communism. On the contrary, the Prosecutor General wrote in April 1952 that "The lowering of the number of fully investigated cases and the quantity of persons charged in bribery cases is explained to a significant degree by the fact that the struggle against this crime in the organs of the *prokuratura* is still insufficient."[60] There were still republics where bribery went essentially uninvestigated and unpunished. In the Armenian SSR, for example, prosecutors brought only four cases of bribe-taking in all of 1952; ten were brought for offering bribes.[61]

In another report later that year, a certain Aleksandrov, the chief of the investigation administration of the *Prokuratura* General of the USSR expressed dismay that the numbers of cases of bribery investigated by the *prokuratura* had dropped so sharply over the years.[62] "Do these data reflect the state of criminality? Is this criminality really falling, and is substantiated evidence of bribery rare?" He proceeds to answer his own question: "Investigated cases show that there are facts of bribery in a great variety of organizations, institutions, and enterprises." "Skilled and persistent investigation has uncovered in many cases a system of bribery," especially in trade, requisitioning and financial organizations. "Similar facts testify to the fact that the comparatively rare launching of cases about bribery is explained not because cases of bribery are not numerous, but because they are rarely uncovered thanks to a struggle against bribery that is clearly unsatisfactory."

Conclusion: a "campaign spasm"

This research points to a substantial degree of continuity between elements of corruption in the late-Stalin period and the corruption endemic to the Brezhnev period. Far from indicating a sharp break between the Stalin and post-Stalin periods, the above investigation into late Stalinist bribery highlights elements of continuity. The 1946 campaign against bribery represents what I call a "campaign spasm," a brief, intense, though ultimately failed, attempt by the party-state to eradicate some sort of unpleasantness afflicting Soviet society. The anti-bribery campaigns anticipated a surge that was much larger, yet similar in key ways, to that resulting from the draconian 4 June 1947 *ukazy* creating harsh punishments for the theft of socialist and personal property. The 4 June 1947 *ukazy* were issued within a year, and they resulted in the arrests of hundreds of thousands of people. The secrecy, low level of arrests and lack of publicity surrounding the bribery campaign, however, stand in marked contrast to the exuberance of the 1947 campaign against theft of state and personal property. In the final analysis, the campaign against bribery was created and carried out without conviction. As the proposed campaign made its way through the legal agencies, both the seriousness and scope of the problem were played down, becoming relatively narrow and muted as it emerged from the bureaucratic process. A quiet campaign protected institutional interests. Agencies were willing to engage in a limited amount of *samokritika* in private forums or in closed letters to the Central Committee, but they hesitated to do so publicly. Institutional sparring and defensiveness from the legal agencies was a major reason for the absence of a public campaign. Agencies wanted to protect themselves, as they tried to deflect blame for criminality. A quiet campaign also protected the state from embarrassment, as the reality of widespread official corruption did not fit the positive public image the Soviet Union was trying to project to its own citizens and to the rest of the world.

From the outset, the campaign against bribery was hardly a fully-fledged attack that addressed the root causes of the phenomenon – hyper-centralized planning, acute shortages of goods and lack of access to housing and justice in the legal system, all of which provided tremendous opportunities for officials. As Bauer, Inkeles and Kluckhohn noted 50 years ago, the informal relationships that linked state functionaries and the population had become fundamental to the functioning of the state and economy.[63]

Notes

1 GARF, f. 9492, op. 2, d. 44, ll. 227–29.
2 Managerial bartering was first described in detail by Joseph Berliner, based on interviews made during the Harvard Project on the Soviet Social System (better known as the Harvard Interview Project), mostly concerning managers' experiences in the 1930s. Joseph Berliner, *Factory and Manager in the USSR*, Cambridge: Harvard University Press, 1957), 160–230. See also David Shearer, "Wheeling and Dealing in Soviet Industry: Syndicates, Trade, and Political Economy at the End of the 1920s," *Cahiers du Monde Russe* 36:1–2, 1995; Paul Gregory, *The Political Economy of Stalinism*, Cambridge: Harvard University Press, 2004.
3 V. I. Kozlov, "Denunciation and its Functions in Soviet Governance: A Study of Denunciations and their Bureaucratic Handling from Soviet Police Archives, 1944–1953," *Journal of Modern History* 68:4, December, 1996, 867–898; Sheila Fitzpatrick, "Signals from Below: Soviet Letters of Denunciation of the 1930s." *Journal of Modern History* 68:4, December 1996, 831–866.
4 On the campaign against theft of state and personal property, see Peter H. Solomon, Jr., *Soviet Criminal Justice under Stalin*, Cambridge: Cambridge University Press, 1996, 408–445; and Yoram Gorlizki, "Rules, Incentives and Soviet Campaign Justice after World War II," *Europe-Asia Studies* 51:7, November 1999, 1245–1265.
5 *Fondy* of the USSR *Prokuratura*, Ministry of Justice, Supreme Court and Ministry of Internal Affairs contain important documentation.
6 Mark Harrison and John Barber are interested mainly in the economic aspects, including industrial and agricultural production, distribution and trade, in *The Soviet Home Front, 1941–45: a social and economic history of the USSR in World War II*, London: Longman, 1991. They discuss briefly unofficial and illegal markets, but their sources did not allow them to explore the official corruption that often made these crimes possible. Julie Hessler has opened up the legal markets for study in her important work. *A Social History of Soviet Trade: Trade Policy, Retail Practices, and Consumption, 1917–1953*, Princeton: Princeton University Press, 2004. Yet, although the markets may have been legal, the process by which things made their way to the markets very often was not. Goods were often stolen from state warehouses, for example. Bribes often enabled this theft, or were paid to cover it up. Elena Zubkova has written on the social history of Soviet society during and after the war, but the place of official corruption and discussions of official criminality are not explored. Although her work examines popular perceptions of a crime wave, the focus is on citizens' fear of rising hooliganism in the cities (89–92). She also writes little about the black market in scarce goods and services that provided the soil for official crime. Zubkova, *Poslevoennoe sovetskoe obshchestvo: politika i povsednevnost', 1945–1953*, Moscow: Rosspen, 2000.
7 GARF, f. 9492, op. 2, d. 49, l. 245.
8 On the famine and provisions crisis that accompanied the end of the war, see V. F. Zima, *Golod v SSSR 1946–47 godov: proiskhozhdenie i posledstvie*, Moscow: RAN, 1996; Zubkova, Ibid., 61–74.
9 GARF, f. 9492, op. 2, d. 49, l. 17. Similar concerns about raw, untrained party cadres were also expressed by the party leadership. See Rittersporn, *Simplifications staliniennes et complications sovietiques: Tensions socials et conflits politiques en U.R.S.S.*, Paris: Editions des archives comtemporaines, 1988. Cynthia S. Kaplan, *The Party and Agricultural Crisis Management in the USSR*, Ithaca: Cornell University Press, 1987.
10 Not until 1948–1949 was the pay of judges and other court employees under the auspices of the Ministry of Justice raised significantly. This pay raise occurred amidst major bribery scandals involving judges and other court employees in a number of Moscow courts. GARF, f. 9492, op. 2, d. 58, ll. 3–38. See the Postanovlenie of the Council of Ministers, "O povyshenii okladov sud'iami I rabotniki organov Ministerstva Iustitsii."

11 GARF, f. 9415, op. 5, d. 95, str. 53.

12 GARF, f. 8131, op. 39, d. 299, ll. 3–3 ob.

13 See, for example, GARF, f. 8131, op. 37, d. 4216, l. 191; and GARF, f. 9474, op. 16, d. 294, ll. 27–28. For cases of corruption in the housing administration in the city of Rostov, see See Jeffrey Jones, "'In My Opinion this is All a Fraud': Concrete, Culture, and Class in the Reconstruction of Rostov-on-the-Don," unpublished PhD dissertation, University of North Carolina, 2000, especially chapter 5.

14 See, for example, report by D. Salin, Prokuror of the Lithuanian SSR, to Mokichev, of 14 April 1947. GARF, f. 8131, op. 38, d. 449, ll. 8–14.

15 OBKhSS reports in GARF, f. 9415, op. 5, d. 98, str. 43.

16 See, for example, GARF, f. 8131, op. 37, d. 3137, l. 46.

17 For the case of Sprimon, see GARF, f. 9492, op. 1, d. 15, ll. 289 ob.–290. See GARF, f. 9492, op. 2, d. 44, ll. 113–18, for a bribery scandal in the organs of the military *prokuratura* in late 1946.

18 See, for example, GARF, f. 8131, op. 38, d. 282, l. 62. The final version of the 1946 Prikaz strengthening the struggle against bribery referred to bribery as an "especially dangerous crime." See also M. P. Karpushin, *Vziatochnichestvo – pozornyi perezhitok proshlogo*, Moscow, 1964.

19 See, for example, the 20 May 1946 report on bribery in 1944–1945, written by Beldiugin of the Ministry of Justice's consultant from the administration of general courts, which states that the "the quantity of people convicted for bribery (the accepting of bribes and the giving of bribes) in certain republics and in the USSR is extremely insignificant on the whole." GARF, f. 9492, op. 1a, d. 478, l. 33.

20 Statistics found in GARF, f. 9492, op. 6s, d. 14. These documents were recently reprinted in *Istoriia Stalinskogo Gulaga: Massovye repressii v SSSR*, tom 1, Iu. N. Afanas'ev *et al.*, eds, Moscow: Rosspen, 2004, 633, 636.

21 See report sent to Stalin and Molotov on the state of crime in the Soviet Union, 1940–1947, signed by Minister of Justice Gorshenin. Report undated, but most likely produced early in 1948. GARF, f. 9492, op. 2, d. 49. Figures exclude infanticide. These statistics are essentially confirmed in *Istoriia Stalinskogo Gulaga*, tom 1, 633, 636.

22 GARF, f. 8131, op. 38, d. 449, l. 91.

23 For such an assertion by the *prokuratura*, see GARF, f. 8131, op. 38, d. 299, l. 42.

24 GARF, f. 9492, op. 2, d. 44, l. 227.

25 GARF, f. 9492, op. 2, d. 38, ll. 119–22.

26

	1943	*1944*	*1945*
Bribe taking:	528	685	981
Bribe giving:	992	1,404	1,858

Sources: GARF, f. 9492, op. 2, d. 38, ll. 119–22; GARF, f. 9492, op. 2, d. 49, l. 277; GARF, f. 8131, op. 37, d. 2817, l. 3. These figures do not include those tried by military tribunals. In 1945, military tribunals convicted 1089 for accepting bribes and 211 for giving them.

27 GARF, f. 8131, op. 37, d. 2817, l. 2.

28 GARF, f. 8131, op. 37, d. 2817, l. 6.

29 To support his allegation of party toleration of bribery, Rychkov recounts an example with facts established by the *Prokuratura* USSR. In June 1946, information came to the Administration of Passenger Services of the Ministry of Transportation that at the Paveletskii station in Moscow workers of the railroad police were taking bribes for giving tickets for passage. A *prokuratura* auditor determined that every day at the station 15 tickets were reserved for employees of the MGB and MVD. Employees of the police, however, sold these unused tickets for bribes to bystanders. The matter

stopped when the deputy chief of the Central Administration of Passenger Service, Terminasov, called the chiefs of the station to his office and merely "suggested" that "they cease with this scandal." No information about this criminal behaviour was passed on to the *prokuratura*.

30 An internal directive letter issued by the People's Commissariat of Justice in December 1944 noted that many judges had "weakened the struggle against bribery" by assigning overly lenient sentences for bribery, an "especially dangerous type of crime." GARF, f. 9492, op. 1a, d. 314, l. 32. Directive letter of 26 December 1944. See also Rychkov's telegram of 7 June 1946, demanding that local justice officials implement the conditions of the directive and report back to Moscow with the results. GARF, f. 9492, op. 1, d. 148, l. 1.

31 GARF, f. 9492, op. 1, d. 478, ll. 30–33; see also GARF, f. 9492, op. 1, d. 514, l. 7.

32 The figure for 1944 was 63.8 per cent.

33 This tendency to punish the givers of bribes less severely is reflected in many documents, including the Minlust survey of bribery for the year 1945, GARF, f. 9492, op. 1a, d. 478, ll. 30–33. Dated May 20, 1946.

34 The *prokuratura* made this point. See GARF, f. 8131, op. 37, d. 4216, ll. 188–91. For their part, the police (OBKhSS was charged with using informants to root out bribery) was critical of its own poor job with its informants, and of wasting time with investigations of official complaints that the *prokuratura* should have undertaken. See the 1946 annual report of OBKhSS, GARF, f. 9415, op. 5, d. 95, str. 51–54, and 1947 annual report of OBKhSS, GARF, f. 9414, op. 5, d. 98, str. 41–44.

35 In December 1946, Rychkov was privately writing that the number of people charged by the police and *prokuratura* is still "extremely insignificant." GARF, f. 9492, op. 2, d. 38, l. 117. Rychkov blames the *prokuratura*, and writes that "I have informed the General Procuror tov. Gorshenin about the unsatisfactory work of the organs of the police and *prokuratura* in exposing bribe-takers."

36 GARF, f. 8131, op. 37, d. 4216, ll. 188–91.

37 GARF, f. 8131, op. 37, d. 4216, ll. 188–91. All regional prosecutors, receiving official or anonymous information about a crime allegedly committed by someone they are supervising, should immediately send a special dispatch to their superior. Given the special complexity of bribery cases, only prosecutors with proper qualifications and experience should be assigned to investigate them. For a case of prosecutors failing to turn in one of their own, see the letter from Salin to the Prosecutor General of Lithuanian SSR, regarding the case of Kondrat'eva, of 10 November 1947. GARF, f. 8131, op. 38, d. 449, l. 15. (Gorshenin replaced Rychkov as Minister of Justice USSR in January 1948.)

38 GARF, f. 8131, op. 37, d. 2817, ll. 1–29.

39 Solomon notes that memoranda on controversial legal issues were often composed jointly by these three agencies. *Soviet Criminal Justice Under Stalin*, 406.

40 GARF, f. 8131, op. 37, d. 2817, l. 2.

41 GARF, f. 8131, op. 37, d. 2817, l. 10.

42 GARF, f. 8131, op. 37, d. 2817, l. 10.

43 GARF, f. 8131, op. 37, d. 2817, l. 11.

44 GARF, f. 8131, op. 37, d. 2817, l. 2.

45 The draft *ukaz* also forbade a person convicted of bribery from ever again occupying an administrative or economic job in any state, cooperative, or social organization or enterprise. GARF, f. 8131, op. 37, d. 2817, l. 17.

46 The Ministry of Justice drafted at least two versions of an *ukaz*. The first is dated 25 June 1946, and the second is dated 4 July. GARF, f. 9492, op. 1, d. 148, l. 2; f. 8131, op. 37, d. 2817, l. 7. The first draft called for a three-year minimum sentence for any official (*dolzhnostnoe litso*) guilty of accepting a bribe. If the guilty party occupied a "responsible position," was taking bribes "systematically," or was extorting bribes, a five-year minimum punishment with the confiscation of property was prescribed.

Under particularly aggravating circumstances, the draft called for the death penalty and confiscation of property. The giving of bribes, or acting as an intermediary in giving bribes, should be punished by a three-year minimum term. In a tradition repeated with the 1947 *ukaz* on theft of state property, the minimum penalties were raised in the subsequent draft, possibly after the intervention of the party leadership. On the 1947 *ukazy* on theft, see Solomon, *Soviet Criminal Justice*, 410–412.

47 GARF, f. 9492, op. 2, d. 38, l. 122. Letter of Rychkov to Zhdanov, 23 May 1946.

48 Although no direct consultation on this question has been found in the archives, the *prokuratura* also apparently opposed creating an *ukaz*, as evidenced by the fact that they crossed out a paragraph that called for issuing a special *ukaz* from the draft Central Committee *postanovlenie*. GARF, f. 8131, op. 37, d. 2817, l. 14.

49 GARF, f. 9474, op. 16s, d. 294.

50 For the final version of the secret *prikaz* "Ob usilenii bor'by so vziatochnichestvom," no. 036/0210/126s, of 15 July 1946, see GARF, f. 8131, op. 38, d. 299, ll. 1–2. For a discussion of the growing use of regulations and laws classified as secret to express legal norms, see Solomon, *Soviet Criminal Justice*, 418–426. All the archival material located in GARF cited in this article is contained in the formerly secret parts of the *prokuratura*, ministry of justice and supreme court archives.

51 For example, Rychkov urged the placing of articles connected with trials for bribery in *Pravda* and *Izvestiia*, and in local newspapers. GARF, f. 9492, op. 2, d. 38, ll. 119–22.

52 The *Prokuratura* General pointed out that prosecution of bribery cases was erratic around the Union; prosecutors worked most vigilantly in the RSFSR and the Ukrainian SSR and did very little in most other republics. GARF, f. 9492, op. 1, d. 514, l. 36. During the first three-quarters of 1946, the courts of the Armenian SSR prosecuted a grand total of four people for bribery; the Kazakh SSR convicted only five; while in those nine months the Turkmen SSR found only 13 people who took bribes.

53 GARF, f. 9492, op. 2, d. 49, l. 277.

54 GARF, f. 9492, op. 2, d. 44, ll. 113–116.

55 Thus, in the first three-quarters of 1946, 322 employees had been fired from the organs of justice and *prokuratura* for various "amoral infractions, among which bribery enjoys a prominent position." During the same period, 210 judges in the RSFSR had been removed for bribery and abuse of position. Yet only 120 of them (57 per cent) were charged with a crime. In the *prokuratura* organs, 249 people were charged with various crimes, with 87 being accused of bribery.

56 GARF, f. 8131, op. 32, d. 58, l. 115. "Dokladnaia zapiska o rabote organov prokuratury po delam o vziatochnichestvo za 1951 g." Dated 7 April 1952.

57 GARF, f. 8131, op. 32, d. 66, l. 2. "Dokladnaia zapiska," addressed to G. N. Aleksandrov, from Khivtsov, Procuror of the investigations section. Dated 17 April 1951.

58 GARF, f. 8131, op. 32, d. 58, l. 19. "Doklad o rabote prokuratury Moskovskoi oblasti po bor'be so vziatochnichestvom vo vtorom polugodii 1951 goda." Sent to General Procurator Safonov, February 1952.

59 The majority of these cases involved housing problems, typically centring around obtaining help to falsely obtain permission to live in subsidized housing.

60 GARF, f. 8131, op. 32, d. 58, l. 116. "Dokladnaia zapiska o rabote organov prokuratury po delam o vziatochnichestvo za 1951 g." Dated 7 April 1952.

61 GARF, f. 8131, op. 32, d. 58, l. 116. "Dokladnaia zapiska o rabote organov prokuratury po delam o vziatochnichestvo za 1951 g." Dated 7 April 1952.

62 GARF, f. 8131, op. 32, d. 58, ll. 132–33. "Dokladnaia zapiska," Dated June 1952.

63 Raymond Bauer, Alex Inkeles and Clyde Kluckhohn noted in 1956 the persistence of certain types of "adjustive, informal mechanisms," including theft, bribery and the state's apparent willingness to abide them. Raymond A. Bauer, Alex Inkeles and Clyde Kluckhohn, *How the Soviet System Works*, Cambridge: Harvard University Press, 1956, 89–93.

7 A darker 'Big Deal'

Concealing party crimes in the post-Second World War era

Cynthia Hooper

On 7 July 2005 in his pioneering account of the opulent yet sinister court life of Ethiopia under Haile Selassie, Polish journalist Ryszard Kapuscinski elaborated on the set of unspoken rules that underlay elite behaviour in a country where it was not uncommon for regional leaders to own palaces filled with gold and roses, even as they were technically obliged to win the Emperor's personal approval for any expenditure of more than ten dollars. As one servitor explained, the dictator allowed, even encouraged, corruption in exchange for unswerving allegiance:

> Thanks to his unequalled memory and also to the constant reports, our monarch knew exactly who had how much. But as long as his subject behaved loyally, [the Emperor] kept this knowledge to himself and never made use of it. But if he sensed even the slightest shadow of disloyalty, he would immediately confiscate everything and take the bird of paradise away from the embezzler. Thanks to that system of accountability, the King of Kings had everyone in his hand, and everyone knew it.[1]

In Nazi Germany, relations between Adolf Hitler and his inner circle were similar, characterized by camaraderie, leadership cliques and legal immunity. NS higher-ups for the most part kept quiet about the corruption that pervaded their own ranks, barring the press from reporting on crimes involving Nazi party members without special permission and insisting on the right of the party to resolve cases of economic 'impropriety' internally, outside the courts. Investigators proceeded according to the informal understanding: 'The small ones we hang, the big ones we let go'.[2] From its inception, the Soviet regime had always cast itself in opposition to such models of 'degenerate' governance. Under socialism, so the propaganda went, those at the bottom of any particular hierarchy would have the right to evaluate the actions of those at the top; constant monitoring by vigilant members of society at large would thus keep the ruling apparatus honest and prevent it from turning into a moneyed stratum of self-interested office-holders. Rhetoric, of course, differed from reality, for regional Soviet elites tended to create the same kinds of insular power networks and 'family circles' typical of most centralized dictatorships.[3] Nevertheless, during the 1920s and 1930s, Stalin periodically attacked the complacency of provincial leaders, most memorably in 1934, when he labelled them 'appanage princes' who thought Politburo decrees were written only 'for fools'.[4] With the dictator's backing, forms of public surveillance – self-criticism sessions, show trials, complaint bureaus, unannounced inspection raids – were fused with severe police sanctions to counterbalance the authority of local cliques. Embezzlement and arbitrary power existed, but any one person's hold on that power was tenuous, and never guaranteed.

This chapter argues, however, that the nature of the Soviet dictatorship changed in the wake of the Second World War, given the gruelling demands of post-war reconstruction and the escalating animosities between East and West. The war had left the USSR in a state of devastation and disorder. It had become a country plagued by poverty, lack of living space and crime, unable to absorb all its veterans or provide for all the widows, orphans and invalids left behind – a place where women outnumbered men in many villages by four to one, where outlaws roamed the provinces, and where more than one million died in famine from 1946 to 1947.[5] Though the regions and republics of the USSR had maintained a largely united front against the Nazis, the country had also disintegrated into a patchwork of diverse communities governed largely from the grassroots over the course of the war-effort years. And while many citizens expected Moscow to continue to tolerate a degree of local experimentation and variation during peacetime, central leaders instead fought to restore political orthodoxy and civic order, simultaneously. After 1945, they embarked on a new, this time domestic, battle, both to rebuild a shattered economy and to re-impose a standardized shape of centrally defined Bolshevik power over a fragmented land. They succeeded, in part, by forging a type of 'tacit concordant' between the central government and local *nomenklatura* elites willing to promote Moscow policies. Vera Dunham, one of the only scholars to locate the beginning of a substantive change in the nature of Soviet political power in the post-war years, coined the term 'Big Deal' to refer to the Kremlin's deliberate cultivation of a host of Soviet middle-class 'organization men' in the provinces. These cadres were, in her opinion, Babbitt-esque figures whose qualities included 'apolitical conformism' along with 'loyalty to the leader, unequivocal nationalism, reliable hard work and professionalism'.[6] Dunham links their rise to a larger cultural turn inside the post-war USSR towards an acceptance of traditionally middle-class, material values of diligence, acquisition and, above all, stability.

This chapter ventures further, arguing that during the post-war period, Moscow came to allow the same kinds of opportunities for illicit nest-feathering and pocket-lining to members of the Soviet *nomenklatura* as those Haile Selassie dispensed to members of his Ethiopian court or Adolf Hitler tolerated among his party faithful. For the 'rapprochement' between Moscow and the middle bureaucratic classes that took place after 1945 was more sinister than the one Dunham describes, grounded as it was in a qualified indulgence of corrupt activities on the part of Soviet elites and a redefinition of the rules of party and state control sufficient to protect these elites from overly vigilant public scrutiny or unauthorized prosecution. In consequence, one can glimpse, despite a continued censorship of relevant archival materials, a melding of organized criminal structures with those of the party-state over the course of the post-war era and, above all, a gradual silencing of public discussion in regard to the problem of wrongdoing within officialdom. This chapter will explore the mechanisms of this more insidious type of 'Big Deal'. In so doing, it will focus on the change in attitudes towards upper-level corruption and on the shift in practices of Soviet control.

Silencing scandal

Secrecy and silence had long been features of Soviet public life prior to the Second World War. So, too, however, had been a tradition of 'speaking out' in defence of Communist Party values. During intervals such as the Great Terror, for instance, Moscow exhorted ordinary citizens to denounce the conduct of officials inside the party/state bureaucracy and even went so far as to criticize those who sought to insure their own well-being by keeping quiet. 'Abuse of authority' thus became a frequent topic at party, election and union meetings. Although there were always limits and risks associated with criticism, during the height of the USSR's hunt for hidden enemies, it was not uncommon for regional potentates to be replaced, thrown in jail, or killed four or five times in succession owing to continual surveillance from below and to the elite intrigues which such surveillance allowed.[7] After 1945, in contrast, the actions of those in positions of authority grew far more protected from the scrutiny of their subordinates. At the same time, government officials hushed up facts about crimes committed by party members and homeward-bound soldiers, both social categories of crucial importance to the Soviet regime. Instead, Communist propaganda cast the long process of demobilization as a way of renewing the party/state apparatus by infusing it with honest and loyal veterans. One-time *frontoviki* starred in a mass-produced mythology of the 'virtuous Communist', as men who had successfully defeated the fascists and returned home to continue to battle for good, often in the bureaucracy and especially in the operations of Soviet security.[8] Images of the upright, virile veteran–policeman–hero who turns down enormous bribes or risks his life to catch a thief (often one who was portrayed as having preyed upon a hapless woman or child) filled both the popular press and papers internal to the security organs.[9] Post-1945 articles deploring the prevalence of banditry and black-market criminality almost always blamed these phenomena on clear-cut groups of 'enemies' left over from the wartime era, former Nazi collaborators and wartime speculators rather than Soviet citizens who had contributed to their country's hard-won victory.[10] Even movies about the post-war period made in the 1960s, such as the cult favourite *The Meeting Place Cannot Be Changed*, a picture based on a series of crimes committed by a notorious Moscow gang from 1950–1953, recapitulate such symbolism. The film pits a 'good' Communist soldier against an iconic 'bad' one, which is to say one who did not, in actuality, serve as a soldier at all: the former is a young but much-decorated veteran sent to revitalize Moscow's anti-speculation police unit after the war, while the latter, his chief antagonist, is a ruthless killer who poses as a demobilized officer by wearing a stolen uniform.[11]

Certain government documents, however, attest to the fact that such lines between 'good guys' and 'bad' were far less clear-cut, although any evidence of their blurring was carefully hoarded. Agencies such as SMERSH, set up to fight spies during the war and afterwards directed towards battling corruption inside the police and judiciary, and the Commission of Party Control, which dealt with wrongdoing among members of the Communist *nomenklatura*, shrouded their work in the highest levels of classification and secrecy. But in their records, as in the Politburo's 'Special Files', all still off-limits to most researchers today, leaders mention widespread collaboration between disgruntled Soviet soldiers and criminal gangs (as well as soldiers' tendencies to form gangs themselves) in the immediate post-war period, especially in the western territories. In Estonia, prosecutors attributed a crime wave in the capital directly to servicemen, 'often officers', claiming that it was 'commonplace' for troops to jump off trains transporting them back to their native villages, 'in order to stay in Tallinn to commit robberies and theft'.[12] One lawmaker claimed that bandit groups, including soldiers, had grown so powerful they were able to 'terrorize the population', but that no authorities wished to acknowledge the problem. 'It's gone so far that employees at several factories have refused to work the evening shift. But if you look at the official accounts of crime statistics for Tallinn over the last three and a half months, it sounds as if everything is fine.'[13]

Reports from the republic's Ministry of Internal Affairs in 1946 tell of demobilized troops looting houses, killing civilians, selling stolen goods to speculators and engaging in frequent shootouts with local police.[14] One typical document describes Russian-speaking soldiers marauding through the forests and high-jacking factory shipments of food and consumer goods – in one case, 275 kilograms of butter – transported via stagecoach along the roads.[15] Higher up the military hierarchy, Red Army battalion leaders commandeered trains and cars, to ship contraband Western goods back to their families. In the early 1960s, the Commission of Party Control infuriated First Secretary Nikita Khrushchev by charging the head of the KGB (an unpopular man and former deputy of executed Lavrenti Beria, but also a close friend of Khrushchev from years of joint service in the Ukraine) with having stolen two million marks worth of German property during the Soviet occupation of Berlin. Khrushchev reportedly demanded the accusations be stifled, exclaiming, 'You are forbidden to make a lot of noise about this. After all many generals committed similar sins during wartime'.[16] The actions of many non-combatant party officials inside Russia towards the end of the Second World War also fell outside the law.[17] Control files list numerous infractions, including the ubiquitous 'stealing and selling of alcoholic spirit from enterprises' on the part of Communists. Investigators noted, as just one example among many, that 'in six factories of the Ulianovskii alcohol trust 325,000 litres of spirit disappeared from 1946–1948 ... at a cost to the state of approximately 65 million roubles.'[18] The Soviet Anti-Speculation

Unit estimated that, of the gifts delivered from the United States after V-E Day, central administrators stole 22,423 items of clothing and footware, 15,467 unspecified 'precious objects', and more than 2.46 million roubles worth of other valuables.[19] In Rostov-on-Don, an inspection found that the Communist director of the ration card bureau had, for a city with about 450,000 residents, printed out some 795,829 extra cards, worth well over half a million roubles, which then were either stolen or sold.[20]

Finally, the activities of many law-enforcers during the early post-war years were similarly dubious, although, once again, this fact was only reported in the most exclusive circles. In 1946 factory directors continually complained about the railroad police, recounting such incidents as a train loaded with 20 tons of coal arriving at its destination with a cargo of only one ton, 880 kilograms in tow.[21] Political police files, briefly available to researchers but now reclassified, include the disgruntled impressions of a new regional Minister of State Security, who arrived in Novosibirsk in 1946 to find what he claimed was a culture of drunkenness, debauchery, stealing and hooliganism among Siberian operatives and their families. The chief pointed out that although officers lived in the most prestigious, 100-unit apartment building in the city, the complex was full of violence and dirt, with gangs of children throwing stones and beating up outsiders. 'On the buildings of our houses are painted fascist swastikas; young people – both boys and girls – curse in filthy language. And this complex houses our best personnel, our best Communists'.[22]

Despite the public mythology of Communist and Red Army virtue and the secrecy surrounding most investigations of illegal activities on the part of elites, many ordinary people were not blind to such abuses. In a number of communities, the first years after the war saw heated, angry discussions about local corruption and a flurry of complaints about what was perceived to be a widening divide between bosses and employees, haves and have-nots. In Samara, at election meetings in 1946 and 1947, members of the rank-and-file hurled numerous accusations against local dignitaries – faulting factory directors, for example, for diverting electricity from street lamps into their own apartments. Citizens spoke frankly about both their own material difficulties and the ways many higher-ups seemed to be enjoying a more comfortable life at their subordinates' expense, often by siphoning food and fuel earmarked for workers into their own hands.[23] Railroad employees in Stalingrad protested gross inequalities among the supplies offered at three separate stores, one for ordinary depot workers, another for railroad administrators and a third for only 80 of the city's most prominent officials. Investigators confirmed that leaders included on a special list routinely received such luxuries as eight to ten kilograms of sturgeon at a time, and that several top railroad executives, together with the Stalingrad MVD chief, had appropriated most post-war 'gifts' from the Americans for their personal use, including 12 trophy pianos.[24] A collective of mechanics and conductors from Iaroslavl protested that, while working in Moscow, they and their families were forced to rent tiny rooms or even corridors for 300 roubles a month, when the director of the railroad, who owned a house and a dacha in Iaroslavl, also kept four apartments at his disposal in the capital.[25] In Moscow, the files of the Central Commission of Party Control contain letters vividly describing a Soviet society made up of a mass of terribly poor workers, ruled over by a small, self-interested, mutually protective, and, by comparison, extraordinarily wealthy clique. One from 1946 began:

> The moment of elections has arrived and in the city of Vodsk we have so many legitimate complaints and misfortunes among the people that it is hard to describe ... In the city an epidemic of typhus ... has broken out, the hospitals are overflowing, apartments also, in the hospital there is nothing but contagion, it is cold ... dirty, teeming with insects, the sick lie in their fur coats, caps, there are no sheets, there is no hot water even for the ill, the food is exceedingly bad, delivery of medical supplies has been interrupted, and yet the head of the city health department, Kurnikova, has received a medal for her heroic work [in helping battle the epidemic]. No one can touch the head of the hospitals, Ermolaeva, in regard to the chaos there, because she is the wife of the head of the NKVD, her protection is strong (*zashchita krepkaia*).
>
> In the city from early morning all the people are on the search for water, the pumps don't work, we take water from open man-holes wherever they are, and on the streets people take turns collecting water from broken pipes. For a population of more than 50 thousand we have only one functioning bathhouse ... there are huge lines to get into the bath and they are only made up of the damned directors of the city ...[26]

The three authors (who did not identify themselves) subsequently described how the head of the city council (*gorsoviet*) together with his close friend, an official in charge of the city's alcohol supply, organized drunken revels for a tight-knit circle of comrades who, the authors claimed, 'will carry secrets to their graves'. One kept an office safe full of quality US suits, intended for families of veterans; another, the head of the city health department, allegedly drew full salaries and benefits from six or seven different postings. According to these writers, city and regional (*oblast'*) leaders were occasionally forced to initiate investigations into wrongdoing for the sake of appearances, but at such moments they typically punished lower-level officials for less serious offences in order to mask larger and more incriminating crimes. In 1945, after leaders of the city *gorsoviet* and Party committee allegedly stole 36,000 roubles worth of food products from the Vodsk supply centre, in order to deflect the attention of Moscow higher-ups away from the full extent of the theft (*dlia otvoda glaz*), they had a state trade inspector file an anodyne report referring vaguely to an illegal appropriation of goods and received a pro-forma warning from the city prosecutor, in consequence. 'Meanwhile', the authors wrote, 'the director of the City Food Production ... continues to make house visits [delivering bribes] to all those who are useful to him, and, as for himself, he continues to feed even his pigs the finest grade of grain'. Ordinary people, the writers concluded, were reluctant to speak out against members of such an insular elite, especially as investigators themselves often hesitated to make waves:

> In the city council you don't meet laborers who have anything to say, and if there were no one would listen to them ... the sessions of the *gorsoviet* are strictly for show, the *aktiv* never speaks out ... In November during 43 sessions only one deputy, a three-time medal winner, dared to criticize the kind of abuses described here, and now no one in the *gorsoviet* will speak to him ... Communists among the workers' *aktiv* have also forgotten how to speak out, and very many Communists are without work. [You] need to assign an investigation [of the charges contained in this letter] to comrades who are capable of uncovering all this decay. Signals we relayed earlier were handed over to middle-level bureaucrats in the regional party committee for verification and these people were encouraged [by Vodsk officials] to turn a blind eye, they were given lots of presents and sent very lovingly on their way.[27]

In this case, the Party Control Commission did launch an investigation which generated some change, although hardly the type of severe punishments one might have expected, considering that virtually all of the charges levelled in the letter were confirmed. Ultimately, the *oblast'* party committee sanctioned the removal of the head of the city council and one of his deputies; however, none of the incriminated officials was arrested, expelled from the party or tried.[28]

Even such partial successes in protesting against the conduct of local *nomenklatura* appear to have grown far more infrequent over time, due, at least partially, to a renewed focus on 'labour discipline' and workplace hierarchy at the expense of criticism from below. In this regard, the years following the end of the Second World War resemble those of 1939–1941, as Soviet authorities struggled to reign in practices of mass denunciation that had contributed to the unpredictable and self-consuming aspect of the Great Terror. During both these intervals, a gradual stabilization of existing power relations and the suppression of unscripted rank-and-file opinion can be observed. In 1939, for example, all the military organizations in the Soviet Union, including the NKVD, revoked point-blank the right of subordinates to critique the performance of their commanding officers. Officials announced that 'in the interest of discipline, criticism of defects in a commander's and [political] commissar's performance would ... instead be considered and discussed at the next higher level of command'.[29] Communist officers in the regions further encouraged their colleagues to hold their tongues, pronouncing, as in 1940 Gorkii, that 'none of us has the right to discuss orders given by the Chief of Police'. Meanwhile, workers fell under uncomfortable scrutiny from above, as documented in party meeting transcripts and newspaper headlines. At the height of the Terror in 1938, editors of the Gorkii *oblast'* police newspaper published an article demanding that leading regional NKVD and party officials be punished for the poor construction of a police dormitory and the abject living conditions residents had consequently been forced to endure. In 1940, in contrast, a similar article about the dormitory blamed not authorities but 'ungrateful' residents for the squalor, charging them with spoiling a quality space by sleeping on their beds in dirty boots, failing to throw away their empty vodka bottles and allowing portraits of Stalin and other leaders to collect dust.[30] As the regional police chief stated that same year, while he was not opposed to 'criticism from below' in principle, he failed to see why this criticism should focus primarily on leaders when there were 'so many insufficiencies within the rank and file'.[31]

Similarly, in the post-war years, and particularly after 1947, employees who dared to complain about their superiors were more often than not subject to intimidation and harassment. Party and state officials inside the same organizations frequently joined forces to silence criticism, prompting local control officials and workers alike to note the inclination of bosses 'not to hang each other's dirty laundry out to air' or 'not to let arguments out of the family hut'.[32] In 1951 the Moscow City Party Committee referenced an incident where the chief of police and the head of the city militia party cell had worked together to remove entire press runs of successive editions of the police newspaper *Na boevom postu*, because they included critical observations about the metropolitan MVD administration.[33] Press coverage, in general, grew increasingly circumspect. In contrast to both the 1930s and the Khrushchev era, leaders during the post-war Stalin years were almost never mocked in cartoons or satires. Articles about wrongdoing within the Communist elite, even following judicial convictions, were only selectively approved for publication. At a Moscow meeting to discuss the role of newspapers internal to the police, a few editors complained about these rules of silence. One recounted how, after he printed an unsigned feuilleton about corruption inside his district, Soviet central police authorities in Moscow phoned his *oblast'* party committee one week later, to announce that the author had 'made a mistake'. As a result, a party meeting was called, allegedly to discuss the feuilleton, but according to the editor, in actuality to discredit the people involved in its publication. 'They spent most of their time talking about the dubious affairs of the editorial board, about the editors, but not about the article', the speaker claimed. He said he was ultimately issued a party rebuke for 'immodest behaviour' in relation to some manufactured, unrelated offence and advised to avoid such controversy in the future.[34]

Another editor from Kazakhstan described how he had tried to publish a satirical paper exclusively for police workers in Alma-Ata, which had included, as part of a discussion of militia efforts to battle debauchery, a sketch of a policeman embracing a vodka bottle and jokes about officers consuming confiscated spirits and then scrambling to hide their sins by adding water to whatever small portion of their 'catch' remained. The editor said that he had printed only two issues of forty copies each, before running into problems from the Central Committee of Kazakhstan. Members had objected to his ironic tone, forbidden future publications, and issued him a rebuke. Such attitudes, the officer remarked, made it impossible to compose anything but sycophantic articles:

We here all speak about criticism, that it is necessary to find space for sharp critical material in papers and it is absolutely correct that ... in our paper in the current year there have been very few critical articles, but the truth is our bosses don't especially want to air their disputes outside the home (*vynosit' sor iz izby*); they say, 'wouldn't it be better to focus on something positive?'[35]

The head of the Political Division of the central Soviet militia administration responded by stating firmly that no publications, however exclusive their audience, could be permitted to discredit the socialist regime. Communists, he said, 'cannot allow' unflattering portraits (such as that of a policeman holding a vodka bottle), even in the interest of loyal self-criticism. Echoing a tenet of socialist realism, the Political Division chief urged editors not dwell on insufficiencies as they existed in the present and commanded them, instead, to show lives and careers in their ideal form, as they should be, while 'talking about such evils [as alcoholism] as relics of the past'. Grim representations of everyday life must be carefully hoarded, he continued, explaining that, as part of a USSR-wide temperance campaign, the Central Committee had gone so far as to approve one naturalistic poster for publication, depicting an intoxicated citizen prostrate in front of his apartment, as his wife, with two children, shuts the door in his face. But even this poster, the speaker continued, was 'only being printed for closed circulation, for hanging in prison sobriety cells, in laboratories, and other appropriate places, but not for broad distribution'.[36] Such comments reflect the secrecy with which the late Stalinist regime addressed any kind of persistent social problem and illustrate leaders' ambivalence about whether the interests of the state were better served by acknowledging troublesome facts or denying their existence.

Closing the door on control from below

This gradual muffling of public discussion of elite corruption was matched by new practices of surveillance inside the regime. This shift is evident in the evolution of the Ministry of Soviet Control, an agency that in the first months after the Second World War strove zealously to uncover scandal. In 1946, its leader, Lev Mekhlis, took the extra step of sending an exasperated letter to the top party and state officials in the republic of Ukraine, excoriating his own inspectors for their laxity in pursuing administrative wrongdoing. His lengthy litany told of controllers appropriating leather coats, boots, alcohol, and food from organizations they were supposed to be investigating, organizing night-time orgies and systematically receiving bribes. It ended by demanding the removal of the Ukrainian Minister of State Control.[37] But Mekhlis' attitudes, however, would change over time. In urging his deputies to be more vigilant, he appears to have antagonized a host of powerful officials. As one historian explains, his Ministry:

> managed to dig out and bring to Stalin's notice within two or three years [of the war's ending] 'cases' against many bosses in Moscow and in the

republics. Quite a few ministries – finances, railways, and defence – were offended. Mekhlis even boasted that 'in the central administration of the armed forces, the prestige of the ministry of state control is great, in short, our controllers are feared.' Perhaps more than anyone else, Mekhlis went too far in his desire to be useful for Stalin. He antagonized influential people in the power-elite. A commission was launched to examine 'cadre work' in his ministry, in other words, to find out weaknesses in his domain . . .[38]

Subsequently, the place of control work plummeted in prestige. Investigators' secondary status manifested itself in low pay, overcrowded offices and miserable living conditions, far below those of colleagues inside the ministries they purported to oversee.[39] At a Moscow meeting in 1952, one senior inspector mentioned his shame at having to take two managers from the Ministry of Fish Production to eat lunch with him, after they had spent the morning going over figures in his office. 'Knowing our buffet I grew terribly embarrassed . . . Fortunately, it turned out we had some new bread that day'. He said that, in comparison, the Fish Production Ministry boasted two luxurious cafeterias. 'Not only do they have a wide assortment of food, but there it is actually pleasant to eat, there it smells of freshness. But with us things are still the same as they were in 1943, during the war'.[40] Above all, many of the Ministry's employees deplored what they saw as a burgeoning number of regulations designed to undercut the authority and limit the mandate of controllers inside state organizations. Their own Ministry, they said, had become far more hierarchical; issues to investigate had to be assigned from above, and inspectors on the ground were not allowed to initiate inquiries, follow-up on accidental discoveries of dubious activities or even write down facts pertaining to directors and department heads without special government permission.[41] As a result, bosses inside factories and trade organizations did not hesitate to resist control workers' appeals for information or thwart their verification campaigns, insisting that Moscow sanction their every action in advance. Inspectors bemoaned the fact that they had no way of holding top managers accountable and could not answer such resistance. 'What was left for [us] to do? Turn around and leave', said one, recounting a fruitless meeting with a deputy director.[42]

The Ministry was forced to abandon former tactics of popular control, which had involved mobilizing members of the rank-and-file in efforts to expose top-level abuse. Taking quite a different tone than he had in earlier years, in 1948 Mekhlis met with prominent Kremlin leaders in order to apologize for the performance of the newly appointed head of Soviet Control in the republic of Azerbaijan. Upon arriving in Baku, this deputy had been assigned to proof a specific accounting problem; however, instead of restricting himself to the investigative parameters assigned by Moscow, he had allegedly begun 'on his own initiative, without any instructions', to receive petitioners and read through community complaints. Although such practices had been ubiquitous in the 1930s, in 1948 Mekhlis blamed them for fomenting an intolerable degree of chaos. To make matters worse, the minister had – shamefully, Mekhlis implied – begun to trust the 'scurrilous' accusations of the dubious riff-raff that besieged his office more than the party/state officials who surrounded him and who had rendered the Soviet government many years of loyal service:

> The head of inspection Emelianov and other controllers from the Ministry set out on the wrong path, using politically damaging methods of inspection, which included showing special attentiveness to unreliable petitioners (*somnitelnye zhalobshchiki*). Through this they continued on the anti-party path of discrediting the most senior party and soviet workers of Azerbaijan, preparing and organizing materials against them, which created an unhealthy atmosphere around the inspection.[43]

This Emilianov had, Mekhlis concluded, 'intentionally shown mistrust towards the leaders of the Central Committee and Council of Ministers of Azerbaijan and conducted himself in relation to them, as if to put people on trial', forgetting that controllers were obliged to treat fellow bureaucrats not as aliens or enemies, but as a trusted comrades, 'our people, ones of us' (*rodnye nashi liudi*). In another violation of post-Second World War procedures, the renegade Minister had himself hired a number of 'unofficial' informants to collect further evidence against an array of local potentates – prompting Baku elites to join forces and contact Moscow, urging a halt to investigations and the Minister's dismissal.

In describing the actions of his subordinate, Mekhlis' comments dripped disdain for the strategies of popular control that had prevailed during the First Five Year Plan and Great Terror years (and which Mekhlis himself had once endorsed). As he concluded:

> The fact that the inspection team had begun to listen to complaints quickly became known throughout Baku. Lines began to form. It is typical, that initially, for the first eight to ten days there were *no* complaints from the populace, but as soon as [news of] this heightened official concern for [whiners] began to circulate, as soon as some people began to obtain residence permits and medical treatment and work reinstatement, complainers began to flock to the Ministry in droves ... I must say that to this day these complainers still persist in congregating ... they won't go away, they keep banging on the windows of the Ministry of State Control in Azerbaijan.[44]

This case clearly emphasizes the most important ingredients of the late-Stalin era 'Big Deal'. These included a deepening alliance between officials at all levels, a willingness to turn a blind eye to self-enrichment, if kept within certain bounds, and a deep suspicion of little people and their motives – all perhaps a consequence, in part, of *nomenklatura* determination to avoid another 1930s-style purge. Denunciation, once lauded as a virtuous civic activity, the preserve of 'daring' people willing to risk their own comfort, began to be described inside the bureaucracy in far more contemptuous terms. Whereas the rhetoric of the

Great Terror had celebrated the role of every Soviet citizen as a voluntary informant, by 1948 agencies shied away from 'amateur' or community-based surveillance activities. In Azerbaijan, Mekhlis claimed that the organization of a cadre of unverified, untrained spies had 'enabled gossip, careless conversation, bias, and the spreading of all kinds of sensational stories'.[45] In other incidents, inspectors were faulted for trusting the 'wrong' people, and for taking the word of even 'gossipy women' over those of tried-and-true state servitors.[46]

Even organizations such as the political police that continued to rely on informant activity over the course of the late-Stalin period came to do so in a much more structured and organized way. In meetings within the central police administration to discuss this issue, officers argued that the 'excesses' of the Terror had resulted from an over-reliance on volunteer vigilantism, and they mocked the hiring practices of the 1930s, when they claimed hordes of housewives had been taken on as undercover NKVD agents after writing on applications nothing more professional than statements such as: 'I could help ascertain those who speculate in bread, because I spend a lot of time standing in lines'.[47] Officers inside the Anti-Speculation Unit advocated, instead, the cultivation of cadres who would be dependent on and subordinate to their minders, provided with fixed assignments rather than set loose to comment on whatever activities they deemed suspicious. 'Politically reliable' informants, they contended, were generally capable only of fingering the most petty of criminals; the most seemingly dedicated volunteers were often the most useless, 'deadweights' who clung to their security service connections and accompanying material benefits, often fired from the rolls as 'ballast', only to return to another police station and offer their services again.[48] Officers called for a 'fundamental re-conceptualization' of agent work, demanding the creation of a spy network made up of 'criminal elements and personages tied to them' in order to penetrate the machinations of an increasingly complex black-market underworld.[49]

In order to achieve this aim, officers repeatedly urged that more and better use be made of 'compromising materials' in order to recruit agents, and that a more efficient pay scale be introduced to reward them (with compensation not constant, but linked to the quality and quantity of information provided). Both memoirs and archival documents suggest the accelerated evolution after 1945 of strategies for forcing cooperation from sources the Ministry of Internal Affairs deemed worthwhile.[50] The political police developed a virtual science of blackmail in the postwar years, touting in MVD textbooks as exemplary cases of investigative work organized around carefully supervised schemes of entrapment, seduction and monitoring through wiretaps or other technological devices. One such example was entitled 'Orientation in Methods' and began with the case of one agent 'L' who found suspicious evidence surrounding a family's apartment, indicating that its members were illegally printing labels from the Stalin State Chemical Factory in Moscow for bootleg containers of homemade dye:

> Monitoring of the house showed that in the apartment of the Ianulinasov family was living their maid 'R', with whom, as it turned out, agent 'L' was

acquainted. In connection to this, the agent was assigned ... to entangle R in an intimate relationship, to gain her confidence, to incline her to him, and, through her, to discern the criminal activity of the [family]. In one of her conversations with the agent, R said that she was living at the Ianuli-nasov's without a residence permit. After a study of the personal character-istics of R we decided to use this circumstance for her recruitment. Operational workers planned and carried out a secret summons and interro-gation of R. At the interrogation, R confirmed that she lived at the Ianuli-nasov's as a maid without a residence permit and simultaneously gave detailed accounts of several facts of their criminal activity. On the basis [of that evidence] R was recruited as an agent with the goal of further investi-gating the designated suspects. The recruitment of R turned out to be a success. Soon she had determined that Ianulinasov and his wife possessed three counterfeit label-making machines, with the help of which every day they prepared from two to three thousand packages of counterfeit dye and earned 130–200 roubles for every thousand pieces.[51]

Such developments in surveillance practice after 1945 did not mark an abrupt break with those that had preceded them; in some ways, rather, they legitimated tactics that had been in use for decades. Yet at the same time, they resolved a tension that had characterized most of the 1930s and particularly the Great Terror years. That decade had seen parallel and interrelated developments towards collective, vigilante activism together with outspoken, often strident criticism, on the one hand, and secret, hidden forms of surveillance, on the other. Post-Second World War authorities, however, cultivated the latter at the expense of the former. They abandoned the more horizontal principles of popular control, with its emphasis on voluntarism, universal amateur involvement and fanatical enthusiasm for party aims, in favour of a vertical, far more calculated, system of information supply, in which official 'handlers' manipulated desig-nated 'sources' in order to achieve precisely delineated, predetermined opera-tional 'objectives'.

Two-tiered justice

For ordinary people, the post-war period was one of tremendous want, brutally hard labour, and subordination to rigorous central decrees designed to discipline the workforce and check any outbursts of possible disaffection.[52] Conditions compelled virtually all citizens to engage in some form of small-scale subver-sive activity (above all stealing state property and selling it on the black market) in order to survive.[53] Such actions, however, carried a not insignificant degree of risk. In 1947, authorities initiated a campaign against theft that required judges to level astronomical penalties against even first-time offenders, who suddenly faced up to ten years instead of a maximum three months in jail.[54] Yoram Gorl-izki has shown that while this campaign resulted in only a fleeting rise in numbers of theft cases heard before the courts, the sentences imposed on the approximately quarter of a million people convicted annually were far higher after 1947 than before.[55] This increasing severity, however, seems to have most affected those lowest down on the social scale, citizens bereft of a protection network, or what in mafia slang is termed a 'roof'. In contrast, Communists in positions of authority who fell under investigation could generally count on both the party and their relevant ministries to take on something of this 'roof' function, in shielding them from prosecution. Elite solidarities, combined with the changing role of primary party organizations inside the workplace, contributed to the development of a two-tiered, status-based system of justice in the post-Second World War Soviet Union – cushioning the *nomenklatura*, but providing ordinary people with little defence from the actions of those above them. During the 1920s and 1930s, party cells had functioned in a very different fashion: from the earliest days of the Revolution, they had been assembled inside workplaces in order to counterbalance bureaucratic authority, monitor elite performance and, on occasion, terrorize those in positions of administrative power. In theory, if not always in practice, party secretaries and state leaders inside the same organization were required to maintain a certain distance from one another, even when those leaders were also members of the Communist Party.[56] After the war, however, party officials actively collaborated with administrators in limiting the autonomy of the rank-and-file, carefully scripting who spoke and what they said in meetings and often hampering investigations of management wrongdoing. As Peter Solomon has noted, 'a party secretary was more likely to be reprimanded by his superiors for failing to anticipate and prevent trouble than to be praised for discovering it'.[57]

Even at the very top of the Communist hierarchy, officials likewise often chose to limit scandal. In a typical case in 1947, the central Commission of Party Control received a protest from a supply agent in a county (*raion*) trade department, who claimed to have been scapegoated by county leaders, after they threw a banquet in a restaurant and wasted 7,000 roubles of state money. To cover up the affair, he said, members of the county party committee (*raikom*) had ordered him to sign a fictitious document attesting to their procurement of 3,200 roubles worth of potatoes. For this, he had been sent to jail, while the *raikom* organizers of the illicit feast had gone untouched.[58] Although the Commission acknowledged the agent's complaints could be legitimate, it refused to pursue the matter. In other incidences of alleged corruption that involved the word of one Communist official against another, the Commission simply decided to abandon cases without either resolving who was to blame or apportioning any punishment.[59] 'We consider that in these conditions it will not be possible to confirm which of them is saying the truth', one file concluded. 'We suggest, as a result, to close the investigation and consign the report to the archives'. During the Great Terror, the Central Committee had issued a set of secret party rules mandating that no Communist could be arrested or handed over to the courts without prior written permission from the appropriate party committee.[60] However, these decrees had been meant only to ensure that party organizations would be able to expel Communists facing imminent arrest from their ranks before the individuals were taken into custody. After the war, in contrast, party leaders came to use these same secret rules to delay, if not outright prevent, unwanted judicial actions against CPSU members.[61] In 1951, the Procurator of Kazakhstan bitterly criticized the consequences of these prerogatives in a letter to the head of the Republic's Central Committee. He claimed that a number of heads of *raion* and *oblast'* party committees took Communists accused of crimes 'under their protection', giving them party 'rebukes' but forbidding their arrest:

> All of this creates lengthy red tape in the investigation of criminal cases, especially those involving groups where several people are accused. Citizens who are not party members and who are involved in such cases are arrested, but Communists who have committed crimes remain at liberty only because the party organs do not allow them to be arrested and tried, which cannot help but arouse justified condemnation among the surrounding population . . .[62]

Moreover, on those relatively rare occasions where party committees broke with routine and recommended criminal prosecution, a number of elaborate protective mechanisms and lengthy procedures still had to be overcome in order finally to get a case involving high-placed Communist officials to court, especially those inside powerful ministries such as Internal Affairs. In 1949, for instance, the head of an *oblast'* MVD division was accused of beating suspects who refused to confess to stealing grain, including a girl whom he hit so badly that she suffered a seizure and was hospitalized for a week. Although three of the officer's subordinates, doctors and a number of additional eyewitnesses confirmed these attacks, a variety of authorities strenuously advocated the officer's exoneration. (Their efforts to save him from prison, however, did seem to have met with equally determined attempts on the part of the Soviet judicial apparatus to continue to push for a trial.) Even once the USSR Minister of Internal Affairs consented to the officer's arrest, in December of 1950 his regional party committee again tried to block prosecution by issuing the officer a 'strict rebuke', but concluding that 'to try him for the violations of legality he committed would not be fruitful, considering a great deal of time has passed, and he has more than 20 years of service in the organs'.[63] In late 1951, the Moscow Party Control Commission reversed the *obkom*'s recommendation and agreed to allow the officer to face charges; however, the final outcome on the ground remains unknown.

In another case, a former county MVD chief was accused by a member of his own police force of summarily shooting several citizens without investigation (including one juvenile whom the chief dragged from the boy's home and executed on his doorstep) during a hunt for a gang of bandits in February of 1945; following a subsequent shootout with the gang, he ordered his men to set fire to a small village in retaliation, destroying, among others, several homes belonging to families of Red Army soldiers. An internal investigation by the MVD confirmed these allegations. Nonetheless, the case still had to work its way up to the level of the USSR General Prosecutor, who was obliged to request formal permission from the Ministry of Internal Affairs to take up the scandal. The officer in question was eventually sentenced in March 1947 to ten years' imprisonment, but even then he was allowed to file a protest against regional lawmakers, who were consequently examined, first by the Soviet judiciary, then by the Party Control Commission, before being cleared.[64] Still, these more or less successful prosecutions stand out as exceptions. In an analysis of the vast amount of stealing taking place from alcohol and food industries, party officials in Moscow concluded that members of the local judiciary and MVD did not attempt to verify the overwhelming majority of citizen complaints they received but rather 'sent them back to the very same organizations against whom the authors were complaining'.[65] Meanwhile, inside those organizations, officials acted to protect their own. The Ministry of Food Production furnished one case in point. There, analysts noted, in those rare cases where an accusation was upheld:

> even then when the facts of stealing were confirmed and it was necessary to take the guilty to court, the Ministry ... and the main departments confined themselves to transferring the administrative workers of various enterprises and trusts to other enterprises, where they continued to live lawlessly (*tvorit' bezzakoniia*). For example, the former director of the Moldavian trust, Stepanov, who earlier had been fired from work and expelled from the party for black-market selling of vodka and other abuses, was chosen to head a division of Glavspirt in October of 1947; Slepsov, removed from his position as director of Khovrinskii liquor warehouse for releasing a large amount of alcoholic spirits without sanction, in 1948 was promoted to the position of Glavspirt senior inspector.... State prosecutors during their investigation of the stealing of alcoholic spirits have not taken the main thieves to court – the directors of factories – but have instead arrested the little people who executed the thievery – rank-and-file factory workers. In all of the cases, prosecutors have violated Soviet legality, covered up material about hidden crimes, and a number themselves have illegally received alcohol [as bribes].[66]

This division of Soviet society into two groups – those subject to the law and those whose networks of protection helped to place them outside it – is reflected in almost any corruption investigation during the post-war period. In 1947, controllers in one ministry succeeded in uncovering, over the course of the year, 'damage to the state' totalling more than 200,000 roubles, embezzlement of 1.138 million roubles and illegal stockpiles of surplus materials worth 4.124 million roubles; yet during this same time they turned only three party members over to the courts.[67] An analysis of state control around this same time mentioned the tendency of factory administrators to condone a degree of economic subterfuge on the part of their deputies. 'Upper-level directors often encourage and cover up the illegal activities of their subordinates', the document reads.

Directors allowed such surreptitious machinations as the 'making of repairs of their personal apartments and cars at state expense' or the 'release at wholesale prices or sometimes entirely for free' of food products and construction materials 'for building personal dachas or houses for qualified workers under their command'. The report faulted ministry officials for their excess liberality in cases of more egregious abuse, contending that, 'They don't even like to transfer workers, and instead . . . will limit themselves to leveling an internal disciplinary rebuke'.[68] An officer in the Uzbekistan police, during an investigation of the militia's anti-speculation unit in 1952, similarly observed that those who broke the rules inside the force generally received nothing worse than 'a pat on the head'. For systematic drinking, one officer got a warning; for beating citizens, a five-day internal garrison arrest; for taking bribes, dismissal. He added:

> In the Fifth Police Division of Tashkent for seven years the boss of the central market was for all intents and purposes the police officer Rasulov, to whom everyone paid an 'offering'. He organized trade in meat at the market . . . On the side, he worked as a chef at all the national festivals . . . Through criminal methods he procured two houses and one dacha. For those crimes he was fired from work but not handed over to the courts.[69]

Conclusion

From 1917 to 1953, one can trace a series of shifts in the rules governing investigation of bureaucratic wrongdoing across the Soviet Union. In the early 1920s, for instance, hosts of amateur inspectors were exhorted to show initiative in uncovering crime inside the Soviet administration. According to Peter Holquist, the political police during the Soviet Civil War went so far as to stress to informants that 'it was not enough merely to describe attitudes; they should also "indicate what explains" them'.[70] During the Great Terror, leaders such as Nikolai Antipov, head of the Commission of Soviet Control (until his own arrest in 1938 and subsequent execution), raised the demands on informant-investigators even further. Antipov contended that they could not be allowed to limit themselves to abstract analyses of workplace shortcomings, but must be required to point fingers, assign guilt and name names in their findings. Such duties, he noted in 1936, made it difficult for his organization to find good low-level cadres, for many people were unhappy with central expectations:

> It is obvious that it is highly unpleasant [for a person involved in control work] that in order to enforce the fulfillment of one or another decrees, he must thoroughly expose shortcomings and reveal the reasons for these shortcomings, and when these shortcomings have been exposed and these reasons revealed, then, of course, the matter is not an abstract one, and [a controller] is required to point to concrete people who directed the work; he is required to draw the appropriate conclusions, and of course, this is not a pleasant thing to have to do . . . [Inspectors] are dissatisfied . . . some

[inspectors] send letters here, some send letters to me and to the Central Committee, asking to be released from their [control responsibilities], even for just six months, asking to be left alone ... and they ask not because we have heavily presumed upon them, but just simply because control is not an especially pleasant thing.[71]

During the post-war period, however, many officials inside the control apparatus expressed frustration about the opposite condition – their inability to conduct more than superficial investigations or to confront local authorities. On the ground, they were blocked from reporting many of the things they saw and repeatedly warned not to discredit representatives of the Soviet regime. When their investigations, in the eyes of superiors, 'went too far', they were chastised for what Moscow termed their tendency 'to heap together a pile of negative facts and hurl them all onto the head' of one or another hapless department or individual.[72]

In the months just after Stalin's death, former members of State Control actually waxed nostalgic for the 1920s and 1930s, including the Great Terror years, recalling them as a time when their own organization had been truly powerful due to its ability to mobilize millions of ordinary citizens in the battle against clandestine bureaucratic corruption. Such attitudes were particularly in evidence at a spectacular assessment meeting convened by the Ministry of State Control in early 1954, to which organizers invited former control work alumnae to speak with current employees. All those who made comments deplored the state of Soviet control in the post-Second World War years compared with what they described as its '1930s glory'. What these alumnae recalled with greatest fondness was their agency's past ties to the Soviet people. 'Now our attitude towards [volunteer] signals is a formal one', commented one speaker. 'We act as if we only trust [officials], as if we're surrounded by alien and hostile people ... But back then, everything very much rested on ordinary people, on the masses, on their signals, which pointed to disorders and shortcomings in the work of the state *apparat*'.[73] Guests urged their younger colleagues to return to what they described as the unparalleled civic energy of the Great Terror years. 'Why is there no active citizenry now?' a senior inspector demanded. 'Where is the *aktiv* hiding? What happened to conscientious, good-faith (*dobrosovestnye*) people? Conscientious Soviet people are everywhere and we have many shoulders we could rely on. But we must admit that we ourselves are to blame for having alienated this *aktiv* in recent years and driven these good citizens away'.[74]

These points of view are, of course, sentimental and highly idealized, never once mentioning the violence and debasement of the era they recall. But they also reflect the fact that many controllers experienced the post-war era as a qualitatively different time than that which had preceded it – one marked by a growing chasm between self-interested leaders and society at large and a rejection, certainly in practice if not in propaganda, of the need for the latter to be able to hold the former to account. The new rules of state surveillance encouraged investigators not to dig too deeply, not to stir up scandal or disrupt hierarchies, and not to publicize the facts of any abuses they might encounter. These rules would shape the future development of the Soviet bureaucracy, laying a foundation for the partnership between organized crime and political dictatorship characteristic of the late Leonid Brezhnev years, and thwarting attempts at reform even today.

Notes

The author wishes to thank the Arts and Humanities Research Council, UK for funding the research upon which this chapter is based.

1　Ryszard Kapuscinski, *The Emperor*, William Brand and Katarzyna Mroczkowska-Brand (trans.), London: Quartet Books, 1983, p. 49.

2　Frank Bajohr, *Parvenüs und Profiteure: Korruption in der NS-Zeit*, Frankfurt am Main: S. Fischer, 2001, pp. 148–163, 166, 171.

3　See, for example, Merle Fainsod, *Smolensk Under Soviet Rule*, New York: Vintage Books, 1963 and James Harris, *The Great Urals: Regionalism and the Evolution of the Soviet System*, Ithaca, N.Y.: Cornell University Press, 1999, pp. 38–70.

4　*XVII s"ezd Vsesoiuznoi Kommunisticheskoi Partii (27 ianvaria – 10 fevralia 1934 g.): Stenograficheskii otchet*, Moscow, 1934, pp. 23, 33.

5　Germans occupied territory on which 45 per cent of the Soviet population had resided; during the war 12–15 million Soviet citizens were evacuated eastward, 27 million soldiers and civilians were killed, and 32,000 industrial enterprises and 65,000 kilometres of railroad track were destroyed. Sheila Fitzpatrick, 'Post-war Soviet Society: The "Return to Normalcy", 1945–1953', in Susan Linz (ed.), *The Impact of World War II on the Soviet Union*, 1985, p. 130. Soviet leaders retrospectively reported the country lost 30 per cent of its national wealth during the war. In *Pravda*, 6 April 1966. See also Michael Ellman, 'The 1947 Soviet Famine and the Entitlement Approach to Famines', *Cambridge Journal of Economics* 24:5, 2000, 603–630.

6　Vera Dunham, *In Stalin's Time: Middleclass Values in Soviet Fiction*, Durham, N.C.: Duke University Press, 1990, p. 17.

7　The NKVD serves as a good example of such successive replacements. Between 1935 and 1941, three people served as heads of the Soviet NKVD, two of whom were shot. In Gorkii, the head of the regional NKVD changed hands five times (transfers which involved one suicide and two arrest/executions) and the head of the regional police force, four (with two police chiefs arrested and shot, one after a brief transfer, and another fleeing town in the dead of night to escape reprisal). See *Zabveniiu ne podlezhit: O repressiiakh 30-kh – nachala 50-kh godov v Nizhegorodskoi oblasti*, Nizhnii Novgorod: Volgo-viatskoe knizhnoe izdatel'stvo, 1993, p. 510; N. V. Petrov, *Kto rukovodil NKVD, 1934–1941: Spravochnik*, Moscow: Zven'ia, 1999, p. 343, 459; Gosudarstvennyi obshchestvenno-politicheskii arkhiv Nizhegorodskoi oblasti (GOPAN), f. 817, op. 1, d. 65, l. 39, 40.

8　Jeffrey Wade Jones, '"In my opinion, this is all a fraud!" Concrete, Culture, and Class in the "Reconstruction" of Rostov-on-the-Don, 1943–1948', Diss. University of North Carolina at Chapel Hill, 2000, p. 252. Although many women also fought on the Soviet side during the Second World War, post-1945 images of the 'virtuous Communist' returning from the frontlines to serve the state further seem to have been predominately masculine ones.

9　An example of such heroic stories is contained in *Post revoliutsii*, 18 June 1949. See also A. M. Beda, *Sovetskaia politicheskaia kultura cherez prizmu MVD: Ot 'Moskovskogo patriotizma' k idee 'Bol'shogo Otechestva'*, 1946–1958, Moscow: Mosgorarkhiv, 2002. Amir Weiner also notes that 'a barrage of popular novels on the post-war countryside celebrated a new hero: the demobilized officer who transferred his zeal from the front to pursue the electrification of the backward countryside. As a rule, the character of the relentless veteran was contrasted with that of a laid-back bureaucrat, most likely one who avoided the front and adapted a "soft" and conservative approach to the tasks of reconstruction'. In *Making Sense of War: The Second World War and the Fate of the Bolshevik Revolution*, Princeton: Princeton University Press, 2001, p. 49.

10 Politburo leaders reproduced this rhetoric, terming bandits 'agents of fascism' even when they acknowledged them to be former Red Army soldiers. Rossiiskii gosudarstvennyi arkhiv sotsialno-politicheskoi istorii (RGASPI), f. 598, op. 1, d. 8, ll. 102, 103.

11 Such tropes also appear in memoirs written in the 1960s about the immediate postwar period. Take, for example, Ivan Parfent'ev, *Proshloe v nastoiashchem: Zapiski byvshego nachal'nika Moskovskogo ugolovnogo rozyska*, Moscow: Sovetskaia Rossiia, 1965. In a note to readers, Parfent'ev writes about his years in the police, saying, 'I worked together with many fighting comrades, friends ... I will never forget them, daring, determined, hardened men. ... I want [in this book] to show the closeness of the Soviet police to the people, its tight connection to them' (p. 5).

12 RGASPI, f. 598, op. 1, d. 16, l. 114.

13 Ibid., l. 111.

14 RGASPI, f. 598, op. 1, d. 8, ll. 30, 108, 109.

15 Ibid., l. 110.

16 Anastas Mikoian, *Tak bylo: Razmyshleniia o minuvshem*, Moscow: Vagrius, 1999, pp. 607, 608.

17 This raises the question, outside the scope of this particular article, of whether *nomenklatura* corruption in the late Stalin period marked a continuity or change in practices that emerged during the war. Certainly many members of the Communist Party and society at large fought the Nazis with a remarkable degree of selfless heroism; however certain documents, again only partially declassified, suggest that those elites who wished to could and did take advantage of the large degree of civic disorder, shortage and lack of oversight to secure their own well-being. One file from the Sverdlovsk Party Control Commission includes a long report from an NKVD informant complaining about bosses having stolen significant quantities of state property, particularly during the evacuation of factories from western Russia across the Urals. This same man accused one worker in 1944 with having remarked to a friend that the region's factory directors have 'fattened themselves up like pigs at the expense of the workers' inside, and they look at workers like cattle, they treat ordinary people worse than the fascists do' (*ozhireli kak svini za schet zheludka rabochikh i na rabochikh smotriat kak na skotinu, obrashchenie s liudmi khuzhe fashistov*). In Tsentr dokumentatsii obshchestvennykh organizatsii Sverdlovskoi oblasti (TsDOOSO), f. 236, op. 5, d. 56, ll. 1, 3, 7.

18 Rossiiskii gosudarstvennyi arkhiv noveishei istorii (RGANI), f. 6 op. 6, d. 1, l. 28.

19 Gosudarstvennyi arkhiv Rossiiskii federatsii (GARF), f. 9415, op. 5, d. 95, l. 60.

20 Jones, op. cit., p. 214. The director was not arrested, but only rebuked for negligence and 'loss of party vigilance'.

21 Rossiiskii gosudarstvennyi arkhiv ekonomiki (RGAE), f. 1884, op. 31, d. 7198, l. 109.

22 A. G. Tepliakov, 'Personal i povsednevnost' Novosibirskogo UNKVD v 1936–1946', *Minuvshee: Istoricheskii al'manakh*, Novosibirsk, 1997, p. 273.

23 Samarskii oblast'noi gosudarstvennyi arkhiv sotsial'no-politicheskoi istorii (SOGASPI), f. 714 op. 1, d. 1149, l. 18.

24 RGAE, f. 1884, op. 31, d. 7201, l. 51. At the urging of the Stalingrad *oblast'* party committee, they ultimately donated one of the twelve to a city cultural centre.

25 RGAE, f. 1884, op. 31, d. 7201, l. 518.

26 Harvard University microfilm collection, Commission of Party Control (Feb. 1946), op. 6, d. 556, l. 204.
27 Ibid., l. 208.
28 Ibid., l. 211.
29 Roger Reese, 'The Red Army and the Great Purges', in J. Arch Getty and Roberta Manning (eds), *Stalinist Terror: New Perspectives*, Cambridge: Cambridge University Press, 1993, p. 212.
30 'Priniat' mery po uluchsheniiu byta', *Na strazhe*, 20.12.38, (51:269), p. 4. Compare with 'V obshchezhitii nado podderzhivat' poriadok' from 21.6.40 (23:354), p. 4.
31 GOPAN, f. 817, op. 1, d. 69, l. 76.
32 Jones, op. cit., p. 311.
33 Beda, op. cit., p. 75.
34 GARF, f. 9415, op. 3, d. 516, l. 79.
35 GARF, f. 9415, op. 3, d. 516, l. 144.
36 GARF, f. 9415, op. 3, d. 516, l. 145.
37 GARF, f. 8300, op. 2a, d. 32, ll. 1–8.
38 N. V. Romanovsky and Zafar Imam, *Russia under High Stalinism: The Last Phase of Stalin's Rule, 1945–1953*, New Delhi: Har-Anand Publications, 1995, pp. 138, 139. Quotation cited from RGANI, f. 386, op. 1, d. 70, l. 18.
39 GARF, f. 8300, op. 1, d. 699, l. 18.
40 GARF, f. 8300, op. 1, d. 701, ll. 21–22.
41 GARF, f. 8300, op. 1, d. 687, l. 20; f. 8300, op. 1a, d. 9, l. 100.
42 GARF, f. 8300, op. 1, d. 701, ll. 20, 23.
43 GARF, f. 8300, op. 1a, d. 9, l. 114.
44 Ibid., ll. 106–107, 115.
45 Ibid., l. 103.
46 GARF, f. 8300, op. 3a, d. 9, l. 15.
47 GARF, f. 9415, op. 5, d. 87, l. 47.
48 Ibid., l. 7.
49 GARF, f. 9415, op. 5, d. 88, l. 6; d. 89, l. 4.
50 See for example, Anatoli Granovsky, *I Was an NKVD Agent*, New York: Devin-Adair Company, 1962, pp. 243, 244.
51 GARF, f. 9415, op. 5, d. 87, l. 144. Ironically, despite the fact that 22 people were ultimately arrested in this operation, the political police never discovered who supplied the dye and how.
52 These include draconian (although arbitrarily enforced) sets of labour laws limiting worker mobility and severely punishing absenteeism, as well as a system of quasi-indentured servitude for young people seeking employment. In Donald Filtzer, *Soviet Workers and Late Stalinism: Labor and the Restoration of the Stalinist System after WWII*, Cambridge: Cambridge University Press, 2002.
53 Julie Hessler refers to 'survivalist strategies' of black-market participation in *A Social History of Soviet Trade: Trade Policy, Retail Practices, and Consumption, 1917–1953*, Princeton: Princeton University Press, 2004.
54 Yoram Gorlizki, 'Rules, Incentives and Soviet Campaign Justice after World War II', *Europe-Asia Studies* 51:7, 1999, p. 1250.
55 Ibid., pp. 1261, 1262.
56 During the Terror, evidence of personal friendship and illicit professional collusion – above all, in the suppression of criticism from below – between party and state officials in the same institution could lead to charges of treason and arrest. In Gorkii, for instance, the police force party secretary was condemned for having played cards at the chief of police's house. GOPAN, f. 817, op. 1, d. 64, l. 89.
57 Peter Solomon, 'Soviet Politicians and Criminal Prosecutions: The Logic of Party Intervention', in James Millar (ed.), *Cracks in the Monolith: Party Power in the Brezhnev Era*, London: M. E. Sharpe, 1992, p. 18.

58 RGANI, f. 6, op. 6, d. 1608, l. 23.
59 RGANI, f. 6 op.6 d. 1586, l. 3; f. 6, op. 6, d. 1600, l. 29.
60 Solomon, op. cit., p. 8.
61 The position of Soviet bureaucrats after the Second World War thus came ever more to resemble that of NSDAP officials in the Third Reich. According to Frank Bajohr, in Nazi Germany, the party had always been anxious to preserve its autonomy and to limit the rights of other organizations to investigate its doings or take its members to court. He writes that, 'Public prosecutors were explicitly forbidden to independently pursue tips [of wrongdoing] as well as to confiscate account books and papers of the NSDAP. The courts in their verdicts were thus to rely completely on documents provided by the NSDAP Reich Protection Minister and the reports of his own accountants'. Bajohr, op. cit., pp. 151, 152, 158.
62 GARF, f. 8131, op. 32, d. 13, l. 15.
63 RGANI, f. 6, op. 6, d. 1576, ll. 30–39.
64 RGANI, f. 6, op. 6, d. 1576, ll. 5–6.
65 RGANI, f. 6 op. 6 d. 1 l. 29.
66 Ibid.
67 GARF, f. 8300, op. 3a, d. 9, l. 17.
68 GARF, f. 8300, op. 1, d. 347, l. 59.
69 GARF f. 9415, op. 5, d. 124, ll. 73, 74.
70 Peter Holquist, 'Information is the Alpha and Omega of Our Work', *Journal of Modern History* 69:3, 1997, p. 431.
71 GARF, f. 7511, op. 10, d. 18, l. 10.
72 GARF, f. 8300, op. 1a, d. 9, l. 95.
73 GARF, f. 8300, op. 2, d. 1033, l. 49.
74 GARF, f. 8300, op. 2, d. 1033, l. 63.

Part IV

New generations

Identity between the yesterday of war
and the possibilities of tomorrow

8 More than just Stalinists

The political sentiments of victors 1945–1953*

Mark Edele

Political loyalties of Soviet Second World War veterans have recently received considerable attention. Historians have proposed three main viewpoints. The first posits that veterans were devout Stalinists, totally dedicated to the leader, regime policies and the system itself. Historians attribute this loyalty to war trauma, social inclusion of veterans in the post-war order, and most of all, the pervasiveness of an official discourse that precluded alternatives.[1] The second viewpoint is that the experience of war liberated the personalities of the frontline generation and sowed the seeds of anti-Stalinism.[2] The third viewpoint, advanced by Elena Zubkova, argues that while the majority of veterans emerged from the war and post-war as loyal Stalinists, this is only part of the story. Within this generally loyal group existed a "liberal faction" of people who were forced by the "facts of the pre-war years, the experience of the war, and observations during the campaign in Europe . . . to reflect, to wonder about the justice of elements of the regime, if not of the regime as a whole." Especially former POWs were a potentially oppositional group because of their precarious social and political position after the war. The general hardships of post-war life, psychological and physical exhaustion, political repression and the impossibility to imagine an alternative to Stalinism prevented the transformation of such opposition into active resistance.[3]

This chapter explores the range of the thinkable, the sources of information and the reasons for discontent among veterans of the Great Patriotic War in the years before Stalin's death. It builds on Zubkova's sketch and revises it mainly in showing that alternatives were in fact possible.[4] There was a whole universe of ideas in circulation, ranging from total support to total opposition, with various shades of gray in between.[5] Not many political positions were outside of the range of the thinkable for veterans in the post-war years.[6]

The backbone of this chapter is formed by a collection of cases of "anti-Soviet agitation" reviewed by the state prosecutor's office of the Russian Federation in the 1950s.[7] These "review files" (*nadzornye proizvodstva*) were produced in reaction to letters of complaint of the prosecuted or their families and as part of the effort to release victims of late Stalinist repressions from the over-full Gulag after Stalin's death.[8] A heated discussion has surrounded the question whether or not we should ignore the evidence about popular moods contained in *svodki* because these were constructed by surveillance agencies. One could easily construct a similar argument against the use of *nadzornye proizvod-stva*.⁹ However, I find it hard to believe that anybody who read even a couple of these files could still maintain that they tell us nothing about political loyalties and beliefs in the post-war Soviet Union. Even if we would adopt a radically positivistic epistemology and read court cases only for what they give direct evidence for – state paranoia and surveillance – we would still have to accept the fact that people who produced these files could think the kinds of thoughts they claimed the veterans in question had expressed. If we are ready to accept this, however (and I cannot see how we could not), why not consider that what the producers of the accusations claimed could actually have been the case?

Possibilities of political thought

Besides full acceptance of the regime, we can broadly distinguish three main political positions of veterans in the post-war years: (1) contest of the legitimacy of the regime on the basis of an alternative utopia, which might or might not express basic values of Stalinism; (2) contest of the legitimacy of the regime on the basis of regime discourse and values; and (3) acceptance of the legitimacy of the regime but misunderstanding of individual policies, or contest of individual policies or people. These different positions shaded into each other and could be held simultaneously or consecutively by individuals. For analytical purposes, however, it is helpful to distinguish between them, in order to get a clearer sense of the range of political ideas during late Stalinism.

   In its most extreme form, the contest of the legitimacy of the regime on the basis of alternative utopias, such as the lands "abroad" or Tsarism, did not reflect basic values of Stalinism itself. Instead, such utopianism drew on antithetical notions such as capitalism, free market economy, basic civil rights such as freedom of speech, and a multi-party system. By "utopia" I mean an idealized world removed in space and/or time from a negatively construed contemporary reality. Utopia in this sense is a counter-image of how the world should be which serves as a basis for criticizing the existing order of things. Such critical utopianism is in many ways the opposite of the official Soviet utopianism of Socialist Realism. In the Socialist Realist mode of looking at reality, the Communist future is used as an excuse for all sorts of shortages, problems and hardships of contemporary reality. Under the gaze of the believer a foundation pit becomes a future Palace of Soviets, and inequality of distribution plus shortages of goods are seen as the start of equality and abundance.¹⁰ Anti-Soviet utopianism, by contrast, constructed an imagined other in order to contrast it with reality. While the dialectics of Socialist Realism used utopia in order to sacralize reality, anti-Soviet utopianism used it in order to put reality's shortcomings into sharper focus.

   The anti-Stalinist utopia came in several versions. First of all it could be removed either in space or in time. Removed in space it became the wonderland "abroad," which could remain unspecified or named as "Yugoslavia," "England," "America" or even "Germany."¹¹ What these symbols stood for was equally differentiated. It could be abundance of material culture – the land behind the border as a generalized and constant feast. It could also stand for culture with a capital C – the Germans or even the Americans as the embodiment of *kul'turnost'*.[12] Further, it could be "democracy" – the possibility to speak one's mind without reprisals, and the possibility to vote in a multi-party system for the government one liked best. One former POW was reported to be "enraptured (*voskhishchat'sia*) by American politics. He said that there are many parties, one can express any thought and point of view, but here it is not allowed to talk about everything, [and] one has to support only one party."[13] The utopia behind the border could also be specified as "capitalism." This word sometimes simply stood for wonderful affluence: in "capitalism" workers labored only three months a year, drove around in cars, and had lavishly furnished apartments.[14] The word could, however, also denote free market structures as a positive alternative to state socialism. "People abroad live well," a veteran reported as saying in 1950, "because there exists private property (*chastnaia sobstvennost'*) and all belongs to the people. In the Soviet Union ... the State is the owner (*sobstvennik*) of everything and the people own nothing."[15] Single farming as an alternative mode of production was another expression of private property as a positive counter-model to state socialism.[16]

These distinct elements of utopia – abundance, capitalism, democracy and *kul'turnost'* – were relatively independent of each other and could be combined in various ways, with one exception: abundance was always part of the picture, with capitalism, democracy and *kul'turnost'* as secondary moments which could or could not be added. Dmitrii Aristarkhovich, for example, was for tiled stone houses, good roads, freedom of speech, freedom of information and a multi-party system. He was against lies, the Stalin cult, the Soviet single-party system, repression of speech and thought, and "speculation" – the latter indicating the absence of "capitalism" from his utopia "abroad."[17] Others did include "capitalism" in their utopia, or concentrated on abundance alone, making no reference to an alternative political system. Thus, utopia did not in every case include a vision of an alternative political system, but even if such a vision was absent, the utopia of abundance abroad could still be used to condemn the Bolshevik project by confronting it with the good life elsewhere. In 1951 one veteran turned metal worker expressed such doubts in front of a store. According to a witness he said that "they suffocated the kolkhoz peasants with work, they built a lot of prisons and the communists have led us to poverty, and further he said that in America and England a good life has been set up for the people."[18]

The second formulation of utopia was removed in time. The golden past was also connected to abundance and could be located simply "in former times" (*ran'she*),[19] before the war,[20] during NEP[21] and under Tsarism.[22] In the latter two cases abundance was usually connected to the absence of kolkhozes.[23] Sometimes, good and bad were unequally distributed over time, as the following historical paradox, formulated by a veteran, suggests: "Despite [the Tsar] little Nikolai (*Nikolashka*) having been an idiot, life was better at that time, while now the leaders are good but life is bad."[24] The comparison with an "other" society removed in space or time could also function as a critical device to measure Stalinism with its own measuring stick. This was the second major position, which contested the legitimacy of the regime on the basis of regime discourse and values. If the utopia was, for example, NEP or even the pre-war Soviet Union, the critique shaded over to a less radical position than that implied by a comparison with capitalism. Similarly, the notion of the wonderland of consumption and *kul'turnost'* abroad was to a large extent a displacement of Stalinist values into an utopian space.[25]

One could also measure really existing Stalinism with Stalinist mental tools by comparing it with Tsarism or fascism – and thus use the regime's own picture of the enemy against itself. In this case utopia became dystopia and social reality was compared not with a positively connoted other, but with a negatively connoted one. One poorly educated (*malogramotnyi*) Ukrainian peasant soldier from Vinnitsia, born in 1915, did not see what the difference between Hitler and Stalin might have been, and why "the people" should fight for the one against the other: "I said, what are we fighting for, it would be better if Hitler and one-our leader of the party Stalin* went and fought [just the two of them]. He who wins, his state should also have victory, in order to not throw the people into the ruin of this war."[26] Another veteran used a court session in 1951 to address the public in the courtroom with the words: "I was in the *gorkom* of the party, there sit only fascists, in the militia organs, in the state prosecutor's office and in general in the investigation agencies also sit fascists. Wherever you go – fascists . . . And this court is not a Soviet, but a fascist, a Gestapo-court."[27]

Or one could deny that the Revolution had made any difference: "In old Russia (*ran'she v Rossii*) only the children of noble (*znatnyi*) parents had good living conditions. The situation today is just the same; "just as in former times (*ran'she*) there were privileged classes, so there are now."[28] Other veterans compared Soviet power with serfdom.[29] Criticism of the existing order also emerged if the claims of the regime were not understood through the dialectics of Socialist Realism but were simply confronted with reality.[30] This subversive method of speaking Bolshevik often resulted in bitter irony. One veteran, born in 1901, who found Bolshevik politics "inhuman and disgusting," wrote in fluent Sovietese to his still serving son:

> Nowhere in the world exists such a situation, where people who are soldiers don't even get leaves. This is inhumanity, not more and not less. I think that they fear to let you go on leave for fear you could learn about life at the home front, so that you don't get infected with the socialist paradise on the Socialist soil.[31]

Apocalyptic teleology is one example of such a subversion of the dominant ideology. Instead of seeing the present in terms of a golden future one could dismiss the official notion that hardships were temporary and life was getting better every minute. The "workers in the USSR live worse with every year, because the wages decrease. Soon we will work completely for free."[32] "At the moment we live poorly, but under communism we will live even worse. Then the people will walk around totally naked (*sovsem razdetyi*)."[33] Instead of marching "from darkness to light" some veterans saw history move in the opposite direction, into apocalypse and a new war "which of course inescapably should be."[34] To use the precise formulation of another veteran: "life does not move towards the sunrise but towards decay (*upadok*)."[35] Such apocalyptic thinking could also be coupled with hopes for a better life in a different society. One veteran waited for the outbreak of atomic war with the USA in which the latter would destroy the Soviet Union. "After this ... they will enact a change and everything will be different."[36]

Others used parts of the revolutionary tradition, such as egalitarianism, and Marxist tools to analyze their own society – a move that certainly would have pleased Milovan Djilas and Leon Trotsky.[37] "I don't believe in Socialism in our country," argued one veteran in 1949, "because one lives good, and others poorly."[38] "I compared the salary of a cleaning woman with the salary of ministers and said, that like that we will not march towards communism."[39] Sometimes such implicit Trotskyism became explicit. One old veteran, born in 1896, remembered that "during the Civil War he saw TROTSKY. He eulogized (*voskhvaliat'*) the Trotskyites and said that they were never enemies of the people."[40] Knowledge of Trotsky was not restricted to the oldest of the veterans, however. One veteran, born in 1919, remarked the following: "Trotsky was a very intelligent person, exceptionally well grounded theoretically, the people respected him, he was a very gifted orator ..."[41]

Other currents of the revolutionary tradition were also remembered. One witness in the case of Aleksei Terent'evich D., a veteran born in 1914, reported: "In 1949 we sat in his apartment and drank. B. drank 100 grams, then D. asked him 'To which party do you belong, to the Mensheviks or the Bolsheviks?' When B. said that he was a Bolshevik, D. said 'this means that you are an idiot.' "[42] But not only the Mensheviks were remembered as a positive alternative to the Bolsheviks. In April 1945 the NKVD arrested a printer who at night used the printing press of the factory *Krasnoe ekho* in the city of Pereslavl' in Iaroslav region to produce fliers for a political group which the authorities described as "anti-soviet" and "pro-fascist." In fact, they seem to have put themselves into the tradition of the Socialist Revolutionaries as the odd text of the confiscated flyer suggests:

APPEAL TO THE INVALIDS OF THE PATRIOTIC WAR!
Enough of your wives' and children's' tyranny!* Enough of the insults by your Soviet butchers (*palachei*).
They mocked you enough, [you,] disfigured by the Soviets.
Kick out the scoundrels – the representatives of the regional committee of the Bolshevik party! Down with the kolkhozes! Down with Soviet power!
Long live our free, brave, great People!
Long live the party of the Socialist Revolutionaries! (the party of the peasants).
COMMITTEE[43]

Drawing on such peasant radicalism of pre-revolutionary origin it was not only possible to "accept the system but reject the regime" (i.e. the particular persons in power),[44] but also to dismiss the entire Bolshevik project of building Communism as an anti-popular movement: One veteran claimed to speak for "the people" who would rise up against "the commune." The strengthening of the kolkhozes, he told one witness "leads to the commune ... soon an uprising will occur, the people are not pleased (*chtoby skoree vosstanie vozniklo, narod nedovolen*)."[45]

The third major position was to accept the legitimacy of the regime but consciously or unconsciously to misunderstand individual policies. This gray zone of ideas did not necessarily express opposition to Stalinism, but still deviated from the propagated model. This middle ground was referred to as "unhealthy moods" (*nezdorovye nastroeniia*) by Soviet authorities. And it was mainly this middle ground, which *svodki* reported on, not the outrightly anti-regime utterances prosecuted under article 58-10. Thus, a reliance on *svodki* does not overstress radical attitudes, as historians have sometimes feared.[46] Rather, such reliance paints an overly *mild* picture of possibilities for discontent under Stalinism. There were several ways to think "incorrectly" or "unhealthily" but not oppositional. One possibility was to reach one's own conclusions about what the correct policy of the Soviet state should look like. Such conclusions were informed by rumor on the one hand, and the reception of propaganda on the other. In September 1947 the *obkom* secretary of Cheliabinsk region reported that

> Among demobilized soldiers and officers in the town of Miasse exists the opinion that in any case in 1948 there will be a war with America, which will be supported by England. Therefore several demobilized voice their dissatisfaction with the results of the Patriotic war. They say: "It was wrong not to destroy the 'allies' after the fall of Berlin."[47]

These veterans, while clearly not in opposition to the Stalinist order, used official discourse together with what they learned through rumor to make sense of the world around them. The extent to which the conclusions they reached deviated from the officially promoted interpretation is illustrated by the fact that in 1951 those who communicated war rumors ("propaganda of war in whatever form") were threatened with prosecution "as severe criminal offenders."[48]

Another possibility was (consciously or unconsciously) to misunderstand the official line. Openly displayed anti-Semitism is a case in point.[49] The visibility of anti-Semitism among veterans is usually interpreted as an effect of Nazi propaganda (G. V. Kostyrchenko),[50] of the "invisibility" of Jews at the front and in reports about heroism (as Mikhoels of the Jewish Anti-Fascist Committee argued),[51] or as a combination of both, making the army "the ultimate incubator for anti-Semitic sentiments" (Amir Weiner).[52] In addition, openly displayed hatred towards Jews also rested on a misinterpretation of the "anti-cosmopolitanism" campaign. As Kiril Tomoff has demonstrated, "cosmopolitan" did not simply mean "Jewish." It was a wider term which included all sorts of "rootless" persons including jazz fans, *stiliagi*, scholars who quoted foreign scholarship, composers who drew on a Western tradition, listeners of foreign radio stations, anybody interested in things foreign and "Zionists."[53] Officially there was no anti-Semitism in the Soviet Union, and agitators stressed its difference to "anti-cosmopolitanism." This is often seen as a cynical move, but it should be taken more seriously, because this complexity of the official line explains why open anti-Semitism was punished. One student veteran at MGU – "an idiot and an anti-Semite" – was obviously unable to understand the fine line between "cosmopolitans" and "Jews" and thus fell victim to the subtleties of the "cosmopolitanism" discourse.[54]

Another case of misjudging the subtleties of allowed discourse was when (allowed and even encouraged) criticism of local circumstances and people (*kritika*) became illegal "anti-Soviet agitation" if this criticism was directed towards the system as a whole, or the central government.[55] One veteran made such a lapse on a local party meeting in Iaroslav region in 1950. In his speech he first criticized the work of the primary party organization as planless – a perfectly legitimate *kritika*. Then he criticized decrees which forced peasants to provide work for timber production as "screwed up" (*golovotiapski*) and noted the clearly forced character of such *corvée* labor. Even worse, he claimed that "our kolkhoz peasants lack any democratic rights." He further criticized the Communist Youth League as a mere mechanism to send kolkhoz youth logging and to coerce them into signing large sums in government bonds (*zaem*). This, he claimed, was the major reason for the lack of popularity of the organization among peasant youth. The problem with all of this was that he did not frame this critique as one of local shortcomings but as a critique of central policy; he did not critique the local Komsomol cell, but the Komsomol as a whole. He "did not name a single name of our local leaders," remembered the chairwoman of the meeting and therefore she concluded "that this was ... not criticism but an anti-Soviet speech against the politics of the party."[56]

Such lapses from "criticism" to "anti-Soviet agitation" did not depend on alternative worldviews but used the hegemonic one to criticize the ruling political order. To do that, veterans did not need to intend to be oppositional. All they needed to be was outspoken and misjudge the boundaries of allowed talk.[57] Such critical attitudes which reproduced the officially sanctioned values are maybe the best evidence for an internalization of Stalinist propaganda discourse: some veterans "spoke Bolshevik" even if they criticized the regime.[58] Other evidence for such internalization includes adoptions of official role models by real-life veterans,[59] or thank-you letters to Stalin or to welfare institutions in which war invalids thanked their benefactors (including Stalin himself) for the wonderful "care" they had received.[60] In such letters, invalids used official language to describe their own situation, inscribed themselves into the officially promoted narrative of the war against fascism as a struggle of light against darkness, interpreted the rather limited welfare they received as a major expression of socialism, reproduced what Jeffrey Brooks calls the "culture of the gift,"[61] and sometimes ended with ovations. Since such letters did not ask for anything but expressed thanks, these feelings cannot be discussed away as instrumental. Memoirs of non-veterans, too, support the picture of demobilized soldiers as staunch Stalinists.[62] Finally, veterans themselves claim that they could not think outside of the parameters of official discourse.[63]

There is little reason to doubt these self-representations as Stalinists by memoirists who in the post-Soviet context would have a lot to gain from retrospectively fashioning themselves as hidden regime critics. However, it would be a serious misinterpretation to take their claims to speak for all veterans as historically accurate. The believers were at one extreme end of the spectrum ranging from an embrace of an idealized version of Western liberal democracy and capitalism to "Stalinism" – with all possible shades of gray between. Moreover, to be a "Stalinist" could mean different things to different people. Film director Grigorii Chukhrai involuntarily demonstrated this point in his memoirs:

> many of us really were Stalinists. ... our multi-national motherland was dear to all of us, as were honor and dignity, ours and that of our parents, our girls, our friends, who did not wish to be slaves of the Germans. We knew how many sacrifices industrialization had cost our parents, and it hurt us when all of this was destroyed."[64]

That "Stalinism" could simply mean loyalty to one's kin and to one's country, or identification with the role of the man as defender of women, had to do with wartime propaganda. It was not "Socialism" or "Stalin" which wartime propaganda called on Soviet soldiers to defend, but polyvalent symbols such as "home," "family," "motherland" or the "honor of women."[65] If "Stalinism" (or regime support) hinged on such polyvalent symbols, it did not necessarily imply loyalty to every aspect of the system or every policy promulgated by the regime. Rather, even the "Stalinists" were likely to lapse into "unhealthy moods" or worse at times.

Sources of information

In exploring the informational base, which allowed veterans to evaluate the world they lived in, it is useful to distinguish between sources and channels of information. By "source" I mean simply where the information originated, while "channel" refers to the way in which this information was transmitted from its source across space and time to be at the disposal of veterans. They used at least four sources of information: official discourse and the revolutionary tradition in general, peasant culture, war experience and foreign propaganda.[66]

The main channel for official discourse was the official network of information, which included propagandists and agitators, and the media network of newspapers and journals, posters, the movies and radio.[67] People did not simply believe official announcements. On the contrary, there was quite some skepticism towards official media among veterans.[68] Based on such skepticism, the information provided through official channels could be read creatively in order to learn something, which was not intended by the senders of the message.[69] It thus allowed a variety of readings, which explain many of the "unhealthy" moods of veterans. Critical interpretation of official announcements and canonical texts, combined with logical inference, is one example for this process. After reading the *Short course* a veteran concluded that Stalin could not have single-handedly authored this text. Instead he thought that "this is a collective work and that the book was written by a commission."[70] Another veteran dismissed Stalin's "genius" on the basis of a close reading of his texts: "If one reads his works one can see that all is taken from Lenin, while there are only very few original ideas."[71]

Such critical attitudes were aided by the structure of official discourse itself, that is, by the information which was transmitted through the official channels. As Karen Petrone and Juliane Fürst have demonstrated, this discourse was no monolith but a complex and multi-faceted entity.[72] Moreover, official discourse had several layers including not only the most up-to-date pronouncements of *Pravda*, but also earlier positions of the revolutionary tradition. For one these earlier layers were kept accessible through books, which did not simply disappear once a certain strand of thought was outlawed. Nikolai Fedorovich D'iakov, who later became an activist on behalf of former POWs, worked for a while in Abkhaziia in 1948. Part of his job as librarian of a sanatorium was to sort out forbidden literature and destroy it. What he did not keep himself he gave to friends – including, for example, a six-volume pre-revolutionary edition of Kliuchevskii, which he gave to a fellow former POW.[73] A group of Trotskyites obtained a "political dictionary" which included biographical information on Trotsky and Bukharin.[74]

Earlier layers of the revolutionary tradition also remained accessible through channels of communication outside of the official network. On the most basic level these were individual memory and face-to-face communication. One could, for example, challenge the Stalin cult as a lie, if one remembered a different story of the role of Stalin in Soviet history: one veteran maintained that Stalin had little to do with the successes of the country, and that "the people learned about him only with collectivization. Before that he was a little-known personality."[75] Face-to-face communication created rumors which allowed the sharing of individual memories and ideas between more and more people, creating a universe of information parallel to the official network.[76] One anti-Soviet veteran explained his knowledge of the coming of war in the following way: "when he worked in the Donbass, one educated person (*odin uchenyi*) told him, that without fail in 1949 will be war, and America will win, because she is stronger."[77] Mediated channels of communication, such as letters, also played a role in this process of informal communication. One veteran was caught because of a letter he sent to his son, who served in the armed forces. Far from cleansed by self-censorship, this letter called the politics of the Bolsheviks "inhuman and disgusting" adding that "really only Bolsheviks are able to such a method of torturing people."[78]

The persistence of these channels – memory, face-to-face communication and rumor – also explains the continued existence of the second source of information – peasant culture, including peasant radicalism. In its origins this culture was pre-revolutionary and while it was certainly transformed by the onslaught of forced collectivization and thirty years of Bolshevik rule, it retained its own characteristics, which cannot be deduced from official policy alone. The widespread anti-kolkhoz feelings nearly two decades after collectivization, the appreciation of single farming, individual initiative and NEP-style markets which peasant-veterans displayed point to the relative independence of this cultural system from state-promoted discourse. So does the persistence of anti-Semitism. After all, Russian peasants and peasants-turned-workers were not known to be philo-Semites before the revolution.[79] Nor did anti-Semitism disappear after 1917.[80]

War experience provided a third source veterans could draw on. According to contemporaries, the dangerous deviation of "Westernness" (*zapadnichestvo*) had increased its influence "during the years of war."[81] Veterans' images of the West rested for one on observations of the effects of lend-lease during the war (especially military machinery and food supplies).[82] Moreover, the exposure to "lands abroad" was crucial. These "lands" looked much different from what soldiers had expected on the basis of inter-war propaganda. There was, for example, Vasilii Ivanovich, a POW who, at least in retrospect, liked what he saw during his imprisonment.[83] He was not alone. Dmitrii Aristarkhovich also saw the West as a POW:

> As they brought us to Germany I saw good, well-maintained roads, good buildings, stone houses with tiled roofs. I talked about that.[84]

> During smoking breaks with the collective P. praised Hitler and said that the Fascist Party admires Hitler and greets even his portrait. ... About his imprisonment he said that he worked in a mine, lived in a barrack. He said that the prisoners stole foodstuffs and lived not badly.[85]

> About the imprisonment P. said that in Germany he lived not badly, because [while] they worked hard there, the food was good, but here (*u nas*) the people at the homefront were fed poorly, worse than prisoners of war in Germany.[86]

Another veteran admitted that he had said "that while I lay in the hospital in Germany, the doctors there treated me better than [would have been the case] in the USSR."[87] As a top political officer of the repatriation administration in Sweden summed up such views: "After they have seen the untroubled life (*bespechnaia zhizn'*), certain individuals among our repatriated [citizens] draw the incorrect conclusion that Sweden is a rich country and that the people here live well ..."[88] Others saw during their war odyssey not only German POW camps, but ran away and fought as partisans in Italy ("in Italy the people live well"), and Yugoslavia ("in Yugoslavia the people live well") and on the basis of this experience concluded that "people abroad live well, because there is private property and everything belongs to the people."[89]

But not only the unsupervised life of the guerrilla fighter, but also the exposure of regular troops to the West during the war was a source for alternative information about life abroad and provided a new frame of reference for judgments about the Stalinist system.[90] This applied even to comparatively poor regions such as those in Eastern Poland, which were annexed in 1939 under the Hitler–Stalin Pact.[91] This turned soldiers, and later returning veterans, into a group of concern for the propaganda organs, who had learned their history lessons well. In February 1945, for example, propaganda officers of the Second Belorussian Front discussed the danger of a new Decembrist movement.[92] Despite the "merciless fight" of the propaganda machine against such potential trouble-makers, veterans continued to remember their experience of countries outside of the Soviet Union and used them as a frame of reference to make judgments about the society they lived in. The authorities associated anti-kolkhoz feelings during the summer of 1945 with the detrimental influence of veterans who had seen the West.[93] In party meetings at Voronezh State University in Summer 1947 the existence of "individual facts of kowtowing before bourgeois science" was linked to those students who had been abroad during the war.[94] The director of the Leningrad State Librarians' Institute noted in a closed party meeting on 10 September that among first year students "exist people, who have been abroad during the war." These suspicious characters thought "German films are better than ours." Even worse they were under the false impression that "abroad, everyday life is better organized."[95] The secretary of the Komsomol *gorkom* of the town of Slobodskoe in Kirov region also linked love of Western comfort with having had a glimpse of it during the war. He criticized the political immaturity of those who were unable to interpret their experiences in the correct Socialist Realist fashion.[96] A similar charge was leveled at demobilized soldiers among the students of Kuibyshev agricultural institute.[97] In a report to the Central Committee the secretary of the *obkom* of Vologodsk region summed up the problem, as it emerged from closed party meetings discussing the KR affair in 1947: in every one of the party meetings held, the point was made that "serious work with demobilized Red Army soldiers, who during the Great Patriotic war have been abroad or who have been POWs" was necessary. "Demobilized individuals," he reported further, "not understanding the reactionary nature (*reaktsionnost'*) of bourgeois culture, eulogize it."[98]

Not surprisingly, oppositional groups, too, saw veterans as critical material: "the front-line soldiers [*frontoviki*]," wrote an OUN fighter in her travelogue and report to her underground cell in October 1945, "are very different from the men of the party and the rear. They, who have seen the West and a different life, express their views much more freely and closer to the truth than the latter [party people]."[99] Foreign propaganda reinforced some of the information, which originated in peasant culture and war experience, and provided additional information. The notion, for example, that "abroad" workers drove cars cannot have originated from first-hand experience during the war. In the 1940s, the only countries where (some) workers drove cars were the USA and Britain.[100] There is a slight possibility that those POWs who had been in the UK had met car-owning workers.[101] More likely, however, these ideas originated elsewhere. Similarly, ideas about free speech, the multi-party system, or a British or American welfare state are unlikely to have originated in personal observations. Such information came from other sources: the propaganda effort of the former Allies USA and Great Britain during the war and in the context of the evolving Cold War. A major channel for this kind of propaganda was face-to-face as well as mediated agitation in the POW camps, taking place between liberation from the Germans and repatriation to the Soviet Union. A second channel was provided by radio stations such as the Voice of America (VOA) and the BBC, but also propaganda journals published by the embassies, such as *Britanskii soiuznik* and *Amerika*.[102] In 1946, 49,000 exemplars of *Britanskii soiuznik* were distributed in the Soviet Union. The majority of the exemplars went to subscribers, but an important section (21 percent) were sold. These copies made the MGB most uncomfortable because they "fall into the hands of accidental readers."[103] Even more problematic were letters to the editor from active service men, who expressed thanks for the interesting material about life in Britain.[104] *Britanskii soiuznik* was not only available in Moscow. More than half of its copies were distributed in the "periphery."[105] *Amerika*, the US equivalent, also reached the provinces.[106] One of the Trotskyites from Kabardinskaia ASSR, Nikolai Vasil'e-vich (born in 1921) got hold of one of the prized copies: "Once I read the journal 'Amerika,' in the journal was written that night work is paid [in addition to the regular wage] in America; I said that here (*u nas*) night work is not paid for."[107] He also learned about the politics of Titoism, and he connected this new information with what he remembered about the Soviet revolutionary tradition: "I said . . ., Tito implemented such policies, which in our country Bukharin and Rykov tried to implement."[108]

While printed propaganda was available only in small print runs and could be controlled comparatively well, more state-of-the-art communications posed a bigger problem for the authorities.[109] On 17 February 1947 the US propaganda channel "Voice of America" (VOA) started Russian language broadcasts, while the BBC seems to have operated earlier. The massive attempts to jam the air-waves had limited results because of counter-measures taken by the radio stations that remained accessible throughout the late Stalin years to those who owned a wave receiver.[110] At least among veterans, these were not necessarily people of intelligentsia background. One of them was Arkadii Ivanovich. Born in 1913, with incomplete secondary education, he worked as an accountant in a kolkhoz in Iaroslavskii region. Up to his arrest in March of 1950 he liked to listen to the VOA and the BBC, which might have contributed to a mishap at a party meeting where he misjudged the fine line between *kritika* and anti-Soviet agitation.[111] One anti-Soviet veteran of peasant background (born in 1922) listened regularly to "radio shows of American and British radio stations, who reported on living conditions in the Soviet Union, on the leaders of the Soviet government and other things. I listened to the foreign radio shows in Novgorod *oblast'*, I owned a radio receiver."[112] Another veteran, a locksmith (*slesar'*) on a kolkhoz in Gor'kii *oblast'* (born in 1923), continuously listened to foreign radio broadcasts, invited acquaintances to listen as well, and functioned as something of an agitator of what one witness called the "Truthful Voice of America" (*Spravedlivyi golos Ameriki*). Functioning as a transmitter from one channel of information (radio) to another (rumor), this veteran passed on what he learned from listening to the radio to the surrounding people. He "often told the kolkhoz peasants, that the peasants in America live well."[113] Moreover, he did this not only in what Jochen Hellbeck has named "profane spaces," but also during public rituals:[114] At an evening meeting (*vecher'*) in the school he said; "the radio station 'Voice of America' reports the truth about the Soviet Union."[115]

Reason for discontent

Overall, veterans lived in a cultural universe that was considerably richer than historians sometimes imagine. This universe included propaganda discourse as well as memory of older times, visions of the good life abroad, information disseminated by word of mouth as well as state-of-the-art communication equipment used by foreign propagandists to infiltrate the minds of Stalinism. The complexity of the available discourses explains why they could think what they thought. It does not explain, however, what they thought, or why they thought what they did.

Simple exposure to certain channels of information played a role. Older veterans, for example, could remember a different regime themselves, while their younger peers had to rely on hearsay which competed with the official view of the world. Some veterans had seen the West themselves, others had not.[116] Access to wireless radio receivers, which allowed listening to foreign broadcasts, and access to Western propaganda journals was restricted as well.[117]

However, simple exposure or lack thereof to sources and channels of information outside of the official network does not explain political sentiments of veterans. First, it was possible to interpret the information given on the VOA as propaganda lies or to read the material situation outside of the Soviet Union in the "correct way." Second, the availability of rumor as a source of information – a source probably as "inescapable" as official pronouncements – meant that the lack of direct access to the primary channel of alternative information (such as a wave receiver, direct exposure to the West and personal memory) could be compensated by hearsay. Therefore, neither a simple generational nor a simple class analysis will do. Lovers of the West, for example, could be found among veterans born in 1893, in 1909, or in 1919; among drivers of combine harvesters and locomotives, or among electricians;[118] but also in social strata beneath and above these skilled workers: neither rank-and-file *kolkhozniki*,[119] nor veterans turned students were immune to such fantasies.[120] The primary problem was thus not lack of information but the creative use of the available cultural forms, which took place in the context of life experiences of individual veterans.

One important variable for whether or not a veteran tended to express hostility towards the regime was how the regime had treated him or her before, during and after the war.

Many veterans from a peasant background never forgot the horrors of dekulakization and collectivization.[121] In general, memories of repression produced lingering resentments. Arrests of parents, their defamation as enemies of the people and subsequent personal experiences of discrimination and stigmatization were all designed to evoke doubts in even the keenest activist. The taint of having been in German imprisonment re-enforced such sentiments.[122] Chukhrai is unlikely to have become the enthusiastic Stalinist, whom he portrays in his memoirs, if his mother had not escaped the purges by pure luck or if he had been taken prisoner and had his life-chances inhibited at every step by the new stigma of having "given himself into imprisonment."[123] One might draw parallels with a veteran who had been a militant Stalinist before the war. He became a convinced anti-Stalinist ("I hate this butcher! I hate him!") once he fell victim to the regime's discrimination against former POWs.[124] Another example for this process is Vasilii Ivanovich, a Leninist-Titoist lover of German culture and the capitalist paradise. His military career stalled because he had been a POW.[125] His contempt for party members ("thieves and swindlers") was obviously the result of having his career cut short and his decline from officer and party member to locomotive driver without political affiliation.[126] His bitterness about his personal situation after the war made him receptive to using his experiences of the outside world in order to criticize Soviet reality.[127]

A veteran's post-war life could also be blighted by membership in a stigmatized national group, or in a group perceived to be stigmatized. Fatikh Fatakhovich, for example, a veteran of Tatar nationality, born in 1906, saw himself confronted with "the Russians," "the Georgians," and "the Jews."[128] He expressed a strong feeling of cultural loss through Russification and longed for a golden past of Tartar cultural freedom (probably located somewhere between the revolution and the 1930s).[129] The most likely characteristic to push veterans towards criticism was, of course, being Jewish.[130] One Jewish veteran reacted to an anti-Semitic remark with a fit of anger, adopting the official enemy description: "From now on I am a cosmopolitan, because in the Soviet Union I did not find a home (*rodina*)."[131] Resentment was not only created by such massive stigmas as being a former POW, the son of a kulak or of an enemy of the people, or by membership in a problematic national group. Simple lack of advancement in post-war life, disappointment of the high expectations generated by victory, and the general post-war disillusionment were sufficient. "After the war they promised us a good life, but in reality they increased taxes, and life became worse and worse all the time. And for what we fought we don't know ourselves."[132] It might not be a coincidence that many of the disgruntled veterans one can meet in prosecution files were living and working in the countryside in the early 1950s – at a time when it had become abundantly clear that the regime would not grant the peasantry any breathing space. While the majority of their peers had made their way out of the countryside in the search for a better life in the cities, those who stayed behind had little hope for betterment, and often reacted with considerable bitterness to the agricultural policies of the regime.[133] One such case was Semen Trifomovich M., a peasant veteran from Riazan' region, who was an apocalyptic teleologist ("life moves towards decay") and a staunch enemy of the kolkhoz order and "the commune."[134] His disaffected statements, for which he got convicted, related mainly to the tightening of the kolkhoz regime in the post-war years. He was opposed to the decrees, which intended to bring an end to wartime laxities in the countryside and amalgamate small collective farms into bigger ones. He also complained about too high taxation and lack of pay. These real-life grievances became a jumping board for his critique of Communism as an apocalyptic movement built on the exploitation of "the people." While some thus read the tightening of the kolkhoz order after the war as a sign that the regime did not intend to pay back wartime service with a better life, others considered the repeal of many of the privileges for decorated veterans in 1947 unfair.[135] For some, this became a jumping board to criticize lack of freedom of speech and democracy.[136]

The tightening of the kolkhoz order, the abolition of special privileges for decorated veterans, and the general hardships of post-war life – at odds with the expectations of victory – were problems touching a large majority of veterans. But there were also more private grievances, which could lead to disgruntlement. One veteran, pushed to explain why he was so critical of the system, declared in court: "I was dissatisfied about the fact that I could not find work for a long time, because I had [received] a severe reprimand from the party (*po partiinoi linii imel strogyi vygovor*). Well, after that I started to express anti-Soviet opinions. I said in conversations that there is no justice in the party."[137] In sum, the perception of having been treated unfairly by the regime, of not having one's war service reciprocated with a "good life", or of generally lacking chances of advancement helped to push veterans towards criticism of the existing order. However, one should be careful not to overstress this point. For one, as the work of Jochen Hellbeck demonstrates, membership in a stigmatized group or victimization by the regime did not necessarily lead to oppositional sentiments. It could also lead to attempts of self-purification fueled by a strong desire to belong to the revolutionary polity on the march towards the radiant future.[138] Thus, while victimization or even lack of advancement *explains* the sentiments of critical veterans, these factors do not *predict* moods in every case. The human mind is too complicated an affair to be captured by simple stimulus–response models of political ideas. Moreover, the inverse proposition – that a good social position after the war led to pro-regime sentiments of veterans – is also not true in every case.[139] The two generals who were shot in 1950 for their (private) criticism of the kolkhoz order and the lack of democracy in the Soviet Union are a case in point. Both belonged to Stalinist high society,[140] were clearly beneficiaries of what Vera Dunham has called the Big Deal, and far from being victims of the kolkhoz order.[141] This did not keep them from having critical thoughts. Ultimately, political sentiments remain impossible to predict. As evaluations of the world they relied on creative acts of interpretation, which cannot be reduced to the necessary, but insufficient, conditions of social position, war experience or access to information.

Conclusion

Were veterans, then, special? Did they constitute a specific group as far as political sentiments are concerned? And what is the role of war experience in all of this? Was the war maybe a watershed in the possibilities to conceptualize society? We can approach these questions from two perspectives – from a comparison with pre-war moods of the population, and from an analysis of the difference between veterans and the rest of the population after the war.

To start with the latter, in many respects veterans were not different from other inhabitants of Soviet society. Anti-Semitism, for example, was by no means restricted to veterans. Especially in the formerly occupied territories of the Soviet Union, anti-Jewish hatred among the civilian population was widespread and could take on violent forms.[142] Many of the grievances veterans expressed were those of their respective social group – such as kolkhoz peasants or workers – and what made them supporters of Stalinism had as much to do with their age, upbringing, social position and life trajectory as with the fact that they were thrown into the war. Former POWs were also not alone in their problematic status as suspicious individuals – they shared this status with former slave laborers and with a wider group of stigmatized persons, and were, thus, part of the normal state of Stalinism.[143] Listening to the VOA, the BBC, or reading *Britanskii soiuznik* and *Amerika* was of course also not restricted to veterans. German radio receivers – which veterans acquired as "trophy goods" during or immediately after the war – were not the only sets which allowed people to listen to non-Soviet channels.[144] Veterans were also not the only group of people who saw the West during the war.[145] They shared this experience with former slave laborers, but also with civil servants and military personnel who went to administer occupied Germany and Austria, or with agents of enterprises who were sent to find much-needed equipment. At the same time, not all veterans had seen the West. What did make veterans different to some degree, was their extremely strong sense of entitlement, which contributed to grievances if it did not coincide with social advancement after the war.[146]

If the political sentiments of veterans are indicative of wider trends in Soviet society, did the war change the possibilities of political thought? To some extent, yes. For a whole decade, from the start of the Stalin revolution until 1939 (the annexation of parts of Poland), the Soviet Union had lived under a very strong form of isolation from the rest of the world.[147] Information about life abroad was scarce and largely mediated through official channels. As the Harvard Project concluded, in the 1930s the system was "quite successful ... in preventing foreign communications from reaching the population."[148] This virtual wall around the Soviet Union burst open during the war, and veterans were one group which transported this new knowledge back to the Soviet Union and kept it alive. The propaganda effort of the former Allies after 1945 helped to keep some windows to the West open, as did the occupation of Eastern Germany. The war thus certainly complicated the cultural universe accessible to Soviet society as it added additional information and channels of information.

However, this does not mean that the picture of the Soviet citizen in the 1930s who only had access to information distributed by the official network is entirely accurate. The persistence of peasant culture as a source of information in the post-war years and the memory of older layers of revolutionary tradition suggest that Sarah Davies was indeed right in proposing the existence of a relatively autonomous popular opinion.[149] Otherwise these elements could not have been transmitted over decades to be at the disposal of critical veterans in the post-war years. Many of the elements analyzed in this article were far from new but had been around in the 1930s as well. To list only a few: anti-Semitism and apocalypse were as much part of the popular lexicon before the war as were contempt for kolkhozes among peasants but also across social divisions, ironic use of official discourse, creative reading between the lines, nostalgia for "good old times," the NEP as an ideal, or notions of free trade and private property as positive forces.[150] There had been discontent about the privileges of the "new class" before 1941 and there had been expectations of war linked to hopes of a possible breakdown of the system.[151] There were people in the 1930s who thought that "the Christ-loving military people will attack from America." Others wanted to "bring back the Tsar," a slogan which sometimes was connected to memories of a different political economy (the Tsar and free trade).[152] Even the notion that the Tsar had been an idiot but life was better under his (presumably idiotic) rule was an idea which had circulated in the 1930s, as had hopes for an assassination of Stalin, or the idea that Trotsky would have been a positive alternative.[153] What seems to be new in the post-war years is that some of these elements were at times combined in a more clearly formulated counter-ideology based on the utopia "abroad" or "America."[154] The emergence of this new counter-ideology sets post-war Stalinism apart from its pre-war incarnation and connects late Stalinism with the decades to come.

Acknowledgments

*I am strongly in debt to Vladimir Aleksandrovich Kozlov who informed me about the declassification of the documents this chapter is based on and gave me access to his database on 58-10 cases. Without his generosity this chapter would never have been written. I also want to thank the members of the Modern European History Workshop, the Russian Studies Workshop (both University of Chicago), and the Institutskolloquium für Geschichte Osteuropas at the Humboldt Universität Berlin for feedback and discussion of earlier drafts. In particular, Alan Barenberg's summary of my rambling argument, critique and suggestions helped immensely. Debra McDougall eliminated many Germanisms and silly jargon. She also suggested a major restructuring, which I adopted. Rosa Magnusdottier and Sheila Fitzpatrick also read and commented on drafts at various stages of gestation. The comments by the students of my 2004 honors class at the University of Western Australia made the restructuring as a chapter easy. Juliane Fürst edited the chapter meticulously and made important suggestions for improvement. Research was made possible by a DAAD Doktorandenstipendium (2001–2002). Writing was supported by a Mellon Dissertation Year Fellowship at the University of Chicago (2003–2004) and by generous leave and travel funds by the School of Humanities of the University of Western Australia in the (Australian) summer of 2004–2005.

Notes

1 Catherine Merridale, "The Collective Mind: Trauma and Shell-Shock in Twentieth-century Russia," *Journal of Contemporary History* 35, no. 1 (2000): 39–55; and Amir Weiner, "Saving Private Ivan: From What, Why, and How?," *Kritika: Explorations in Russian and Eurasian History* 1, no. 2 (2000): 305–336.
2 E. S. Seniavskaia, *1941–1945. Frontovoe pokolenie. Istoriko-psikhologicheskoe issledovanie* (Moscow: RAN institut Rossiiskoi istorii, 1995).
3 Elena Zubkova, *Russia After the War. Hopes, Illusions, and Disappointments, 1945–1957*, trans. Hugh Ragsdale (Armonk, N.Y.: M. E. Sharpe, 1998), 105–107, 25–27; quotation: 26.
4 My analysis is deeply influenced by the analysis of the Harvard Project on the Soviet Social System. See Raymond A. Bauer, Alex Inkeles and Clyde Kluckhohn, *How The Soviet System Works. Cultural, Psychological, and Social Themes* (Cambridge: Harvard University Press, 1956); and Alex Inkeles and Raymond Bauer, *The Soviet Citizen. Daily Life in a Totalitarian Society* (Cambridge, Mass.: Harvard University Press, 1961), esp. 159–188.
5 See Bauer, Inkeles and Kluckhohn, *How The Soviet System Works*, 101; Sarah Davies, *Popular Opinion in Stalin's Russia. Terror, Propaganda and Dissent, 1934–1941* (Cambridge: Cambridge University Press, 1997), 6–9 and passim; and Vladimir Shlapentokh, *A Normal Totalitarian Society. How the Soviet Union Functioned and How It Collapsed* (Armonk, N.Y.: M. E. Sharpe, 2001), 127–152.
6 This implies that not only the historiography on veterans, but also the scholarship on "Stalinist civilization," subjectivity and the problem of belief needs some adjustment. For the developing orthodoxy on this topic see Stephen Kotkin, *Magnetic Mountain. Stalinism as a Civilization* (Berkeley: University of California Press, 1995); id. "[review of Sarah Davies, Popular Opinion in Stalin's Russia]," *Europe-Asia Studies* 50, no. 4 (1998): 739–742; Igal Halfin and Jochen Hellbeck, "Rethinking the Stalinist Subject: Stephen Kotkin's 'Magnetic Mountain' and the State of Soviet Historical Studies," *Jahrbücher für Geschichte Osteuropas* 44, no. 3 (1996): 456–463; Hellbeck, "Fashioning the Stalinist Soul: The Diary of Stepan Podlubnyi (1931–1939)," *Jahrbücher für Geschichte Osteuropas* 44 (1996): 344–375; id., "Speaking Out: Languages of Affirmation and Dissent in Stalinist Russia," *Kritika: Explorations in Russian and Eurasian History* 1, no. 1 (2000): 71–96; and id. "Working, Struggling, Becoming: Stalin-Era Autobiographical Texts," *The Russian Review* 60, no. July (2001): 340–359. For challenges to this paradigm see Davies, *Popular Opinion in Stalin's Russia*; Juliane Fürst, "Prisoners of the Soviet Self? – Political Youth Opposition in Late Stalinism," *Europe-Asia Studies* 54, no. 3 (2002): 353–375; id., "Re-examining Opposition under Stalin: Evidence and Context – A Reply to Kuromiya," *Europe-Asia Studies* 55, no. 5 (2003): 789–802; and Mark Edele, "Strange Young Men in Stalin's Moscow:

The Birth and Life of the Stiliagi, 1945–1953," *Jahrbücher für Geschichte Osteuropas* 50, no. 1 (2002): 37–61.

7 All of the cases are from the state procuracy of the Russian Federation (there are similar cases in the procuracy of the Soviet Union, which I did not review for reasons of time constraints). Based on the exhaustive database of V. A. Kozlov I selected 76 cases to review. I selected these cases based on the principle of diversity of social background and age. I reviewed a little over half of this sample and have detailed noted on 38 cases in my files (I did not take notes on some of the files, which were similar in type to those I had reviewed already).

8 Use of this type of sources is still in its infancy. An important example of work which draws most of its information from procuracy materials is V. A. Kozlov, *Massovye besporiadki v SSSR pri Khrushcheve i Brezhneve (1953 – nachalo 1980-kh gg.)* (Novosibirsk: Sibirskii khronograf, 1999). Two catalogues of these documents are essential for any work with *Nadzornye proizvodstva* of cases of "anti-Soviet agitation" (paragraph 58-10 of the Russian criminal code). One covers the period prior to 1953, and is available as a computer database in GARF. The second catalogue is published: V. A. Kozlov and S. V. Mironenko, eds., *58-10. Nadzornye proizvodstva Prokuratury SSSR po delam ob antisovetskoi agitatsii i propagande. Annotirovannyi katalog mart 1953–1991* (Moscow: Mezhdunarodny fond "Demokratiia", 1999).

9 The interested reader can find a detailed source critique in my dissertation. "A 'Generation of Victors?' Soviet Second World War Veterans from Demobilization to Organization 1941–1956," (Chicago: The University of Chicago, 2004), 442–450.

10 Sheila Fitzpatrick, "Becoming Cultured: Socialist Realism and the Representation of Privilege and Taste," *The Cultural Front. Power and Culture in Revolutionary Russia* (Ithaca: Cornell University Press, 1992), 216–237, esp. 217.

11 For examples see the case files GARF (Gosudarstvennyi Arkhiv Rossiskoi Federatsii) f. A-461, op. 1, d. 1887 (America); d. 110 (Germany and America), which are both described in detail in Edele, "A 'Generation of Victors?'," 432–436. For Yugoslavia see GARF, ibid., d. 152, l. 7; and for England see d. 131, l. 18.

12 See for example the case file GARF f. A-461, op. 1, d. 110; described in detail in Edele, "A 'Generation of Victors?'," 435–436.

13 GARF f. A-461, op. 1, d. 132, l. 12.

14 GARF f. A-461, op. 1, d. 110, described in detail in Edele, "A 'Generation of Victors'?," 435–436.

15 GARF f. A-461, op. 1, d. 152, l. 7.

16 GARF f. A-461, op. 1, d. 1202, l. 13.

17 GARF f. A-461, op. 1, d. 132, passim.

18 GARF f. A-461, op. 1, d. 131, l. 18.

19 GARF f. A-461, op. 1, d. 1114, l. 7.

20 "I do not deny, that I talked about the hard life of the workers in the Soviet Union, because in comparison with the pre-war level, the wage of the workers today is the same, but foodstuffs are several times more expensive." GARF f. A-461, op. 1, d. 125, l. 16.

21 GARF f. A-461, op. 1, d. 110, described in detail in Edele, "A 'Generation of Victors?'," 435–436.

22 GARF f. A-461, op. 1, d. 1887, described in detail in Edele, "A 'Generation of Victors?'," 432–434.

23 GARF f. A-461, op. 1, d. 1119, l. 5, 22; d. 1863, l. 4; d. 1867, l. 4; and Otdel partiinoi informatsii upravleniia po proverke partorganov TsK VKP(b), "Informatsionnaia svodka," (15 January 1947), RGASPI f. 17, op. 88, d. 810, l 81–102, here: 100.

24 GARF f. A-461, op. 1, d. 1755, l. 8.

25 See Catriona Kelly and Vadim Volkov, "Directed Desires: Kul'turnost' and Consumption," in *Constructing Russian Culture in the Age of Revolution: 1881–1940*,

ed. Catriona Kelly and David Shepherd (New York: Oxford University Press, 1998), 291–313; Vadim Volkov, "The Concept of Kul'turnost': Notes on the Stalinist Civilizing Process," in *Stalinism. New Directions*, ed. Sheila Fitzpatrick (London: Routledge, 2000), 210–230; and Julie Hessler, "Cultured Trade. The Stalinist turn towards consumerism," in ibid., 182–203.

* Note the lack of fluency in Bolshevik speech. If forced to repeat a "slanderous" remark about Stalin, prosecutors and witnesses avoided naming the sacred name of the leader and replaced it by "the leader of party and people," "one of the leaders of party and government" or similar formulations. The accused here struggled to execute this kind of speech-act as instructed by his interrogators. He simply was unable to use these rules and produced an ungrammatical sentence which failed to omit Stalin's name.

26 GARF f. A-461, op. 1, d. 1820, l. 11. For information on him l. 2.

27 GARf f. A-461, op. 1, d. 1015, l. 14. The court in question was a *narodnyi sud*, not the court which later sentenced him for anti-Soviet agitation.

28 GARF f. A-461, op. 1, d. 129, ll. 20, 21; d. 125, l. 16.

29 GARF f. A-461, op. 1, d. 125, l. 17.

30 On official discourse as basis of youth resistance see Fürst, "Prisoners of the Soviet Self?"

31 GARF f. A-461, op. 1, d. 1867, l. 4; for the "disgusting politics" quotation: l. 5.

32 GARF f. A-461, op. 1, d. 129, l. 21.

33 GARF f. A-461, op. 1, d. 1114, l. 5.

34 Ibid., l. 5. On the Soviet master narrative inverted in such apocalyptic thought see Igal Halfin, *From Darkness to Light. Class, Consciousness, and Salvation in Revolutionary Russia* (Pittsburgh, Pa.: University of Pittsburgh Press, 2000).

35 GARF f. A-461, op. 1, d. 1202, l. 13.

36 GARF f. A-461, op. 1, d. 152, l. 8.

37 Leon Trotsky, *The Revolution Betrayed. What is the Soviet Union and Where is it Going?* (New York: Pathfinder, 1972 [orig.: 1937]); Milovan Djilas *The New Class. An Analysis of the Communist System* (New York: Praeger, 1957).

38 GARF f. A-461, op. 1, d. 1824, l. 7.

39 GARF f. A-461, op. 1, d. 143, l. 9.

40 GARF f. A-461, op. 1, d. 1863, l. 4.

41 GARF f. A-461, op. 1, d. 1886, l. 12. Another veteran, born in 1921, frequently talked with two friends about Trotsky and Bukharin. They concluded that "they had been sentenced undeservedly." GARF f. A-461, op. 1, d. 1655, ll. 32–33.

42 GARF f. A-461, op. 1, d. 152, l. 22. For his birth year: l. 6. The "idiot" confirmed this story: l. 22.

* sic! "Dovol'no tiranii Vashikh zhen i detei!"

43 Obkom secretary A. Larionov to Malenkov (5 May 1945), RGASPI f. 17, op. 125, d. 310, l. 14. Unfortunately the file does not give further information about the group. Thanks to Masha and Sophie Steynberg for comments on my translation of this flyer.

44 Such personalization of discontent while accepting major features of the system has been found to be a major phenomenon among Soviet DPs who left during and after the war. Bauer, Inkeles and Kluckhohn, *How The Soviet System Works*, 29–35, esp. 34; Inkeles and Bauer, *The Soviet Citizen*, 291–295.

45 GARF f. A-461, op. 1, d. 1202, l. 12.

46 This refers largely to reports collected by security organs. Sheila Fitzpatrick, *Stalin's Peasants. Resistance & Survival in the Russian Village After Collectivization* (New York: Oxford University Press, 1994), 327; Davies, *Popular Opinion*, 10–14. Party reports have the opposite tendency. See Olga Velikanova, "Berichte zur Stimmungslage. Zu den Quellen politischer Beobachtung der Bevölkerung," *Jahrbücher für Geschichte Osteuropas* 47, no. 2 (1999): 227–236, here: 234; Bauer, Inkeles and

Kluckhohn, *How the Soviet System Works*, 42–43; Lesley A. Rimmel, "*Svodki* and popular opinion in Stalinist Leningrad," *Cahier du Monde Russe* 40, no. 1–2 (1999): 218. On the problems of *svodki* as a source see also Lynne Viola, "Popular Resistance in the Stalinist 1930s: Soliloquy of a Devil's Advocate," *Kritika: Explorations in Russian and Eurasian History* 1, no. 1 (2000): 45–69; and Zubkova, *Russia after the War*, 7.

47 RGASPI f. 17, op. 125, d. 518, l. 12.

48 Zakon "O zashchite mira," (12 March 1951), *Sbornik zakonov SSSR i ukazov Prezidiuma Verkhovnogo Soveta SSSR (1938g. – noiabr' 1958 g.)* (Moscow: Iuridicheskaia literatura, 1959), 143–144.

49 There are many reports on anti-Semitism among veterans. See for example RGASPI f. 17, op. 125, d. 190, l. 16. First quoted by G. Kostyrchenko, *V plenu u krasnogo faraona* (Moscow: Mezhdunarodnye otnosheniia, 1994), 16; and id., *Tainaia politika Stalina. Vlast' i antisemitizm* (Moscow: Mezhdunarodnye otnosheniia, 2001), 243. See also Amir Weiner, *Making Sense of War. The Second World War and the Fate of the Bolshevik Revolution* (Princeton: Princeton University Press, 2000), 289.

50 Kostyrchenko, *Tainaia politika Stalina*, 242.

51 Letter of Mikhoels to Shcherbakov, 2 April 1943, RGASPI f. 17, op. 125, d. 127, l. 175–175ob. Mikhoels did not claim that Jewish decorated soldiers never appear in the press (which would be incorrect), but that Jewish heroism is not stressed enough to counter the rumors about their lack of engagement in the war effort.

52 Weiner, *Making Sense of War*, 219–222, 293–294 (quotation).

53 Kirill Tomoff, "Creative Union: The Professional Organization of Soviet Composers, 1939–1953," PhD diss. The University of Chicago, 2001, chap. 5.

54 Liudmilla Alexeyeva and Paul Goldberg, *The Thaw Generation. Coming of Age in the Post-Stalin Era* (Boston: Little, Brown, and Co., 1990), 43–44.

55 See Merle Fainsod, *Smolensk under Soviet Rule* (New York: Vintage Books, 1958), 378; Zubkova, *Russia after the War*, 144–145; and Robert W. Thurston, *Life and Terror in Stalin's Russia 1934–1941* (New Haven: Yale University Press, 1996), 186, 192. On the rules of "Stalinist democracy" see also Alexei Kojevnikov, "Games of Stalinist Democracy. Ideological discussions in Soviet sciences, 1947–52," in *Stalinism. New Directions*, ed. Sheila Fitzpatrick (London: Routledge, 2000), 142–175.

56 GARF f. A-461, op. 1, d. 1436, l. 9.

57 For another example see GARF f. r-9401, op. 2, d. 134, ll. 182–83. [Osobaia papka Stalina], also quoted by Zubkova, *Russia after the War*, 75.

58 See Bauer, Inkeles and Kluckhohn, *How The Soviet System Works*, 29–35; and Kotkin, *Magnetic Mountain*, 198–237.

59 Weiner, *Making Sense of War*, 46–49.

60 RGASPI f. 17, op. 88, d. 470, ll. 44–45, 47–49.

61 Jeffrey Brooks, *Thank You, Comrade Stalin! Soviet Public Culture from Revolution to Cold War* (Princeton, N.J.; Princeton University Press, 2000).

62 See Alexeyeva and Goldberg, *The Thaw Generation*, 29–55, 43–44, 54–57; and Mikhail Gorbachev, *Zhizn' i reformy*, 2 vols, vol. 1 (Moscow: Novosti, 1995), 66–67.

63 Grigorii Chukhrai, *Moia voina* (Moscow: Algoritm, 2001), 21–22; id., *Moe kino* (Moscow: Algoritm, 2002); 142. Zubkova, *Russia After the War*, 26, 32.

64 Chukhrai, *Moia voina*, 281–282. Chukhrai likes to present his "generation" as a generation of Stalinists who could not but believe in the leader until the Twentieth Party Congress opened their eyes. Chukhrai, *Moe kino*, 142.

65 Richard Stites, ed., *Culture and Entertainment in Wartime Russia* (Bloomington: Indiana University Press, 1995); Mark Edele, "Paper Soldiers: The World of the Soldier Hero according to Soviet Wartime Posters," *Jahrbücher für Geschichte Osteuropas* 47, no. 1 (1999): 89–108; Brooks, *Thank You Comrade Stalin!*,

159–194; and Lisa A. Kirschenbaum, "'Our City, Our Hearths, Our Families': Local Loyalties and Private Life in Soviet World War II Propaganda," *Slavic Review* 59, no. 4 (2000): 825–847. Given the behavior of Wehrmacht, SS and *Einsatzgruppen* on Soviet soil, the internalization of Soviet hate propaganda also needs little explanation. See Omer Bartov, *The Eastern Front, 1941–1945: German Troops and the Barbarisation of Warfare* (New York: St. Martin's Press, 1986).

66 This discussion owes a lot to Inkeles and Bauer's model of the "system of communications" in the Soviet Union as made up of two parallel but interconnected systems – the official and the unofficial. See *The Soviet Citizen*, 159–165.

67 Generally on the official network see Inkeles and Bauer, *The Soviet Citizen*, 159–161; Alex Inkeles, *Public Opinion in Soviet Russia* (Cambridge, Mass.: Harvard University Press, 1967); and Peter Kenez, *The Birth of the Propaganda State: Soviet Methods of Mass Mobilization, 1917–1929* (Cambridge: Cambridge University Press, 1985). On propaganda during the war see Stites, ed., *Culture and Entertainment in Wartime Russia*. On newspapers see Matthew Edward Lenoe, *Closer to the Masses. Stalinist Culture, Social Revolution, and Soviet Newspapers* (Cambridge, Mass.: Harvard University Press, 2004); and Brooks, *Thank You, Comrade Stalin*. On posters see Victoria E. Bonnell, *Iconography of Power. Soviet Political Posters under Lenin and Stalin* (Berkeley: University of California Press, 1997); and Frank Kämpfer, *"Der Rote Keil." Das politische Plakat. Theorie und Geschichte* (Berlin: Gebr. Mann Verlag, 1985). On poster propaganda during the war see Edele, "Paper Soldiers: The World of the Soldier Hero according to Soviet Wartime Posters."

68 "in the radio they chatter (*boltat'*) and blow about (*trepat'sia*) . . . everybody watches foreign movies with delight, but there's no point in watching ours, it's just agitation." (GARF f. A-461, op. 1, d. 129, l. 20); ". . . all they write in our newspapers about America and England is a lie." (GARF f. A-461, op. 1, d. 131, l. 18.)

69 Inkeles and Bauer, *The Soviet Citizen*, 159–188.

70 GARF f. A-461, op. 1, d. 132, l. 10.

71 GARF f. A-461, op. 1, d. 1886, l. 11.

72 Karen Petrone, *Life has become more Joyous, Comrades: Celebrations in the Time of Stalin* (Bloomington, Ind.: Indiana University Press, 2000); Juliane Fürst, "Prisoners of the Soviet Self?"

73 Nikolai Fedorovich D'iakov, *Mechenye. Dokumental'nye zapiski byvshego soldata* ed. N. Mitrokhin (Moscow: Indformatsionno-ekspertnaia gruppa "PANORAMA", 1999) (= *Dokumenty po istorii dvizheniia inakomysliashchikh* vyp. 11, September 1999), 143, 256, 145–146.

74 GARF f. A-461, op. 1, d. 1655, l. 33.

75 GARF f. A-461, op. 1, d. 132, l. 14.

76 See Inkeles and Bauer, *The Soviet Citizen*, 185; and Fitzpatrick, *Stalin's Peasants*, 67–91, 286–296.

77 GARF f. A-461, op. 1, d. 1069, l. 29.

78 GARF f. A-451, op. 1. d. 1867, l. 4.

79 On working class anti-Semitism before the Revolution see, for example, Charters Wynn, *Workers, Strikes, and Pogroms. The Donbass-Dnepr Bend in Late Imperial Russia, 1870–1905* (Princeton, N.J.: Princeton University Press, 1992). For a concise overview over the history of anti-Jewish policies and anti-Semitism before the Revolution see Heinz-Dietrich Löwe, "Antisemitismus," *Lexikon der Geschichte Rußlands. Von den Anfängen bis zur Oktober-Revolution* ed. Hans-Joachim Torke (Munich: Verlag C. H. Beck, 1985), 36–38.

80 For a concise overview: Gerhard Simon, "Antisemitismus," *Historisches Lexikon der Sowjetunion 1917/22 bis 1991* ed. Hans-Joachim Torke (Munich: Verlag C. H. Beck, 1993), 24–25. On peasant anti-Semitism in the 1920s see, for example, Fainsod, *Smolensk under Soviet Rule*, 157. On anti-Semitism in 1930s in urban

settings see Sheila Fitzpatrick, *Everyday Stalinism. Ordinary Life in Extraordinary Times. Soviet Russia in the 1930s* (New York: Oxford University Press, 1999), 130, 168, 169, 186–187, 207, 215, 252 n. 64. See also Terry Martin, *The Affirmative Action Empire. Nations and Nationalism in the Soviet Union, 1923–1939* (Ithaca: Cornell University Press, 2001), 389–390.

81 A professor of Voronezh State University speaking at a closed party meeting discussing the KR affair in 1947. RGASPI f. 17, op. 122, d. 273, l. 40.

82 GARF f. A-461, op. 1, d. 129, l. 18, 20, 21.

83 GARF f. A-461, op. 1, d. 110, l. 7.

84 GARF f. A-461, op. 1, d. 132, l. 9.

85 GARF f. A-461, op. 1, d. 132, l. 10.

86 GARF f. A-461, op. 1, d. 132, l. 11.

87 GARF f. A-461, op. 1, d. 125, l. 17.

88 Zam. po politchasti predstavitelia upolnomochennogo SNK SSSR po delam repatriatsii v Shvetsii A. Beliaev, "Politdonesenie," (28 April 1945), RGASPI f. 17, op. 125, d. 314, l. 51.

89 GARF f. A-461, op. 1, d. 152, ll. 7, 9, 22. Note that he also listened to VOA: ll. 6, 7.

90 Zubkova, *Russia After the War*, 18. On traces of this process in letters of soldiers see E. Sherstianoi, "Germaniia i nemtsy v pis'makh krasnoarmeitsev vesnoi 1945 g.," *Novaia i noveishaia istoriia*, no. 2 (2002): 137–151. The much-quoted Belov might have over-stressed the case slightly when he wrote that the war, the German occupation and seeing the West in 1945 "nearly nullified all previous efforts of Soviet propaganda." Fedor Belov, *The History of a Soviet Collective Farm* (New York: Praeger, 1955), 72.

91 I am indepted to Paul Stronski for making this point during a lunch break discussion in RGASPI. For a description of Red Army soldiers on a buying frenzy in the annexed Polish territories in 1939 see Jan T. Gross, *Revolution From Abroad. The Soviet Conquest of Poland's Western Ukraine and Western Belorussia. Expanded edition with a new preface by the author* (Princeton: Princeton University Press, 2002), 28–29, 45–50. See also Mark von Hagen, "Soviet Soldiers and Officers on the Eve of the German Invasion: Toward a Description of Social Psychology and Political Attitudes," in *The People's War. Responses to World War II in the Soviet Union*, ed. Robert W. Thurston and Bernd Bonwetsch (Urbana: University of Illinois Press, 2000), 186–210, here: 199–200.

92 Seniavskaia, *1941–1945. Frontovoe pokolenie*, 91–92.

93 Zubkova, *Russia after the War*, 63.

94 RGASPI f. 17, op. 122, d. 273, l. 42.

95 RGASPI f. 17, op. 122, d. 261, l. 213–213ob. The director suggested to put such ideas into the right perspective: The "comfort of the average American is not an easy thing," he claimed, not denying the existence of such comfort. However, the poor American is "a slave of comfort and not able to feel the deep feelings, which we have become used to consider normal among Soviet people."

96 RGASPI f. 17, op. 122, d. 273, l. 135.

97 RGASPI f. 17, op. 122, d. 275, l. 44.

98 RGASPI f. 17, op. 122, d. 273, l. 32.

99 Weiner, *Making Sense of War*, 369, 378 (quotation). Weiner dismisses her notion of *frontoviki* as "Glossing over the reputed hostility of Red Army troops toward the nationalist movement and its cause."

100 Such prosperity confused European leftist radicals considerably. In 1929 the German anarchist Ernst Toller met car- and house-owning workers in the US and went to considerable pains to discuss away this apparent prosperity as objective exploitation. Ernst Toller, *Quer Durch. Reisebilder und Reden* reprint of the 1930 edition with an introduction by Stephan Reinhardt (Heidelberg: Verlag Das Wunderhorn, 1978), 14–19, for the cars 16. Eric Hobsbawm recounts his own surprise when

one of his comrades returned from a party assignment to wartime Coventry "open-mouthed" and asked his friend if they were aware that "up there the comrades have *cars?*" Eric Hobsbawm, *The Age of Extremes. A History of the World, 1914–1991* (New York: Michael Joseph, 1996 [orig. 1994]), 305.

101 On POWs in Britain and their repatriation see Pavel Polian, *Deportiert nach Hause. Sowjetische Kriegsgefangene im "Dritten Reich" und ihre Repatriierung* (Munich: R. Oldenbourg Verlag, 2001), 93–94.

102 Vladimir Pechatnov, "The Rise and Fall of Britansky Soyuznik: A Case Study in Soviet Response to British Propaganda of the Mid-1940s," *The Historical Journal* 41, no. 1 (1998): 293–301. Thanks to Rosa Magnusdottier for pointing me towards this article.

103 RGASPI f. 17, op. 125, d. 436, l. 28.

104 RGASPI f. 17, op. 125, d. 436, l. 29.

105 See table 6.1 in Edele, "A 'Generation of Victors?'," 480.

106 Pasha Angelina, the star tractor driver, was a subscriber to both in 1947. Sheila Fitzpatrick and Yuri Slezkine, eds, *In the Shadow of Revolution. Life Stories of Russian Women from 1917 to the Second World War* (Princeton, N.J.: Princeton University Press, 2000), 306.

107 GARF f. 461, op. 1, d. 1655, l. 33.

108 GARF f. 461, op. 1, d. 1655, l. 29.

109 On the importance of foreign radio broadcasts for the "second public opinion" in Nazi Germany see Detlev Peukert, *Inside Nazi Germany. Conformity, Opposition, and Racism in Everyday Life* trans. R. Deveson (New Haven: Yale University Press, 1988), 54.

110 Inkeles, *Public Opinion in Soviet Russia*, 251; id., *Social Change in Soviet Russia* (Cambridge: Harvard University Press, 1968), 346–379; Walter L. Hixson, *Parting the Curtain: Propaganda, Culture, and the Cold War, 1945–1961* (New York: St. Martin's Press, 1997), 35–37; Brooks, *Thank You, Comrade Stalin*, 210; and Mark Edele, "Strange Young Men in Stalin's Moscow: The Birth and Life of the Stiliagi, 1945–1953," *Jahrbücher für Geschichte Osteuropas* 50 (2002): 51–52.

111 GARF f. A-461, op. 1, d. 1436, ll. 8, 10.

112 GARF f. A-461, op. 1, d. 125, ll. 16 (quotation), 22.

113 GARF f. A-461, op. 1, d. 131, ll. 17 (quotations), 29.

114 Jochen Hellbeck, [answer to Sarah Davies] *Kritika. Explorations in Russian and Eurasian History* n.s., 1 no. 2 (Spring 2000), 440.

115 GARF f. A-461, op. 1, d. 131, l. 18.

116 Those who had been abroad often felt superior to those who were not in the know: "There you have culture," commented one veteran to a swearing driver, "it seems you have not been in Germany, and don't know cultured life." GARF f. A-461, op. 1, d. 1169, l. 10.

117 Most Soviet radio sets were cable-receivers, which only allowed listening to Soviet channels transmitted over a network of land-lines. For a discussion of the extent of the availability of wave receivers see Edele, "Strange Young Men," 51–52.

118 GARF f. A-461, op. 1, d. 1887; d. 110; and d. 1114. For a detailed description of these three cases see Edele, "A 'Generation of Victors?'", 432–439.

119 GARF f. A-461, op. 1, d. 1128, l. 7; for biographical information on him ll. 4, 35–36.

120 This fact became a topic in many of the party meetings which discussed the KR-affair in 1947: RGASPI f. 17, op. 122, d. 273, l. 42; d. 261, l. 213–213ob; d. 273, l. 135; d. 275, l. 44.

121 See for example the case of Ivan Vasil'evich R. GARF f. A-461, op. 1, d. 1867.

122 This, in a nutshell, is the biography of D'iakov, *Mechenye*, passim.

123 *sdalsia v plen* was the official formulation for "having been taken prisoner," implying active desertion from the own side. On Chukhrai's war and post-war life see his *Moia voina*; and *Moe kino*.

124 D'iakov, *Mechenye*, 145, 146. On repatriates as a "category of people, aggrieved by the regime" whose humiliating and insecure social status made them lean towards opposition see Zubkova, *Russia after the War*, 107.
125 GARF f. A-461, op. 1, d. 110, ll. 46, 49, 54. The file is contradictory here. According to one document he was a POW between 1941 and 1943; according to another, he was freed in 1941 "under questionable circumstances" and lived in the occupied territory until 1944.
126 GARF f. A-461, op. 1, d. 110.
127 For an artistic representation of the bitterness among POWs, which was produced by the combination of having put one's life on the line for the regime during the war and getting stigmatized for this in post-war life see Chukhrai's movie *Chistoe nebo* (Moscow: Mosfilm, 1961). See also Ludmilla Alexeyeva's description of her bitter "Uncle Borya." Alexeyeva and Goldberg, *The Thaw Generation*, 42–43.
128 GARF f. A-461, op. 1, d. 129, ll. 16, 18, 20, 21, 36.
129 GARF f. A-461, op. 1, d. 129, l. 18.
130 GARF f. A-461, op. 1, d. 1169, l. 10.
131 GARF f. A-461, op. 1, d. 1854, l. 28; for biographical information see l. 27.
132 GARF f. A-461, op. 1, d. 1863, l. 4.
133 On outmigration from the countryside see Edele, "A 'Generation of Victors?'," 328–335.
134 GARF f. A-461, op. 1, d. 1202, ll. 17, 22.
135 See Edele, "A 'Generation of Victors?'," 127–134.
136 GARF f. A-461, op. 1, d. 125, l. 16.
137 GARF f. A-461, op. 1, d. 143, l. 9.
138 Hellbeck, "Fashioning the Stalinist Soul;" and id., "Speaking Out."
139 On the extent of social mobility among veterans see Edele, "A 'Generation of Victors?'," Chapter 4.
140 V. N. Gordov was a Hero of the Soviet Union. He held the rank of colonel-general and had been the commander of the Privolzhskii military district. F. T. Rybal'chenko held the rank of major-general and had worked as the Chief of Staff of the district commanded by Gordov. R. G. Pikhoia, *Sovetskii Soiuz: istoriia vlasti 1945–1991*, 2nd edn (Novosibirsk: Sibirskii khoronograf, 2000 [orig.: 1998]), 38.
141 Dunham, *In Stalin's Time*.
142 Mordechai Altshuler, "Antisemitism in Ukraine toward the End of the Second World War," *Jews in Eastern Europe* 3, winter (1993): 40–81.
143 See Fitzpatrick, *Everyday Stalinism*, chap. 5.
144 See Edele, "Strange Young Men," 52.
145 See Zubkova, *Russia after the War*, 72.
146 On the assertiveness of veterans see Weiner, "Saving private Ivan," 317.
147 Fitzpatrick, *Everyday Stalinism*, 5; Kotkin, *Magnetic Mountain*, 225.
148 Bauer *et al.*, *How the Soviet System Works*, 26.
149 Davies, *Popular Opinion in Stalin's Russia*.
150 See for example: Fitzpatrick, *Stalin's Peasants*, 9, 45–47, and passim; Lynne Viola, *Peasant Rebels under Stalin. Collectivization and the Culture of Peasant Resistance* (New York: Oxford University Press, 1996); Bauer *et al.*, *How the Soviet System Works*, 38, 114, 119; Inkeles and Bauer, *The Soviet Citizen*, 159–188; Fainsod, *Smolensk*, 248, 252. On anti-Semitism see fn. 79–80 above.
151 Fitzpatrick, *Everyday Stalinism*, 105, 205.
152 Fitzpatrick, *Stalin's Peasants*, 66, 68, 69.
153 Elena Osokina, *Our Daily Bread. Socialist Distribution and the Art of Survival in Stalin's Russia, 1927–1941*, trans. Kate Transchel and Greta Bucher (Armonk, N.Y.: M. E. Sharpe, 2001), 157.
154 On *Amerika* as a myth in post-war society see Rosa Magnusdottier, "American Myths or Soviet Realities? State, Society, and Social Control, 1945–1959" (PhD dissertation in progress, UNC, Chapel Hill).

9   Children's lives after Zoia's death

Order, emotions and heroism in
children's lives and literature in the
post-war Soviet Union

Ann Livschiz

The Great Patriotic War was a struggle for survival and the ultimate test of the
Soviet system, one for which the whole country was ostensibly preparing for
decades. In the years leading up to the German invasion, concerns about Soviet
children – their moral fibre, work ethic and physical endurance – reached new
heights, reflecting the anxiety of a maturing system about the loss of revolutionary
fervour and the fate of the Soviet project, which believed its survival depended on
the young generation. But the wartime record of the majority of the Soviet population, young and old, was seen by the state as vindication of all its pre-war policies.[1] Specifically, it affirmed the success of the Soviet way of upbringing
(*vospitanie*) by showing that the vast majority of young people proved loyal,
capable of hard work and sacrifice and exhibited correct moral judgement. The
state was justifiably proud. There was an element of relief and surprise as well. In
the words of Olga Mishakova, a Komsomol Central Committee secretary, "yesterday's pampered, spoiled young person (*iznezhennyi, izbalovannyi podrostok –
beloruchka*) – today gives 2–3–4 times the norm of production per shift."[2] There
was even a twinge of guilt – "all these years we have criticized these children, but
just look at their performance in this war. Has the war transformed them, or have
they just been this great all along, and we did not notice, and did not realize what a
great job we have done with them?"[3] Another way of looking at it was to see the
war as the solution to the many problems with labour education and moral training
plaguing the Soviet system. As the head of the Children's Publishing House
(*Detgiz*) Liudmila Dubrovina triumphantly put it – "all of the pre-war problems
with [children's] upbringing were corrected by the war."[4]

However, in addition to the pride, joy and relief, there was also concern. The
war not only showcased the successes of the Soviet system of upbringing, but it
also exposed its weaknesses. Most of these problems, such as insufficient discipline and lack of practical skills, were already noted during the experience of the
Russo-Finnish war, and a little more than a year was not enough to eliminate
them.[5] But the Great Patriotic War was much longer, more intense and "total,"
and the experience of occupation as a test added new problems. "War on the one
hand elevated the people (*podniala narod*), but a part of it rotted (*razlozhilas'*),
and in some way people's old habits (*perezhitki*) were refreshed," commented
Olga Mishakova in 1946.[6]

If the experience of war brought out the best in people, it also brought out the worst. This was particularly the case, from the Soviet point of view, in the areas occupied by the Germans. The state placed part of the blame on the Germans and their methods.[7] "They purposefully tried to morally corrupt (*rastlit'*) our youth by the illusion of an easy life, erotic books, pictures, pornographic poems and pictures."[8] Yet the reasons why the Germans could have any success were considered to lie in the shortcomings in the Soviet system of upbringing, which now needed correcting.[9] The experience of occupation, when people were left without the benefit of Soviet propaganda for a few years and were thus prone to lose their Soviet outlook, fortified the state's belief in the need for perpetual propaganda directed at various population groups, especially children. The quest for moral purity would dominate the post-war years, but steps were taken during the war to correct perceived problems. The experience of war simultaneously confirmed the success of the creation of Soviet citizens and established that the process of transformation was far from over. Indeed, with some adjustments, it would have to continue in the post-war years. The focus of this chapter is the way in which post-war need for rebuilding and normalization shaped both the content and nature of indoctrination of children and the expectations placed upon them by educational establishment and cultural figures.

On the problem of order

It was not just the content of the children's minds and souls that troubled officials and parents, but very concrete discipline problems, which were at least partly the result of reduced supervision owing to wartime conditions. The war also brought with it a plethora of serious social problems – an increase in *beznadzornost'* and *besprizornost'*, juvenile delinquency and crime, poverty, broken homes and loss of fathers – not to speak of the impact of violence as both physical and psychological trauma. For the state, it was crucial to regain control and re-establish order, particularly in light of the realization that parents were unable to cope with post-war disorders.[10] In addition to the various decrees on reducing *besprizornost'* and *beznadzornost'*, Rules of Conduct, detailed guidelines for children's behaviour in public, in school, and at home, that were promised in the mid-1930s, were finally approved and promulgated in 1943. The final product emphasized order, obedience, politeness and deference to authority figures. To help institute order in schools, from the autumn of 1943 onwards, boys and girls in big cities would be attending separate schools – another pre-war measure that was implemented during the war.[11]

Criticism of the moral shortcomings of Soviet children could be taken too far, of course, particularly when discussed in public, rather than in conferences behind closed doors. Thus the publication of an article by the head of the Moscow Department of Education, Orlov, in the journal *Moskovskii bolshevik* in March of 1944 caused a minor scandal reaching all the way to the Party's Central Committee Sector of Schools. In his article entitled "On discipline and the conduct of our children" Orlov "demand[ed] the reintroduction into the schools of punishments similar to those in the Red Army Statutes, and to use similar methods in general schools as are used in the [military] schools." We can see what exactly the party found troubling in the article by consulting the notes drawn up for CC Secretary Shcherbakov. Orlov wanted to reintroduce the *kartser* – a punishment cell used in pre-revolutionary schools – and to leave punishments to the discretion of the teachers. Orlov made "slanderous" (another word for criticism gone too far) statements about the Soviet schools, suggesting that "instead of training the young generation for harsh reality, and bringing them up as disciplined, trained and willed people, we, because of a sentimental pity for them, train them from an early age for moral dissoluteness (*k ras-pushchennosti*)." The problem with the article was further exacerbated by the fact that it was published in an official journal of the Moscow party *gorkom* and signed by the head of the Moscow Department of Education, which meant it was taken as a directive and used in some schools.[12] Apparently, not all comparisons between Soviet schools and their pre-revolutionary predecessors were meant for public consumption.

Another consequence of the war was an increased independence on the part of the children and young people, who in many ways had to manage on their own, and became more assertive and self-sufficient.[13] However, these features did not fit into the model of post-war society the state had in mind. Consequently, the post-war vision for model childhood behaviour became even more narrowly prescribed and called for even more state regulation in the sphere of education. While this was proclaimed as a sign of the party's care for children and their upbringing in time of crisis, behind the scenes the rhetoric was different. Measure of control was indicative of the degree of concern the state felt over this matter.[14] The war gave the state both the opportunity and the means to launch a more fervent attack on the problem of child discipline. The post-war period would not see a relaxation, but rather a tightening of the controls in the hope of fashioning moral citizens out of its charges.

On the usefulness of wartime heroes

The youthful heroes of the Great Patriotic War, who had died in the fulfilment of their patriotic duty, could be safely frozen in their moment of triumph, with either real or manufactured life-stories of near perfection leading to the heroic deed, which was presented for other children's emulation. Children and youths who survived, proved much more problematic for the state. The much-lauded "rapid maturity" during the war, which resulted in their contribution towards the war effort, came at a cost. The war led to a loosening of social controls over their behaviour, resulting an increase in hooliganism (still a broad concept covering a wide range of behaviour, from mildly annoying to criminal), loss of discipline problems in schools and other related phenomena. War forced children to be more independent. Circumstances forced them to learn to manage on their own. Not surprisingly they became more assertive, more self-sufficient and less prone to defer to adults.

After a most protracted, bloody and costly war, both state and population had a desire for a "return to normality." The problem lay in the definition of that term, which had different meanings to different population groups and certainly for the state. But if there was one issue, over which state and the general population were in agreement, it was the fact that re-establishment of public order was an essential component of normality. This meant in particular the re-establishment of control over children. The post-war years were characterized by a myth of a shiny, happy society. An important part of this shiny, happy society meant returning children to their appropriate place – namely schools. Thus, the post-war years saw a shift in the models for emulation presented to children.

Fear of losing control was one of the overriding themes troubling Soviet authorities in the post-war years. While the state was able to get most of what it wanted and needed from the population during the war, the degree to which the state depended on the will and willingness of the people to obey became glaringly clear. The lesson learned from the war by the Soviet state was not that Soviet people could be trusted, but that more control was needed. The first step towards regaining control was firmly establishing a monopoly on war memories and commemorative practices. The legacy of the Great Patriotic War – its interpretations and uses both by the state and the people – was a complicated matter, not least so when it came to dealing with the problem of rearing the young generation.

The experience of the war brought changes to the heroic pantheon. The child hero of the 1930s Pavlik Morozov was not removed from the pantheon of heroes after the war.[15] Yet the war produced new heroes – boys and girls, young men and women – who also sacrificed their lives, but whose feats were not complicated by adversarial and ambiguous relationships with their fathers. If anything, they represented the unity of generations in the face of a common outside enemy. They may have taken the place of their fathers, or fought alongside them, united by their love for Stalin and their common Soviet motherland. While there must have been children of traitors who turned on their fathers and proved their loyalty and dedication to the state, none of them made it into the heroic pantheon.

Furthermore, in the newly biologized hunt for enemies during the war and in the post-war years, there was no room for Pavlik-like behaviour.[16] There was no campaign encouraging children of bourgeois nationalists, Chechens or later Jews – "cosmopolites" to denounce their fathers and save themselves. The Soviet family, while in practical terms decimated by the war, in symbolic terms had closed its ranks and presented a united front against the enemy, justifying the faith placed in it by the state, and further redeeming itself as a thoroughly Soviet institution. The "fathers" did their part in the war, and their survivor's reward would be a return of obedient, polite and studious children to the family hearth. They did not need to fear denunciation from their children because they already proved their loyalty on the battlefield. "They are united in the face of the common enemy. The parents are proud of their children; they inspire and support them."[17] Children were now more than ever reminded of all the sacrifices their parents made for them, and how much they did for their children in daily life (they "raise you, feed you, clothe you").[18] These trends persisted and flourished, with the glorifications of veteran father figures tempered (albeit severely) by the cult of Stalin.[19]

But the usefulness of the war cult for children in the post-war years had a number of limitations. The new heroes were not always the best models for obedience – one of the key characteristics for the post-war Soviet child. The type of children who became heroes during the war at times had a pre-war life that did not make them suitable models. On the one hand, they served well as examples of the transformative power of war in which everyone united against the enemy and when even former delinquents found a way to contribute. This was quite a popular trope in the wartime literature.[20] The most famous "former hooligan" was one of the heroes of the Komsomol underground group *Young Guard* – Sergei Tiulenin (though it should be noted that his mother strenuously objected to this characterization of her son by Aleksandr Fadeev, the author of the eponymous novel which told the story of the heroism of young men and women in an underground organization in Krasnodon, most of whom were eventually arrested, tortured and executed).[21] Tiulenin was perceived as one of the most successful characters in Fadeev's book – "the most charming image" as one of Fadeev's colleagues put it.[22] He was definitely the least "perfect" and most human male protagonist, in stark contrast to, for example, the saintly Oleg Koshevoi – underground organization's commissar. The dilemma about such imperfect heroes is captured in one of the notes received by Lev Kassil' at a readers' conference – "Don't you think that when Sergei Tiulenin jumped from a window, he was a hooligan? And then he became such a hero!"[23]

While wartime heroes offered an example of behaviour during extreme situations, after the conflict it became important to demonstrate to children how to behave in daily peacetime life. Children may have dreamt of performing heroic deeds on the battlefield, but it did not necessarily translate into proper public behaviour or good grades. Daily routine did not appear to offer the glamour or possibility of heroism to the children. One way to apply the lessons of war to peacetime was to have classroom discussions of letters from army units to children, which raised questions such as "What would have happened to our Motherland, if we had the [low] level of discipline that you have in your school?"[24] It is difficult to assess the effectiveness of this technique, though it gives some clues as to how the cult of war could have lost its potency.

Another approach was not to rely on the war cult to address all necessary propaganda needs. In a February 1949 conference on pioneer work, a list of biographies that seventh graders should be studying included adult state and party leaders, as well as revolutionary and Civil War heroes. No Great Patriotic War heroes were listed by name, though they could have been covered under "etc." Even the cult of Zoia Kosmodemianskaia – arguably the most developed cult around an ordinary person as a war heroine – was considered to have its limitations:

We need to familiarize children with the surrounding life in school and at home, to show models of good human work on concrete examples and materials; give them an understanding of honesty, patriotism and real work. We can read about Zoia and it is necessary to read about her, but that is not enough. We must demonstrate fortitude and heroism with examples from life around us.[25]

The search for the right balance between war heroes and labour heroes would preoccupy education and Komsomol officials for years to come.[26]

The experience of war showed that it was not always the people who made the best speeches who turned out to be true heroes.[27] However, this went against the prevailing trends of public ritual (both in the adult and in children's realm), which attributed a great deal of significance to public and highly ritualized expressions of loyalty. The state also did not like to rehabilitate former villains or social rejects. This attitude is captured perfectly in a passage from O. Gonchar's *Znamenostsy* (1946–1948), a passage that seems to begrudge former *detdomovtsy* and delinquents their moment of glory during the war, while reasserting the traditional social hierarchy:

By the way, have you noticed, what groups produced the greatest number of heroes in battle? – From former *besprizornye*, from *belomorkanal'tsy*? Not at all … The best fighters are yesterday's workers, miners, metalworkers, collective farmers, and in general, members of honest working professions. Because war is first and foremost work, the most difficult of all jobs known to man, without days off, without vacations, 24 hours a day. . . .[28]

Thus, the unadulterated record of heroism was unacceptable for use in establishing post-war order. The war needed to be safely sanitized for children's (and adult's) consumption. It was used widely, but also very selectively. It was critical to explain the proper significance and context for heroic actions and their applications to children's daily life, namely getting good grades and incessantly expressing gratitude. While the images of dead heroes could be fairly easily adjusted, living "hero" children presented a bigger problem – those who served in partisan or regular army units during the war and returned to schools or trade schools. It was crucial to depict them as not resting on their laurels, but rather working hard to make a peacetime contribution as well.[29] For example, in the prize-winning book by I. Vasilenko *Little Star (Zvezdochka)*, a boy who was a decorated member of a partisan detachment realized how important it was for him to acquire the practical skill of learning how to make machine parts ("little stars" of the title) to make a contribution to the rebuilding process after the war. Until Stalin's death, the teachers, Komsomol officials and cultural figures who wanted to make use of the moral lessons of the war had to walk a tightrope to ensure they did not "treat the war in passing or as a distant past," while not choosing the wrong war episodes to promote.[30]

Post-war culture

Similar to the education and party officials, children's writers – those engineers of little human souls – saw children's wartime behaviour as either a confirmation of the writers' good work or an indictment of the quality of writing for children. At the meeting of the children's section of the February 1944 Plenum of the Writers' Union, the discussion explicitly turned to the nature of children's experiences during the war, what children's behaviour said about the quality of their upbringing, and the direction literature for children should take in the postwar years. Children were reaffirmed as the purest element in Soviet society, captured by the statement from Dubrovina that even under occupation "children remained faithful to their motherland."[31] In a long speech by Ilya Ehrenburg, the author proclaimed that the goal of writers was to

> fortify ethical norms ... [Before the war] we did not devote a lot of attention to feelings, high and low (*vysokie i nizkie*). We did not give images of evil in a setting familiar to the reader. Our "evil" was conventional and remote from the readers. Young men and women did not know that evil can be right next to them, and it happened that when they saw it, they did not recognize it as evil.... It is through [works of literature] that young souls learn how to tell good from evil, beautiful from base.[32]

Thus, the inability of some young people to make correct ethical decisions was attributed directly to the failures of Soviet literature, and by extension – Soviet writers.

Samuil Marshak, one of the deans of Soviet children's literature, continued the discussion in a similar vein the next day:

> We should not hide the horrors of war from children – after all, they were witnesses, participants and victims of it. But alongside the cruelty that war brings, and maybe even more importantly – we should show the children those best and purest qualities that opened up in people at the front – generosity, self-sacrifice, faithfulness, devotion to duty, patience, endurance, and selfless courage.[33]

Literature was supposed to show "generosity of feelings." Marshak underscored the importance of poetic literature, because "only poetic literature can cultivate in children feelings and high morals, free from boring didacticism.[34] Leonid Sobolev took the discussion even further:

> A new generation will grow up after the war. All that was tragic, frightening, majestic, and beautiful that our generation lived through in this war should become a fruitful soil on which these new generations will grow. To show the soul (*dushu*) of everything, first and foremost to children and young people. Nowhere else has literature acquired this much power and

significance as at this time, right before a victory, before the new way of life for a great multi-million people. Many young men and women have been crippled by war. Their souls have been crippled. But the breath of war, the breath of victory, the images of Soviet people, who gave their life, strength to their motherland and victory, this is an image for those, who will be growing up [after the war]. Only then will the precious blood spilled in battle, blood of the defenders from the mange of fascism on this globe, will not be wasted, and then and only then, if we use our entire force of writers' duty to bring up young human souls, generations of our sons and grandsons, our future human society.[35]

The nature of the discussion is noteworthy because the focus was on the intangible and ephemeral qualities, rather than the concrete and practical skills – a 1930s preoccupation. Thus, it was literally the hearts and souls of children the writers were concerned about, and it was these hearts and souls and the moulding of them that the writers saw as their provenance. This was clearly a continuation of a pre-war debate, which I have characterized elsewhere as "tanks versus feelings." Though it ran as an undercurrent in most discussions about children's literature through the 1930s, the debate came to the forefront in the aftermath of the Russo-Finnish war, where some party officials and cultural figures explicitly stated (in behind-the-scenes discussions) that the blood of young people who died in the snow was on the hands of writers who did not sufficiently prepare them for adult life and for war by dwelling on "damned feelings" rather than skills and "tanks." Not everyone agreed with this formulation, and a number of prominent children's writers fought tenaciously for their right to focus on feelings, which they saw as preparing children both for life and for war. Somewhat unexpectedly, the war appeared to have strengthened the position of the "emotions" camp, which included a number of prominent and popular children's writers, such as Samuil Marshak.[36] Behavioural problems of children and young people, whether under German occupation or at the homefront, were seen to be due not to insufficient military or labour training, but rather to insufficient "emotional training." It was far easier to acquire labour and military skills than it was to acquire humanity or an understanding of the concept of right and wrong. It was this that shapes the direction of the development of children's literature in the post-war years.

The Soviet state's commitment to maintaining the prominent place of literature and writers in Soviet society has been sufficiently noted and explored in several historical studies. As Vera Dunham put it, "literature was the repository of regime's myths."[37] While serving as a source of legitimacy for the regime, the war was also a source of empowerment for different groups within Soviet society. The degree to which these groups acquired empowerment correlated to their state-recognized contribution to the war effort. Writers (and other cultural figures) were no exception. Their contribution to the war effort, and thus their importance in the Soviet social and cultural hierarchy, was ultimately determined by the recognition the state granted them.[38] With their social and ideological importance confirmed, many children's writers were eager to do their part to help the state in the way they thought was necessary, deeming themselves to be in possession of sufficient Soviet credentials to be able to make the determination of what was necessary on their own, without waiting for instructions or guidelines.

Any discussion of post-war culture is impossible without considering the 1946 *Zvezda-Leningrad* decree, which essentially constituted a programmatic statement on culture, education and morality, as well as a lesson in proper expression of patriotism and Russian nationalism in the post-war setting.[39] The Soviet writers' excitement in the immediate post-war years was subdued by the devastation this decree brought upon the cultural community. It both demoralized writers, particularly those resistant to standardization in their writing or with "cosmopolite" tendencies, but also reaffirmed the ideological importance of literature in the Soviet Union.[40] In the aftermath of the decree, the imposed cultural conservatism even exceeded that of the 1930s. Despite an obsession with "remaining true to reality" and with facing and conquering, weight and scope of post-war troubles were too much to admit and acknowledge. Thus, generally, the main conflict, according to Katerina Clark, was "between the good and the as-yet-less-good."[41] There emerged a limited list of problems, which were officially acknowledged as existing, and were repeatedly and ritualistically attacked, conquered and eliminated.[42] Attempts by writers to stray from the list and raise other issues of concern in their works left them open to often vicious attacks for distorting Soviet reality, blackening the good name of Soviet children/families/schools/parents/teachers, or for giving prominence to atypical problems rather than showcasing the typical and the uplifting.[43]

The remarkable thing about this time period, however, was that despite the straightjacket of Zhdanovite cultural politics, there still were some confident children's writers determined to get their visions in print and their voices heard (or read).[44] Some children's writers had enough clout to have even their "questionable" visions published, and even praised, before a critical consensus could be reached. Another avenue for "questionable literature" was through regional and cooperative publishing houses. Looking at three cases of such explorations that were met with a general wave of criticism and occasional praise helps illustrate some fascinating aspects of post-war Stalinist culture.

A small book by Kornei Chukovskii called *Doggie Kingdom* was published by the Ministry of Agricultural Machinery's publishing house in 1946. News of this publication eventually reached the Sector for Agitation and Propaganda of the Central Committee in December. The work was cited as an example of why there was need to control small publishing houses, whose publications were often of "low ideological and artistic/aesthetic quality." The offending book told a story of two boys who enjoyed sadistic attacks on a dog. The dog escaped and went to the "Doggie Kingdom" to complain to its ruler – Ulialiai Eighteenth – who in turn brought the boys to his domain, where he forced them to experience the kind of treatment they had subjected the dog to. This proved to be an important formative experience for the two delinquents, who promised never to torture any living creature again (because there was apparently also a "Kittie Kingdom," a "Duckie Kingdom," etc.). They subsequently returned to the human world.[45] In light of the increase in violence among children after the war, the book seems to be addressing an important issue, demonstrating the importance of a humanitarian approach to all living things and condemning animal cruelty. However, the work was roundly condemned in *Kul'tura i zhizn'* in an article "Vulgarity (*poshliatina*) under the flag of children's literature" as a "glaring example of vulgarity":

> The author took a mocking (*izdevatel'skii*) approach to the important goals of raising our children. Due to the labour-corrective politics of a dog-autocrat (boys in doghouses, wearing dog-collars on chains, eating smelly rotten leftovers and having dogs ride on them), the kids become sensible, cultured, kind little boys (*painki*), and are sent home. There, in human society, they begin a new life following all the rules of dog moral code. Chukovskii is trying to pass off this lampoon as a fairy tale, and the myopic editors and publishers are distributing this composition, which offends children's feelings, and their conception of humanity. Children must be protected from works that preach zoological morality.[46]

A number of features of this attack were standard fare. By 1946, Chukovskii had become a favourite whipping boy for writing the "wrong kinds of fairy tales" after his wartime publication of an anthropomorphic allegory of the battle between Germany and the Soviet Union. Yet he remained a favourite among children and parents. For a publishing house interested in turning a profit, books by Chukovskii were very attractive.[47] Hence, the big dilemma – to publish or not to publish.[48] Yet, despite its humanitarian aspiration, if in a somewhat twisted form, the book was deemed unacceptable in the post-war cultural and political climate – "offensive" to the honour of Soviet children, despite frequently mentioned concerns about violent (and abusive) behaviour of children, especially boys during that time.

Shortly after returning from working on frontline newspapers, Ruvim Fraerman finished his book *A Long Voyage*, which was published in 1946. Prior to publication, the book was characterized as being "permeated with the sense of victory, the bright future of our school youth."[49] Advance praise came even from the usually hard-to-please Dubrovina, who talked about the character of the teacher in the book as being "on a pedestal" and able to see the potential in students.[50] This positive assessment changed after the book was published. The book told a story of an all-girls school in the year after the war ended, when the girls' favourite young male teacher returned from the front with a disfigured face. The main character – a popular and good student – cannot deal with her formerly favourite teacher, and refuses to come to his class or study, all the while agonizing about her future path in life (she wants to be a teacher) and about the relationship between morality and aesthetics. While her classmates are appropriately horrified at her behaviour, the teacher seems to understand, and uses the anti-pedagogical device of giving her a good grade despite her lack of attendance with the goal of jogging her conscience. In a review published in *Literaturnaia gazeta*, Elena Kononenko (frequent contributor on topics of morality and gender) condemned the book for lacking a reasonable "drama" and for the unjustified popularity of the main protagonist.[51] The book evoked a great deal of protest from the educational establishment – so much so that at the 1947 meeting to discuss the work of the Children's Publishing House *Detgiz*, Lev Kassil' felt compelled to defend it as an attempt to do what writers were supposed to do – talk about schools. While he personally liked it "least of all of Fraerman's books," apparently the book did resonate with the intended readership, provoking "a great spiritual movement among children" and a "stream of letters." Thus, Kassil' advocated that instead of putting the book in the "criminal" category, the novel should be argued about and discussed:

> The book is full of big feelings and now, when the post-war period has entered our literature, when we need books about morality, when we need to master the experience of war but not by descriptions of battles, but through conversations about what kind of transformation took place in our consciousness, in the consciousness of our youth, these events produced. This book has a right to exist. You do not have to think it is successful, but it should not be treated as simplistically, as it is done in some circles. It is an attempt to talk about the creation of the soul of a Soviet young woman. It is a book that forces one to think about issues that are of interest to our youth.[52]

The phrase "this book has a right to exist" is an interesting choice of words. Three years later, the book was still condemned by the educational establishment for its negative depiction of teachers.[53] Tellingly, Kassil's more public defence of Fraerman on the pages of *Literaturnaia gazeta* was couched in more conditional and critical terms.[54]

So how could one depict heroism and models for behaviour in daily life? What traits were supposed to be cultivated in children? This question emerged as a contributing factor to the long-standing and ongoing conflict between writers and the educational establishment. This was just one of many institutional disagreements between organizations with overlapping jurisdictions. In such instances it was the party that would be looked upon by both sides to determine who was right and punish those who were wrong. However, often it took some time for the party to get involved. In this case, writers viewed the educational establishment as stifling, and often enough created unflattering images of teachers in their works. Not surprisingly, such negative and disrespectful depictions irked teachers. On the matter of writers v. teachers, the May 1941 decree "On the Work of the Children's Publishing House *Detgiz*" was supposed to form a solution, setting guidelines how to depict teachers and schools in children's literature and ensuring that writers did not put any "anti-pedagogical" ideas into their books. In particular, writers were not supposed to create literature that would "take children into the sphere of emotions, that did not correspond to the feelings of Soviet children."[55] Instead, they were supposed to produce "school-themed literature" with heroic teachers and model students.

Significantly, the writers' interest in feelings, or the so-called "question of love and friendship" was met with a great deal of annoyance by the pedagogical establishment. They believed it would distract them away from socialist ideals into a "world of petty-bourgeois illusions."[56] Thus, any literature that brought such issues to the forefront became suspect. Apparently, the replacement for "tanks" was to be "morality" and not "feelings." Yet it was not made clear how to achieve this change in topos.[57] For example, at a June 1946 conference on the work of the children's journal *Murzilka*, the representative from the Komsomol Central Committee emphasized that his "mistake consists in the fact that he aspires to limit himself to the narrow, little personal world (*mirok*) of a person."[58]

Part of the problem was the fact that the literature of emotions, by necessity focusing on the inner world of people, brought out issues of psychological trauma inflicted by war – trauma that was officially, following Stalin's commandment about healing wartime scars and wounds, not supposed to exist. Exploring feelings without dealing with the costs and scars of war was ostensibly what the writers were trying to do. In light of this, the quest by a number of children's writers to inject feelings and humanity into cultural products aimed at children – and in the case of the more prominent children's writers, using their power to get them published and later be publicly flogged for them – underscored both the writers' feeling of self-importance but also their faith in the transformative power of literature.

Perhaps no book published in the immediate post-war period attempted to deal more explicitly with the scars of the war than N. Kal'ma's *Diary of Andrei Sazonov*, published in 1947. When Andrei's father was killed at the front, the boy decided to keep a diary – just as his father had done. Yet it turned out that his father had not died in battle, but rather lost his hearing and ability to speak, partially lost his sight and had his hands amputated. While all his mother's friends thought they should leave him in the convalescence home, Andrei and his mother decided to bring the father home to take care of him. The book, taking the form of a diary, deals with Andrei's attempts to take care of his father and the house, while going to school. (Andrei's mother works full time and is trying to finish her dissertation.) The book alludes to psychological trauma (Andrei's father went to the movies and when they showed a newsreel with German trophy weapons, he began to scream, but refused to leave the theatre). Andrei is essentially on his own trying to decide how to cope with his father's disability. At one point, he decides that perhaps his father can hear and tries to read and talk to him, keeping it a secret from his mother so as not to upset her. There is a happy ending of sorts. A visit from a friend, who was rescued by Andrei's father, prompts the invalid to begin recognizing first his friend and then his wife and son. His healing and reintegration process is furthered by a pair of prosthetic arms, which he receives towards the end of the book.

Kal'ma also explicitly tackles the question of "everyday heroism." One of Andrei's classmates, Lesha, wants to be a hero, but is rude to his mother, who wants him to do prosaic things like buy kerosene or bread. When Lesha behaves rudely towards an old woman on the street, she laments, "What will become of you?" When Lesha replies "A hero," the old woman countered "of you will not come a hero, they are not like this (*ne vyidet iz tebia geroi, ne takie oni*)." Lesha's indignation ("How would you know?") is interrupted by the appearance of the old woman's son – a captain in the Guards, who just came back from some prosaic errands – buying bread and getting his mother's felt boots mended. Here was a clear message of how wartime heroism was to be transferred to everyday life, and that heroic masculinity was not incompatible with politeness and respect.

The author addressed this theme in many of her works. For example, she published an article in *Literaturnaia gazeta* in 1947 "Cultivating good qualities," in which she noted that while during the war the emphasis had been on instilling public virtue in children, it was crucial to remember the importance of cultivating private virtue.[59] Though Kal'ma's novel received a positive review in *Literaturnaia gazeta* in 1948, a few years later it came under scathing attack as "completely opposing party directives" and therefore "harmful to readers."[60] And yet, according to the reviewer, the book provoked a stream of letters from children, who recognized themselves and their problems.[61]

The fundamental debate about the direction of children's literature centred around the question of what it meant to be a hero in everyday life, about defining proper modes of behaviour for Soviet children, and finally, about reconciling order and heroism. Definitions of heroism had to be adjusted to new peacetime conditions. The adjustments sought by some in the writing community came into conflict with the post-war cultural climate. Evoking emotions in "romantic" literature that would inspire children was fraught with danger. Thus, in the late 1940s and early 1950s, it was love for labour and work that emerged as a key theme and ultimate demonstrator of patriotism and peacetime heroism. Notably, Simonov's condemnation of Fraerman's book explicitly chooses to define "romanticism" as a focus on hard work and reconstruction effort in real life. Anything else he characterized as "false romanticism."[62]

Emphasis on "reality" and heroic labour feats shifted the attention of writers from the inner world of children to their participation in public life, signalling an increasing emphasis on proper appearances – form over content. This fit the needs of post-war Soviet society but contributed to the growing gulf between representation and reality. It was also emblematic of how the state chose to deal with a wide range of social problems – namely, pretend that they did not exist. The straightjacket of school discipline, at least in theory, conflicted with the street and domestic life of the children. In turn, children adapted by perfecting the art of dissimulation and hypocrisy. This disparity between what children said and what children did – "speaking Bolshevik" as opposed to "acting Bolshevik" – was gradually beginning to get noticed with increasing alarm.[63] Yet at least in the children's world, order triumphed. Only after Stalin's death would the problem of children's hypocrisy be examined in greater detail, and responsibility would be attributed to Soviet life itself.

Notes

* I would like to especially thank David Key for his extremely helpful comments and suggestions on earlier versions of this article, and the volume's editor Juliane Fürst.

1 Those groups deemed unfaithful met with an unenviable fate. On the importance of the war for the future of the Soviet system, see A. Weiner, *Making Sense of War: The Second World War and the Fate of the Bolshevik Revolution* (Princeton: Princeton University Press, 2001); E. Zubkova, *Poslevoennoe sovetskoe obshchestvo: politika i povsednevnost', 1945–1953* (Moscow: ROSSPEN, 2000); E. Seniavskaia, *Frontovoe pokolenie, 1941–1945: istoriko-psikhologicheskoe issledovanie* (Moscow: In-t rossi-iskoi istorii RAN, 1995); V. F. Zima, *Mentalitet narodov Rossii v voine 1941–1945 godov* (Moscow: Rossiiskaia akademiia nauk, Institut rossiiskoi istorii, 2000).

2 Rossiiskii Gosudarstvennyi Arkhiv Sotsial'no-politicheskoi Istorii (henceforth, RGASPI), f. M-1, op. 32, d. 64, l. 16.

3 *Sovetskaia pedagogika* (1941), no. 11–12. Fictional characters expressed similar sentiments on pages of children's books. The following exchange between an elderly collective farmer and a member of the military appeared in Valentina Oseeva's prize-winning *Vasek Trubachev i ego tovarishchi* – "Here you look, they are still children, but they already have heroes among them. And why? It's all in their upbringing, that's all there is." – "Yes, a Soviet upbringing. Builders of the future. They are going to live under communism" (V. Oseeva, *Vasek Trubachev i ego tovarishchi* (Moscow: Detgiz, 1961), 499).

4 RGASPI, f. 17, op. 125, d. 281, l. 111. It was not just the high-ranking officials in the centre who commented on this transformation (or revelation). See, for example, Tsen-tral'nyi Gosudarstvennyi Arkhiv Istoriko-politicheskikh dokumentov Sankt-Peter-burga (henceforth, TsGA IPD SPb), f. 24, op. 11, d. 178, ll. 12–13; Gosudarstvennyi Arkhiv Rossiiskoi Federatsii (henceforth, GARF), f. A-2306, op. 71, d. 67, l. 18 (1947); L. Magrachev, *Reportazh iz blokady* (Leningrad: Leninzdat, 1989) 142–143, 151.

5 For an example of such concern, see reports in RGASPI, f. M-1, op. 32, d. 64, l. 17. For a more in-depth discussion of the significance of the Russo-Finnish War, see chapter 3 of my dissertation, A. Livschiz, "Growing Up Soviet: Children in Soviet Russia, 1918–1958" (PhD dissertation, Stanford University, 2006).

6 RGASPI, f. M-1, op. 5, d. 278, l. 7. Praise began almost immediately – June 30, 1941 in RGASPI, f. M-1, op. 32, d. 6.

7 RGASPI, f. M-1, op. 32, d. 23 contains a detailed description and analysis of the nature of German propaganda in the occupied territories.

8 Mishakova in 1945. RGASPI, f. M-1, op. 5, d. 244, l. 24.

9 See, RGASPI, f. M-1, op. 32, d. 23 for a detailed analysis of the shortcomings of the Soviet system of upbringing. See also a report from an underground pro-Soviet organization to the central headquarters of the partisan movement in April 1943, describing problems with youth's behavior under occupation reprinted in B. V. Sokolov, *Okkupatsiia: pravda i mify* (Moscow: Ast-Press, 2002), 9–10.

10 RGASPI, f. 17, op. 126, d. 7, l. 41.

11 The Rules of Conduct and the decree are discussed in greater detail in my dissertation.

12 RGASPI, f. 17, op. 126, d. 13, ll. 23–24.

13 It turned out that exposure to the real-life working environment did not always have a purifying effect on young people – "they learned to swear terribly, became hooligans; they smoke, they even have beginning stages of alcoholism" (GARF, f. A-2306, op. 69, d. 3176, l. 86).

14 See, for example, RGASPI, f. 17, op. 126, d. 13, and f. M-1, op. 32, d. 64, ll. 16–18.
15 For example, RGASPI f. M-1, op. 23, d. 1446; various editions of *Pionery-geroi*. V. Gubarev's 1940 book about Pavlik was republished after the war a number of times in the period from 1940s through 1970s (and possibly even later). The book's transformation into a play by Gubarev himself was not until 1953. The Stalin prize- (3rd degree) winning poem *Pavlik Morozov* by S. Shchipachev was not written until 1951. During a 1957 conference on improving the work of the pioneer organization, Pavlik was mentioned as the model on which pioneers should be raised. (RGASPI, f. M-1, op. 5, d. 648, l. 145). When the Pioneer Book of Honor was inaugurated in 1957, Pavlik Morozov became its first entry.
16 The concept of the biologization of enemies is introduced in A. Weiner, "Nature, Nurture, and Memory in a Socialist Utopia: Delineating the Soviet Socio-Ethnic Body in the Age of Socialism," *American Historical Review*, vol. 104, no. 4, October 1999, 1114–1155.
17 *Literaturnaia gazeta* (henceforth, *LG*), 6 April 1946, p. 3.
18 For example, *Narodnoe obrazovanie* (1946), no. 7: 2. As part of the continued glorification of all things Russian, the proverbs and sayings about respecting the elders and parents were trotted out.
19 Not surprisingly, after Stalin's death the ritualized homage to veterans really took off.
20 See, for example, Arkadii Gaidar's *Timur's Oath* (June 1941). A story written by Evgenii Shvarts for Leningrad radio in October 1941, called "Leningrad Kids," described the transformation of two children, who "were not very good boys, and were not particularly liked in their building for being mischievous, but behaved heroically once the war came. ("*Leningradskie rebiata*" as cited in L. Magrachev, *Reportazh iz blokady*, 146–147.)
21 Fadeev went on location, interviewed the surviving members and parents of the executed young-guardists, and produced the novel, which received the highest Stalin prize (100,000 roubles) in 1946 and went on to become one of the most popular books of the post-war period.
22 RGALI, f. 631, op. 15, d. 822, l. 69.
23 RGALI, f. 2190, op. 2, d. 468, l. 85.
24 RGASPI, f. M-1, op. 5, d. 243 (1945), l. 13.
25 RGASPI, f. M-1, op. 5, d. 385, l. 14. Zoia – a young woman who was (according to the official version of events) tortured by the Germans and executed by them – seemed to have captured the hearts and minds of Soviet citizens, particularly young people. In 1942 Zarechnaia's *Goriachee serdtse* (a small book about Zoia) ranked third in the number of letters it received from readers (behind a book about Suvorov and a compilation *Be A Hero* (*Bud' geroem*)). (Rossiiskii Gosudarstvennyi Arkhiv Literatury i Iskusstva (henceforth, RGALI), f. 630, op. 1, d. 91.) However, according to Alexeyeva, the initially potent cult of Zoia grew stale through repetition by 1950, if not earlier. (L. Alexeyeva and P. Goldberg, *The Thaw Generation: Coming of Age in the Post-Stalin Era* (University of Pittsburgh Press, 1993), 61.) Given the self-referential nature of Soviet cultural production, few works of fiction for children written about the war fail to include at least one reference to Zoia as a motivation and inspiration to boys and girls alike, while some even make the whole plot revolve around obeisance to her. Even members of the Young Guard organization (allegedly) referred to her as their inspiration.
26 See, also, GARF, f. A-2306, op. 69, d. 3175, ll. 44–45 (1945); RGASPI, f. M-1, op. 5, d. 475, l. 36 (1951).
27 One can see how this idea had a certain amount of appeal to writers, particularly those who were already concerned with the hypocrisy in the population, especially in children.
28 O. Gonchar, *Znamenostsy* (Moscow: Sovetskii pisatel', 1952), 154.
29 See, for example, I. Vasilenko, *Zvezdochka* (Moscow: Detgiz, 1963).

30 For criticism on forgetting the war too quickly, see RGASPI, f. M-1, op. 32, d. 369, l. 84 (1946) or not using the war enough – RGASPI, f. M-1, op. 32, d. 367, l. 27 (1946).

31 RGASPI, f. 17, op. 125, d. 281, l. 113.

32 RGASPI, f. 17, op. 125, d. 280, ll. 170–171.

33 RGASPI, f. 17, op. 125, d. 281, l. 62.

34 RGASPI, f. 17, op. 125, d. 281, ll. 66, 68.

35 RGASPI, f. 17, op. 125, op. 282, ll. 103–4.

36 This topic is discussed in greater detail in my dissertation. A number of pre-war "tank" proponents died during the war, often at the front. This is not meant to suggest that there were no casualties among the "feelings" people, but simply that it is possible that someone like B. Ivanter would not have changed his position on the tank–feelings issue.

37 V. Dunham, *In Stalin's Time: Middleclass Values in Soviet Fiction* (New York: Cambridge University Press, 1976), 29.

38 See, for example, a long list of writers nominated for medals and decorations for their creative work, especially during the war, in RGASPI, f. 17, op. 125, d. 306. The list included writers such as Kornei Chukovskii, who would suffer sustained attacks in the immediate post-war period, and Lev Kvitko, who would have this very medal taken away and be executed in 1952 along with other members of the Jewish Anti-Fascist Committee.

39 A. Zhdanov, *Doklad o zhurnalakh "Zvezda" i "Leningrad"* (Gospolitizdat, 1952).

40 Once again the empowerment aspect of the decree should not be forgotten – after all, it offered ammunition and set the parameters of rounds of mutual incrimination in the cultural community as well.

41 K. Clark, *The Soviet Novel: History as Ritual*, 3rd edn (Indianapolis: Indiana University Press, 2000) 194.

42 Not surprisingly, this led to the decline (or "decay") in the quality of cultural production, and became reflected in lack of popular enthusiasm (empty theatres, etc.). (M. Zezina, "Crisis in the Union of Soviet Writers in the Early 1950s," *Europe-Asia Studies*, vol. 46, no. 4 (1994), 655.)

43 Even prize-winning books could get in trouble. Thus, Oseeva's *Vasek Trubachev*, mentioned earlier in this chapter, was awarded the 25,000 roubles Stalin prize in 1952, but criticized for including too many children from broken families – why couldn't the author have chosen some "more typical" Soviet families? (See, *LG*, 7 January 1950, p. 3 for a critique; 28 January 1950, p. 2 for a defence, and 2 February 1952, p. 3 for an attack.) Interestingly, apparently Oseeva's book was attacked from another side as well – as a training manual for traitors (because children are encouraged to report on each other) (RGALI, f. 631, op. 15, d. 1016).

44 The mercantilist motivation, of course, cannot be discounted here. However, if it was purely a question of money, more standard fair would have been a more prudent choice.

45 RGASPI, f. 17, op. 125, d. 464. The copy of the book was included in the file.

46 *Kul'tura i zhizn'*, 10 December 1946, p. 4. Of course, "zoological morality" was a clear reference to Nazis – the most recent example of "zoological ideology" in action.

47 The debate around the book was quite passionate, though often bordering on ridiculous. For example, concern as to why it was "small animals" like rabbits that seemed to be representing the Red Army in the story.

48 What is interesting to note in this case (especially since it is not clear how many people read the actual booklet, at least in this edition) was the fact that Chukovskii wrote the story before the revolution, a context in which having an animal kingdom be a monarchy made perfect sense.

49 RGASPI, f. 17, op. 125, d. 306.

50 GARF, f. A-2306, op. 69, d. 3177, ll. 82–83.

51 *LG*, 3 August 1946, E. Kononenko, "Drama bez pochvy," p. 2.

52 GARF, f. A-2306, op. 71, d. 67, ll. 39–40.
53 *LG*, 31 August 1949, B. Iziumskii, "Uchitel' v knigakh dlia detei," p. 2. In this case the criticism is particularly interesting, since it is coming from a former teacher-turned-writer, author of *Alye pogony* – a book about Suvorovite schools.
54 *LG*, 21 June 1947, L. Kassil', "Ot nas zhdut," p. 3.
55 See explanation by Dubrovina in RGASPI, f. 17, op. 125, d. 281, ll. 110–111. Notably the decree's relevance to the post-war years was emphasized.
56 RGASPI, f. 17, op. 125, d. 424, l. 69 as cited in E. Zubkova, *Poslevoennoe sovetskoe obshchestvo: politika i povsednevnost' 1945–1953* (Moscow, 1999), 138.
57 Year in and year out refrains about insufficient quality of the way questions of morality were raised on the pages of children's newspapers – rather than dealing with "big questions," there was instead fuss about little things. In particular CC VLKSM was blamed for insufficient guidance in the matter (RGASPI, f. M-1, op. 32, d. 365 (1946), d. 369 (1946), d. 401 (1946), 627 (1950), d. 665 (1951), d. 704 (1952); op. 5, d. 372 (1948), d. 512 (1952)).
58 RGASPI, f. M-1, op. 5, d. 278, l. 3. Occasionally, exploration of feelings was met with approval from some teachers, but only in very specific contexts. For example, Fadeev was praised for his depiction of teachers and for "chastely and pedagogically ethically discussing love and friendship" in *Young Guard*. GARF, f. A-2306, op. 71, d. 67, ll. 16–17.
59 N. Kal'ma, "Vospitanie dobrykh chustv," *LG*, 26 April 1947, p. 2.
60 RGASPI, f. M-1, op. 32, d. 623, l. 9. It is possible that part of the venom unleashed against the book also had to do with the fact that the author was Jewish. More generally, I have not yet been able to determine the precise "cosmopolitan" component in the attacks against certain writers, such as Kassil', Fraerman, Kal'ma, etc. (See, for example, RGASPI, f. 17, op. 132, d. 395.)
61 F. Vigdorova, "Povest' o khoroshem syne," *LG*, 19 May 1948, p. 2.
62 *LG*, 28 January 1950, p. 2.
63 GARF, f. A-2306, op. 71, d. 2, l. 147; d. 73, ll. 84–5. The term "speaking Bolshevik" is taken from S. Kotkin, *Magnetic Mountain: Stalinism as a Civilization* (Berkeley: University of California Press, 1995).

10 The importance of being stylish

Youth, culture and identity in late Stalinism

Juliane Fürst

Vladimir Gusarov, 20 years old, a good Komsomol member and son of a high-ranking Party official, described his life in the immediate post-war period in the following way:

> The general mass of people was disciplined, obedient, even demure, but I did not want to think about why this was so and for what price this was achieved.... I had the first girlfriend of my life, we kissed in all sorts of places and my clothes were simply ravishing: I had acquired an American leather coat belonging to the collection of items that came with an American 'Studebaker' ... and when I attracted strange looks then I attributed this to my personal extravagance and the eye-catching manner of my uninhibited lover.[1]

This was hardly the thought process and behaviour expected from a young-ster who had enjoyed a good communist education, was active in public life and had just survived a bruising war that had seen his family evacuated and ripped apart. Vladimir neither fulfilled the expectations of the system, which demanded self-sacrifice and political devotion, nor the assumptions of critical Western observers and émigré scholars, who anticipated a wave of youthful discontent after the upheavals of war. Instead, Vladimir was thoroughly preoccupied with his personal life and appearance. His duties as an election agitator he accepted as one of the unavoidable pains of life. Both his conformist and his apolitical beliefs existed by default rather than by choice. His fellow Soviet countrymen served as the grey backdrop to his colourful lifestyle. His identity was deter-mined by things he consumed privately – his clothes, his girlfriend and his love for the exotic world of the theatre. Yet, as will be demonstrated, Vladimir Gusarov was very much a young man of the late Stalinist mould.

The Great Fatherland War left Soviet society in a state of physical and ideo-logical devastation, confusion and contradiction. On the surface the Soviet Union had weathered the upheavals of war remarkably well. Stalin was the only one of the war leaders who remained in power by the end of 1945. The Soviet system had not only been confirmed, but legitimized by victory. The 1930s had proven themselves as a decade of successful rallying to the Soviet cause and successful construction of a Soviet identity. Consequently, all reconstruction efforts were geared towards the resurrection of the pre-war order. Yet it was clear to both leaders and population that life was not going to return to what it once had been. The system of institutions, internal structures and control organs was exhausted, overstretched and understaffed. The totality with which the regime had thrown all its resources into the war effort left it at loss once victory had been achieved. Many of the country's best young men had ventured deep into alien territories and been exposed to the life and ways of European society.[2] The lack of control had allowed the emergence of non-orthodox behaviour and in some instances led to the partial collapse of authority.[3] After the hardship and horror of sustained warfare on Soviet soil, the Soviet people felt entitled to a new beginning.[4] After years of sacrifice they expected rewards in the form of more liberal policies.[5] After prolonged existence in extreme and crammed circumstances, they yearned for privacy and entertainment.[6] It is thus no surprise that the late Stalinist years were a period of intense renegotiation between individual and system, in which both sides introduced new notions of what it meant to be Soviet in a post-war world.

Young people occupied a central place in the struggles and debates of the post-war period. On the one hand they embodied the hope for a better future. They were to reap the fruits of victory. A pantheon of young heroes served as their inspiration and rendered them eager to prove their usefulness to the motherland. They promised to be dynamic in a society exhausted by war. On the other hand this was a generation whose childhood had been marred by violence, flight, evacuation and hardship. They grew up in a period of intense ideological indoctrination, but also confusion and contradiction. Born to parents who themselves had been young in the halcyon days of the revolution, their generational rebellion could not be subsumed by political or ideological change. Denied wartime glory they were alienated by the superiority of the returning *frontoviki* (soldiers returning from the front) and looked for an independent identity. The influx of Western goods and ideas from occupied Eastern and Central Europe provided the opportunity to experiment with new types of self-fashioning and entertainment and served well in establishing both separateness from the Soviet masses and internal cohesion among smaller groups. With the onset of the Cold War the Soviet state soon considered itself under threat by young people's quest for a new post-war identity and began to narrow its definition of Soviet youth and Soviet youth culture. Young people's behaviour and thoughts became weapons in the ideological struggle that was unfolding on the world arena. In official vision youth culture had to be political, ideologically orientated and useful to the needs of the Soviet collective. The young post-war generation favoured a more inclusive definition of Sovietness. Their Soviet youth culture was to be free from strict political and ideological constraints and allow apolitical and private pursuits.

This chapter will outline three areas of contestation between young people and the state in the post-war years, demonstrating how youth attempted to renegotiate the norms of the system and how, in return, the system hardened its position. First, the vigour with which young people threw themselves into a dancing

craze soon raised concern among the authorities over the exclusive and apolitical nature of such entertainment. Second, the enthusiasm for Western cinematic
musicals led to worries about ideological corruption. Finally, the disposition of
many young people to find identity and differentiation through fashion or other
types of consumption caused a hostile campaign alerting Soviet youth to the
dangers and depravity of an apolitical and self-centred lifestyle. Rather than
communicating through the usual channels with the authorities, such as the
youth press and Komsomol representatives, the young post-war generation
argued its point through action, behaviour and display – often, even without the
conscious participation of many of the actors involved. It was only the response
from the authorities that transformed the debate into a verbal discussion, displayed in the media of the time. The debate about form and content of post-war
youth culture was not limited to a clear-cut duel with the Soviet authorities on
one side and young people on the other. The question of what it meant to be a
young Soviet was also raised by and debated among young people themselves,
pitching those who tried to find direction and ideological certainty against the
mass of more apolitical youth. The following analysis of youthful behaviour and
official responses reveals that the post-war years harbour the roots of many of
the Soviet Union's future generational problems. Yet the examination of youth's
manifold quests for identity and the diversification of youth culture in this time
also challenges our traditional assumptions of late Stalinism as the apogee of
conformity and suppression.

The post-war dance craze

There was not much entertainment for a young person in the war-ravaged Soviet
Union. Very few theatres and cinemas were operating. The cultural work of the
Komsomol had been neglected for years in favour of the war effort. Many buildings that had housed youth clubs or houses of culture were destroyed or occupied by other agencies. Yet there was a kind of youth entertainment that could
be enjoyed anywhere and everywhere as long as a harmonica player was at
hand. Post-war youth danced. At any occasion and in any place young people set
up makeshift dance floors and spent their time turning to the tune of waltzes,
foxtrots and tangos. The post-war dance craze embraced all sections of society.
Working class youth danced in the factory clubs and open dance squares, where
crippled veterans would play waltzes and foxtrots for a few roubles. They
danced at the weekend, after work and even in their lunch breaks. In the
destroyed Stalingrad, all-day dance squares began to appear, which were in use
almost 24 hours a day due to different shift patterns.[7] In the villages, because of
the absence of men, young girls had to take over the role of harmonica players
and male partners. Yet they still danced every free minute of their time.[8] Russian
pliaski (traditional folk dances) were joined by Western styles such as the waltz,
foxtrot and tango taught to the kolkhoz youth by travelling dance instructors.[9]
Schools, universities and their respective Komsomol organizations were
traditionally avid organizers of dance events.[10] In addition to the numerous

school and spontaneous street and *dvor* dances, official cultural establishments put on dance evenings in public places, with the dance hall in Gorky Park enjoying a particularly strong reputation among entertainment-hungry Moscow youth.[11]

An enthusiasm for dancing was not a new phenomenon in Soviet youth culture. Urban youth had adopted a penchant for the foxtrot already in the period of NEP.[12] In the 1930s the Komsomol press embraced the cause of proper and demure dancing attributing to it an aura of *kul'turnost* and judging its mastery an attribute of a good socialist youth.[13] Yet dance never acquired an undisputed place in the officials' vision of Soviet youth culture.[14] As early as 1938 *Komsomol'skaia pravda* noted with concern that some dancers had 'changed their names to Willi and donned jackets with "unbearably" broad shoulders, claret shoes with fringes, and hats with narrow brims'.[15] In 1941 a Komsomol commission discovered to their horror that textile workers in Orekho-Zueva spent almost all of their free time dancing and virtually none on ideologically enlightening activities.[16] Youth danced not only at the special dance evenings, but also before every showing in the cinema, when the assembled audience would switch on the radio and dance without even taking off their coats.[17]

The post-war dance craze thus built on a long tradition of a lifestyle straddling the borders of what was acceptable in the official Komsomol youth culture. Yet the availability of new music from the West – arriving in the Soviet Union in the form of 'trophy records' – and the desire to escape the seriousness of war intensified both the frequency of dances and their 'bourgeois' character. After the war dance squares and pavilions shot up in every *raion*.[18] The dance floor became the romantic meeting point of post-war *iunoshi* and *devushki*, having acquired even more importance since in urban areas schools were gender-segregated.[19] Soon dancing was found even more often on the social calendar than in pre-war times. In 1950, a Komsomol instructor noted that the club of the textile factory Petr Anismiv in Leningrad's Kirov *raion* had dances on four evenings of the week and offered no other entertainment.[20] Similar numbers were reported from Stalingrad, where the club *Kul'tstroia* planned dances on 25 evenings out of 30.[21] Russian dances only survived in the countryside – even in official dance circles traditional *pliaski* were shunned.[22] According to a foreign observer, 'the shuffling foxtrot seemed to be the national dance'.[23] Student youth fought stubborn battles with the authorities to fill their *vechera* with exclusively Western dances and ban the ballroom and traditional ones favoured by officials.[24] A Siberian cultural worker complained in 1947: 'From where does the students' love for foxtrot, tango and various rumba blues come from? Why do they not display love for Russian or ballroom dances?'[25] In some *vuzy* it was not even the conventional foxtrot, tango and rumba that dominated the dance floor, but, in keeping with the latest fashion, the jitterbug and the boogie-woogie.[26]

Despite attempts by the Komsomol to counter the trend of the 'tiring sentimental tango and the Americanized foxtrot'[27] and to urge clubs and Komsomol organizations to concentrate on 'educational' measures, the activists on the ground displayed a stubborn resistance when dealing with the question of dance and ignored messages from above. Some dared to tell officials that youth needed rest and amusement after the hard times and refused to change their programme of daily *tantsy*. Others paid lip service to the need for enlightening activities, but nonetheless continued to organize dance evenings, which they often hid under names such as *vecher molodezhi* or *vecher otdykha*.[28] In many organizations dances were used as a way to lure young people to the less popular lectures and assemblies.[29] The reluctance of the local activist to take note of the criticism from above was also fuelled by practical concerns. Aside from fulfilling public demand, dances provided clubs with a vital source of income, since young people were prepared to spend up to ten roubles for an evening of dance enter-tainment that required little or no preparation. There was also considerable anxiety that, if Western dances were taken from the programme, youth would assemble in 'private flats at friends and dance to more "powerful" and "stylish" music'.[30]

The lure of the trophy films

'The dance fever'[31] was soon joined by another traditional favourite pastime of youth – the cinema. Indeed, to the disgust of officials, the two often went hand in hand with cultural establishments promising '*kino i tantsy*' on an almost daily basis.[32] Cinema was the easiest escape from the daily hardship. Mikhail Gor-bachev remembers that he would spend his last rouble on a film while subsisting on canned beans and other 'dry rations'.[33] A large part of the films on show were wartime and post-war productions with good, educational content, promoting both patriotism and Soviet socialist values. Yet authorities were soon alarmed by the tendency of youth to flock to films that were decidedly apolitical. The worry was compounded by the fact that most of these were foreign, since in the time of the 'film hunger' – as Peter Kenez has termed the period of low productivity in the Stalinist post-war years[34] – the Soviet authorities released films captured by the Red Army in Germany and Eastern Europe in order to bring in much needed revenue. The American films *The Count of Monte-Cristo*, *Sun Valley Serenade*, *Stagecoach*, *The Roaring Twenties*, the German serial *The Indian Tomb* and film versions of several operettas all were immensely popular with young people.[35] Vasilii Aksenov confessed to watching *Stagecoach* no fewer than ten times and *The Roaring Twenties* no fewer than 15.[36] Mass popularity, not to say hysteria, was reserved for two other films: *Girl of My Dreams*, released in 1947, and *Tarzan's New York Adventures*, released in 1951. The bard Bulat Okudzhava remembered about *Devushka moei mechty*:

> it was the one and only in Tblisi for which everyone went out of their minds, the trophy film, The Girl of My Dreams, with the extraordinary and indescribable Marika Rökk in the main role. Normal life stopped in the city. Everyone talked about the film, they ran to see it whenever they had a chance, in the streets people whistled melodies from it, from half-open windows you could hear people playing tunes from it on the piano.[37]

Documents confirm Okudzhava's description about the craze the film triggered among young people across the whole country. From Gorkii to Kiev, Komsomol officials noted that young people watched the film not once or twice, but numerous times.[38] Students from Novocherkassk institutes sent a delegation of their comrades on a train journey of several hours in order to buy tickets for the film in Rostov.[39] The Komsomol secretary of the Simferopol' Medinstitut reported that his female fellow students were prone to imitate Marika Rökk's hairstyle and that the corridors of the institute were filled with the whistling of the film's songs.[40] Four years later the film *Tarzan* (made in the US in 1942, but shown in the dubbed German version) had a similarly strong impact on youth imagination and youth culture. Johnny Weismüller became an instant hero and model inspiring not only the hairstyle *pod Tarzana*, but also a cult of imitation with young people howling ape-man screams and prancing around the corridors.[41] Students were caught transcribing 'vulgar songs' about Tarzan during their seminars and even in remote villages in Belorussia Tarzan, Jane and Boy enjoyed unreserved popularity.[42]

Youth preferred colourful, Western films because they allowed them to escape the harsh and grey reality of post-war life. A young lad, asked by Komsomol officials, explained that youth valued foreign films because of their 'ability to address a light subject that entertains the viewer, gives him rest and instils in him the feeling of something beautiful'.[43] A cultural functionary in Rostov complained that while the film *The Indian Thief* drew 256,200 viewers, the season's top Soviet film *Tales about Siberia* attracted only 186,200.[44] The most successful Soviet film of the post-war period was unsurprisingly an imitation of the colourful and light-hearted Western films – the rural music comedy *Kuban Cossacks*. One of the first colour films of the Soviet Union *Kubanskie kazaki* featured a standard love plot between two kolkhoz chairmen competing at a country fair in a setting that was always happy and sunny. Maia Turobskaia has argued that the film is best understood, not as the cynical beautification of a countryside that was close to death in real life, but as a traditional fable of courtship and marriage that played on traditional narrative patterns. Rather than hiding a political message in a seemingly apolitical story-line, the film was indeed just what it promised – a 'golden dream' for an impoverished audience.[45] One of the 40.6 million viewers, who made the film the second most popular picture in 1949,[46] recalled how the film with its peasant tables bursting under food sustained him and his friends through a time of hunger and hardship. 'When we were hungry, we watched the food buffets in *Kuban Cossacks* – that was our happiness'.[47]

Followers of fashion

While thus a decidedly apolitical and entertainment-hungry culture was on the rise among post-war youth in general, some young people indulged in these trends in an intensified and eye-catching manner that seemed to suggest that indeed differentiation, not pure entertainment, was their driving force. Building on the general craze for all things Western and entertaining, these youths added to an increasingly apolitical youth culture a culture of individualism. While certainly neither the first nor the only one, the youth subculture of the *stiliagi*, which flourished in the larger Soviet cities from the late 1940s onwards, has become the epitome of this trend. Yet, as the following discussion will show, they can only be understood properly in the context of their peer group, which by no means provided an antithetical background to their exalted behaviour. Rather, *stiliagi* culture built on a general trend apparent among a great number of youth. Non-conformist behaviour, probing the borders of what was possible, displayed many shades and variations, most of which have disappeared without a trace due to their ephemeral and temporary nature and the subsequent oversimplification of all alternative forms of behaviour under the term *stiliachestvo*.[48]

The confusion of war had weakened the tight grip Komsomol and authorities used to have on Soviet youth culture – both de facto and, even more important, ideologically. After the war there were simply fewer institutions and personnel available to implement the vision of the perfect Soviet youth. Yet even the vision itself had become confused and fuzzy on the edges. What was to define Soviet youth culture in a post-war Soviet Union? How to incorporate new values, norms and traditions shaped during the war? What to make of concessions granted in the heat of fighting and what to do with the new challenges arising out of the involuntary exposure of Soviet youth to other cultures? This vacuum of ideological direction and concrete directives was quickly filled by young people with a variety of behaviour and cultural markers, which, probing the elasticity of traditional Soviet norms, saw no contradiction between Sovietness and the pleasures offered by Western style and entertainment.

While conscripts and officers stationed in Germany and Eastern Europe brought Western trophies to the Soviet Union, it was young adolescents who had stayed behind the lines during the war who really leapt onto the new dress and entertainment culture. The original bearers of the imported goods either moved on to respectable family life or had brought the Western artefacts only as prestige trophies. Youngsters like Aleksei Kozlov built a good jazz collection by stealing records at school dances to which unsuspecting *komsomolki* had brought their fathers' treasures. Kozlov excused his thieving of real American jazz such as Glenn Miller and Benny Goodman by claiming that for their owners they meant nothing, while for him and many of his contemporaries they meant the world.[49] From 1947 onwards youth's access to American music was fuelled by extensive music broadcasting of foreign stations. The fact that young people listened to these stations (despite a severe shortage of airwave receivers and jamming from 1948 onwards) was borne out in several attacks on the programmes by the press.[50] Multiple personal and archival testimonies confirm the success of foreign radio. The film director Andrei Konchalovskii answered the question of what influenced him most in adolescence with 'the Voice of America, Radio Svoboda and the BBC', whose programmes he remembers as 'frightfully entertaining'.[51] Aleksei Kozlov and his fellow *stiliaga* Victor Kosmodam'ianskyi testified to the enormous popularity of Willis Conover's 'Music

USA', transmitted in English by the Voice of America.[52] While both Kozlov and Kosmodam'ianskyi were dedicated music lovers with some quite fancy equipment, the BBC and Voice of America were well-known entities to students, workers and even kolkhoz youth.[53] The complaints of Komsomol officials and occasional articles in the press indicate that fast Western music, which encouraged fast and stylish dancing, had made it to almost every youth club and was received with great enthusiasm, even by those who otherwise did not break the normative code for Soviet youth.[54] The newly established position of the *massovik*, who was to monitor the programme of official evenings of entertainment, was not quite without its dangers, as the following remark made in an assembly of trade union workers demonstrates: 'There are still incidences, when our masters try to eradicate these Western dances from the dance squares, and they risk virtual stoning'.[55]

Western clothes were another desirable item that made its way along the West–East trading route. Hundreds of thousands of conscripts furnished themselves with Western-style outfits,[56] which with some luck could turn up in commission shops in Moscow and Leningrad, where foreign correspondents and diplomats deposited some of their old garments. Fashionable Soviet youth, imitating Western – or what they believed to be Western – style was spotted not only in Moscow and Leningrad, but all over the country. In the Ukrainian Zaporozhe *oblast'* a *raion* Komsomol secretary deplored the use of lipstick and eye pencil by Komsomol girls dressed in Western clothes.[57] In the Stalingrad *oblast'* girls went to Moscow to obtain the fashionable 'Meningitis hat', a little cap perched on the corner of the head, while boys were keen to obtain trousers and jackets with zips both on the hip and on the breast – a vague imitation of American uniforms.[58] Even in a rural *oblast'* such as Riazan' the Party noted with concern that 'certain groups literally worship everything Western in their appearance and dress code'.[59] Since original items of Western clothing were hard to come by, the second best solution was a tailoring atelier ordered to imitate the desired style. The practice was so widespread that *Krokodil* published in 1948 a spoof diary of an atelier worker, in which not only the outlandish tastes of society women were highlighted, but also those of their spoilt teenage daughters.[60] Less well-to-do youngsters asked their mothers to tighten their trousers and widen their ties.[61]

Young people, however, not only picked up the material goods, which came in the baggage of the returning *frontoviki*. Rather, it was the *idea* that difference was possible and desirable that left the largest impact. The glimpses of an entirely different world, which seemed bright and vivacious, captured the imagination of teenagers and induced them to create a fashionable world on their own. In May 1946 *Komsomol'skaia pravda* reported from the factory club of *Uralmash* in Sverdlovsk that some young workers had taken to the habit of giving themselves trendy names such as Kleka, Mike and Koka, which they had traded against their traditional Russian names in order to be 'up to scratch'. Alongside their strange names they sported clothes of the latest fashion such as wide trousers and broad shoulder pads, strange haircuts including what would today be called a Mohican and, most remarkable, the trend of having one of their front teeth covered by a gold crown.[62] While such fashion had its roots more in the traditional chic of the Russian criminal underworld than in foreign influences, the trend to lap up new ideas and differentiating features is already present. Moreover the girls and boys in question believed themselves to be followers of Western fashion. This was clearly a misperception, yet a very revealing one. A picture even more closely resembling a fully developed subculture along Western lines was painted by *Komsomol'skaia pravda* journalist Garbuzov in his April 1946 article 'An evening in the Gigant'. Again the setting is a factory club, in which a minority of dancers, dressed in 'trousers, which ended above their boots and on the head a cap with an ugly peak not thicker than two fingers', danced something called the '*linda*', which involved 'throwing legs unnaturally to the side', while 'the upper body remains lifelessly rigid'. The female partners of what the author nicknamed the *koroly* were heavily made-up, wore artificial hairstyles, perms and vulgar bracelets around their wrists.[63] A cartoon in *Krokodil* of the same year depicts the urban and more affluent counterpart to the fashion-conscious worker of Uralmash and Iaroslavl, who, while more sophisticated in his choice of clothing, demonstrated nonetheless the same 'unhealthy' preoccupation with 'standing out'. The caption reads: 'I yanked a world class suit – swell, the tie – groovy, and the shoes make you dizzy! Now I'm cultured – vow'.[64] From here it was not far to the *Broadway* culture of the *stiliagi*, whose fame and prominence rests more in their discovery by the Soviet press than in their extraordinariness.

The *stiliaga* was the culmination of a style evolution that drew on both Russian and Soviet sources as well as new impetus from the West. As such he (or more rarely she) represented the trendy pinnacle of a youth society for whom fashion and entertainment were vital ingredients of their self-identity. Contrary to common perception, *stiliagi* culture was not a reserve of the rich and privileged. Most available accounts give no indication that the followers of the subculture came from wealthy backgrounds or that they imitated richer friends. Neither the saxophonist Aleksei Kozlov nor the author Vasilii Aksenov, famous chroniclers of the culture, grew up in privilege. Arkadii Bairon, whose police interview about his life as a *stiliaga* is located in the former Komsomol archive, remembers that he had to quit school in order to support his family – a fact that did not stop him from becoming a style-conscious young worker with the appropriate haircut and clothes from 1952 onwards.[65]

Looking at *stiliaga* culture it is important to remember that it was a subculture in flux with fringes and sub-subcultures and a coherence that was based on nothing other than 'being different'. The fact that *stiliaga* culture lay indeed at the junction of pre-war entertainment culture and wartime and post-war Western encounter is apparent in many of the markers with which they labelled themselves. The word *stiliaga* itself was the creation of the authorities – most likely coined by the *Krokodil* journalist Beliaev – and was judged offensive by the *stiliagi* themselves.[66] Indeed, before the start of a more concerted campaign against *stiliagi* under Khrushchev, several terms seem to have been in circulation. Apart from the aforementioned *koroly*, the term *poprygunshiki* – cannot-sit-stills – was circulating in Leningrad and applied to young people, who, in the words of an activist 'twisted every dance in their own fashion' and were threatened with expulsion from official dances.[67] Self-confessed *stiliagi* refused to adopt a label, identifying via a common appreciation for jazz, a broadly outlined dress code and a special jargon almost incomprehensible to outsiders. They called themselves *chuvaki* and their girls *chuvichi*, they *batsali stilem* (danced with style) at their *khatakh* (parties) and they *bashliali* (bought) their *shmotki* (clothes) from the *faltsovshiki* (speculators) on the market in order not to look like the *zhloby* (normal people).[68] This language was rooted in the slang of professional – and often Jewish – musicians of the revolutionary and pre-war period and interspersed with American names and words indicating the Western influence. The *stiliagi*'s main meeting point, the left side of Gorkyi Street running from Pushkin Square down to *Okhotnyi riad*, where new clothes were paraded and information about music and parties exchanged, was known as 'Broadway', 'Main drag' or simply 'Brod'.[69] A sub-subculture of *Stiliagi*, which seems to have appeared around 1952 – the *Shtatniki* – Americanized the jargon even further with words like *trauzera*, *khetok* and *shuznia*.[70]

The most important marker of the *stiliagi* were their clothes – the right outfit was the ticket into the main crowd. As Aleksei Kozlov remembers, if he 'was not going to change his look, nobody (of the Broadway regulars) would recognize him as one of their own'.[71] Again, *stiliaga* dress style betrays borrowings both from the Russian/Soviet past as well as from the Western present mixed in with true *stiliaga* creativity, which was often the result of botched imitation attempts. The early *stiliaga* in Beliaev's feuilleton still wears the wide trousers associated with the sailor and thief culture which Kozlov evoked in his *dvor* descriptions and which the *Komsomol'skaia pravda* correspondent encountered in the *Uralmazh* factory club.[72] Yet, by the early 1950s, the 'Broadway' look was tight trousers with extremely wide flares measuring up to 22 centimetres. Similarly the colourful appearance of Beliaev's *stiliaga* with an orange-green jacket, canary-pea green trousers and socks 'resembling the American flag'[73], which stood out from 'the drabness and often raggedness of the clothing of the masses'[74] in 1949, was toned down to a more quality-orientated scheme in the 1950s, when the rest of society started to brighten up. When Aleksei Kozlov hit the Broadway, a long jacket with closed pockets and a white silk scarf were the attributes of the in-crowd.[75] In 1954 the *stiliaga* Arkadii Bairon described the current fashion as

> a coat with arms in style called Reglan. At the front there is a double coquette on which buttons are sown ... The most fashionable jackets now have one button and are very long, moreover and broad in the shoulders and tightening lower down. The arms have extended cuffs ... To wear broad ties is not in fashion now, therefore I ask my mother to tailor every tie I buy in a shop down in order to make it narrow. Very fashionable are shoes with thick soles.[76]

As is apparent from these accounts, two of the most significant items of the *stiliagi* dress collection were the elaborately made shoes, which could weigh two and a half kilograms[77] and their ties, which, from the early 1950s, were decorated with jungle scenes featuring monkeys, palms or other exotica – in reference to the cult of Tarzan.[78] The film starring Johnny Weismüller with long hair combed back created not only the *Tarzanet*, young people imitating Tarzan, but also gave the *stiliagi* a name for their trademark long hair, which they now wore 'pod Tarzana'.[79] While Western films, especially the new Italian cinema of the 1950s, were a major source of inspiration for style-hungry youngsters, the possibilities of developing new trends were endless. Observation and minute copying of clothes worn by Western delegations visiting the Soviet Union was a well-known practice. More surprisingly, the official Soviet fashion journals also featured on the style horizon of young *stiliagi* – after all their fashion was inspired by the very same Western couture that was judged so detrimental when aped by independent youngsters.[80]

Alongside jargon and dress, music and dance figured as pillars of *stiliaga* culture. Here again, it is apparent that the Broadway culture of the late 1940s and early 1950s was based on developments among youth at large, but dared to take things a step further. The *stiliagi* insisted on dancing 'with style' and listening to authentic, rather than sovietised jazz. They obtained records on the black market, recorded songs from the BBC and Voice of America and organized underground concerts, whose staging was known only to a select crowd of people by word of mouth.[81] The *stiliaga* Arkadii Bairon in a statement to police claimed an intrinsic connection between jazz music, *stiliaga* style and – implicitly – the return to a peace-time society. Denying the common assumption that it was the film Tarzan (whose impact on youth behaviour has been discussed above) which had shaped the specific hairstyle of his set of peers, he – in the unreflective manner typical for many *stiliagi*[82] – simply points to the simultaneous emergence of jazz and 'long, straight hair, brushed into the forehead' in 1946, which also happened to be the first year untainted by war and witness to the largest number of demobilizations and repatriation. 'This year I consider the year when jazz music blossomed here and hence these (*stiliaga*) hairstyles appeared'.[83] Dance, too, was a marker that was enacted in the framework of mainstream youth culture and then taken to new limits. *Stiliagi* testified that many of their dance evenings were indeed officially organized dances, at which they would assemble and engage in stylish, eye-catching dancing.[84] They pushed the limits of the acceptable by dancing in a stylized, sensual way, annoying the organizers and other dancers. Yet their surroundings were not always hostile and *stiliagi* relied on willing Komsomol officials to grant entry to official evenings (even though the practice of changing clothes inside the venue after having passed the bouncers became common practice in later years).[85] Their faster dances such as the jitterbug and the lindy hop inspired student bands to explore quickly tempered jazz and early rhythm and blues.[86] Thus while, since differentiation was the goal of the game, the *stiliagi* needed a backdrop of ordinary youth, *stiliagi* could only enact their practices if the right music was played and the right kind of crowd admitted. Rather than presenting an exception from mainstream youth, *stiliachestvo* was intrinsically intertwined with it.

It was this diversity of 'stylish' trends that allowed a wide variety of political attitudes and self-perceptions. Clearly the majority of the trendsetters were not only at the pinnacle of fashion, but also represented the height of political indifference. Indeed, their very *stiliaga* existence was based on the fact that they had banished politics to the absolute fringes of their life. Slavkin[87] conceded that the *stiliaga* were, in general, 'common kids, simple boys, the majority of whom were not intellectually gifted and few who could formulate their position on social question or a political opinion'.[88] Yet, while Victor Kosmodam'ianskyi up to this day insists on the absolute apolitical nature of the trend, which he followed happily alongside his membership in the Komsomol,[89] some of his contemporaries were vaguely aware of the rebellious implications of their stance against conformity and an official culture soaked in politics and ideology. Konchalovskii recalls in his memoirs the provocative attitude of his friend Iulik, whose father, a close associate of Bukharin, had spent many years in the camps. At a time when Jazz was unacceptable in public he asked an orchestra in a Black Sea resort to play the Chattanooga Choo-Choo and sang himself in English causing a mighty scandal, in which a Red Army officer shot wildly in the air in protest.[90] Aleksei Kozlov was not only non-conformist in his choice of dress, but extended his risqué lifestyle to literature after having been introduced to forbidden authors such as Remarque, Huxley and Babel by older *chuvaki*.[91] At least one of his 'Broadway' friends was an 'ardent anti-Sovietnik', who had squared his cultural dissatisfaction with his political beliefs.[92] The personage of Boris Pustintsev, who encountered the *stiliagi* scene in St. Petersburg with firmly anti-Soviet views and saw a connection between his dress, his disdain for the Komsomol and his political convictions, was an exception. His friendship circle, with whom he discussed politics and who were to become his co-accused, did not overlap with his peers from the Leningrad Broadway.[93] Ironically it was often privilege rather than external dress code that seemed to inspire anti-Sovietness. Vasilii Aksenov encountered among the *shtatniki* an open disdain for the Soviet Union, which initially shocked the trendy, but naïve lad from Kazan.[94] Yet, even for a politically aware youngster like Aksenov, whose mother had been arrested in the purges,[95] it was only in American exile possible to conclude that 'when you think about it, *stiliagi* were the first dissidents'.[96]

Negotiating identity

Youth officials from Party and Komsomol arrived at Aksenov's conclusion much faster. Always suspicious of apolitical and light-hearted entertainment the authorities soon attempted to reclaim lost ground and force the return of a more collectively minded and ideological youth culture. Apoliticalness, individualism and self-centredness were all charges that started to appear on a regular basis in meetings and press. The issue of youth culture became inextricably linked to the campaigns of the Zhdanovshchina, the Roskin and Kliueva affair, the anti-cosmopolitan affair and the general ideological competition that was ensuing on the world stage. After a lull in ideological work during the war, when propaganda was deliberately inclusive to gather all strengths available, 1946 saw the first return to a campaign championing ideological purity. Zhdanov in his attacks on Akhmatova and Zoshenko was keen to underline the detrimental effect their work had on youth. While the scandal surrounding the supposed selling of medical secrets to the West by the cancer researchers Roskin and Kliueva and the anti-cosmopolitan campaign were primarily directed against the established intelligentsia (especially Jewish intelligentsia) and their alleged liability to worship the West, both campaigns were keen to involve young people. Young people were feared to be victims of Western influence, while at the same time they were expected to take on the role of persecutors of their contemporaries and – more controversial, but also more enticing – their elders.

The discussion of the dance craze in the Komsomol press and among Komsomol officials clearly mirrors both official fears about dance as a catalyst for Western corruption and the desperate attempt to reconcile youth's most popular pastime with the requirements and values of an officially sanctioned Soviet youth culture. Realizing the necessity and inevitability of accepting dancing as a part of a Soviet adolescence, the authorities went to great lengths to attack certain aspects of dancing without ever clearly questioning it as a general form of entertainment. Concentrating on the public spaces in which dancing took place and which were damned for their spontaneous, unsupervised nature, the numerous Party and Komsomol discussions clearly reveal the fear that a dancing youth was beyond their realm of influence – physically and mentally. 'Dances in clubs and open squares – what are they?' asks the Party secretary of the Stalingrad *oblast'*, 'They carry no educational value. . . . In every *raion* we have open-air dance floors, where youth is assembling. And how do our organisations assert their rule there? Not at all. Youth is all on its own. They come and go and nobody is following them'.[97] The adult public fuelled such concerns with letters, in which they warned about the 'inexcusable damage' done to youth by 'daily dancing' and reminisced about the diverse cultural work clubs used to perform in their own youth.[98] In general, dance floors became associated with uncontrolled, hooligan and sexual behaviour,[99] which was viewed as a direct consequence of the pervasiveness of Western dances. 'The choice of records consists of unimaginative (*bezidedeinyi*) products for the performance of the Kozin or Western European dances including the Rumba. It is therefore no coincidence that apolitical attitudes and amorality are wide-spread',[100] reads a Komsomol report from 1952. An official in Leningrad proudly reported that the number of Western European dances had been reduced from making up half of the programme to four per evening and were to be phased out. The sceptical reaction and frantic proposals for correct lessons, control and tight supervision he elicited from his colleagues demonstrate the helplessness officials felt in the face of youth's continued enthusiasm for foxtrot, rumba and the slow waltz.

The director of the marble hall at the Palace of Culture pointed to his experience. When dance evenings were replaced by so-called cultural evenings, consisting mainly of educationally valuable programmes, attendance dropped catastrophically. In December 1952, 800 to a 1,000 people came every day to the marble hall. After the changes the number in March and April plummeted to 300 to 400 and continued downward in the following weeks. Letters and oral complaints expressed general dismay about the banning of the tango and the slow waltz.[101] Young dancers moved in a world to which the Komsomol – as an organization devoted to 'useful' and 'enlightening' activities – had little access. 'We have not yet gained access to the free time (*dosug*) of youth ... we are still so far from youth', admitted a Komsomol official from Leningrad, who also conceded that views from activists lower down the hierarchy did not necessarily overlap with those of the centre, making the implementation of a truly 'Soviet' time even more difficult.[102]

Given such insecurities, it is obvious that the Komsomol could not tolerate any entertainment universe that essentially ran parallel to its own provisions. The popularity of the trophy films was thus another deep source of worry, since they not only induced young people to spend time watching a 'worthless' film, they also instilled them with precisely those apolitical and hedonistic thoughts that were so directly opposed to the Komsomol mentality of hard work and personal sacrifice. However, since the films were sanctioned from above, the press remained stonily silent, with the exception of a collection of readers' letters in *Kul'tura i zhizn'* in March 1947, which complained about the films' 'vulgarity'.[103] The policy perplexed many Komsomol officials, activists and the more politicized part of youth. It was only during the discussion about the Roskin/Kliueva affair, concerning the two cancer researchers accused of having sold secrets to the West, that trophy films came into the firing line. 'In our cultural propaganda work we speak very little about the capitalist encirclement', complained the VLKSM *oblast'* secretary at a discussion of the affair in Zaporozhe before wailing against the lies of American radio and film.[104] Other activists agreed with comrade Ukrainets from Zaporozhe. 'These films are very harmful', concluded the Crimean assembly, while in Kursk it was reported that, 'youth does not understand that *our* films are the best in the world'.[105]

Youngsters who had forgotten about the superiority of Soviet culture became objects of particular official scorn. In order to separate such 'lost souls' from the healthy part of youth, who while not yet fully 'toadying' to Western culture were susceptible to seduction, the young *koroly, stiliagi* and Jacks, Johns and Kletas were ridiculed and portrayed in the press in the worst and most unattractive manner. Already the names given to such budding subcultures were supposed to indicate the silliness of their lifestyle. The term *stiliagi* was a journalistic invention with clear negative overtones.[106] The same was true in all likelihood for the *koroly* (kings) of the club 'Gigant'.[107] The ridicule was intensified through the repeated portrayal of the ignorance and stupidity of young people devoted to dance and fashion. 'They have nothing left in the head', said a young worker about the *koroly* at the Gigant. 'Everything went into their legs'.[108] Young Soviets, who had been brought up believing in the elevating properties of education, were supposed to be deeply repelled by *Krokodil's*

description of the ignorance displayed by Beliaev's *stiliaga*, who believed Hein-
rich Heine to be a French king and the St. Vitus Dance another new addition to
the Western European dance repertoire.[109] Fashion-conscious youth was linked
to a whole host of other nasty attributes. They were portrayed as spoilt and lazy,
exploiting their hard-working parents. They were accused of speculation, a dis-
honest lifestyle and overtly sexual behaviour. At the 1954 Komsomol Congress
they were even suspected to be American informers and spies ready to betray
their fatherland.[110]

However, it was not only the regime that debated the direction and nature of
post-war youth culture and its implication for the Soviet collective. Confused by
the different signals given from above and puzzled by the contradictions
between official demand and social reality, different positions of what it meant
to be a Soviet youth emerged among Soviet youth. While the trend to apolitical
entertainment was widespread, not all young people considered dancing, cinema
and fashion as central to their youthful lives. Indeed, there was a significant
minority who were outright appalled by what they considered to be the devalu-
ation of communist youth culture. Yet rather than siding with the authorities in
the fight against 'hedonists', they considered Soviet officialdom implicated in
the changes. Concessions, laxity and admission of unprepared youngsters in the
Komsomol had, according to their interpretation of events, brought about the
current malaise. Ideologically conscious youths would raise the issue of
'depraved youth' outright in meetings and self-published journals. They attacked
declining ideological standards among young people, deplored vanity and self-
fashioning and ridiculed those who put their private interests above the collective. Instead they called to a return to the values of revolution and war, inspired
by a mixture of ideological fervour and adventurous romanticism.[111]

While, as both memoirs and documents testify, there are numerous examples
of such debates flaring up among Soviet youngsters in the post-war period, the
example of a group of pupils from Astrakhan shall stand for countless other
instances, which involved similar views and visions, but are rarely as well documented. The group of friends consisting of three boys and two girls believed that
youth had veered away from a culture worthy of the Revolution. Youth had to
rediscover patterns of behaviour that would distinguish it from 'bourgeois'
youth. They took specific issue with the habits of their classmates in an open
lecture and discussion – approved and sanctioned by the school directorship and
local Komsomol – on 'Negative phenomena among our youth'. Their scorn was
in particular directed against the post-war popularity of operettas, which they
considered vulgar entertainment unworthy of Soviet people, and the flirtatious
nature of post-war gender relations, which they characterized as silly and superficial. They hailed Maiakovskii and tried to emulate the coarse vocabulary associated with the time. Inspired by the novel *Molodaia gvardiia* they named the
phenomenon of – in their opinion – 'un-Soviet' youth behaviour Filatovshchina
after the novel's character Valia Filatova, who emerges from the novel as a flirt
and a philistine.[112] The main ideas of the lecture were followed up in a self-
produced almanac titled: '*Bonzai samokritiku*'. Several more students

contributed to the publication by responding either positively or negatively to the ideas aired in the lecture.

As is apparent from the contributions to this piece of *samizdat'*, youth was by no means united in its perception of Sovietness. Many youngsters were quite ready to champion individualism over the old-fashioned notion of collectiveness. In a letter to the authors a girl expressed her disagreement that Maiakovskii was essential to a true Soviet youth identity. 'We girls can love Simonov, Esenin and Shchepachev and still be Soviet', she wrote. Another one defended her love for operettas, which she sees as a legitimate place to 'recreate and think of nothing'. Yet, another group of girls took issue with the accusations that bourgeois forms of politeness detract from the revolutionary spirit that should prevail between men and women. 'A girl should not be helped into her coat? She should not be treated to ice cream or the theatre? Why not? There is nothing dirty about this!' Another letter took a practical approach to what was important to contemporary youth: 'All that we pupils need now is a golden medal ... It is permitted not to read what is written about Indonesia. It is better to learn the foundations of Darwinism, so that one knows everything that one needs for the ten-year exam'. Finally, another letter exposed most pointedly the two opposing directions, in which youth searched for post-war identity and culture:

> You think that you run ahead of life, but in reality you are about 25 years behind. Maybe after the revolution one could scream 'Down with the operetta!', 'Build new relationships!', but now? No, this is not the time, my good old people! You propose to spit on the operetta, study only 'Anti-Duhring ??' and discuss politics? How boring are your ideals, your narrow soul and interests! How can a person live without jazz, funny songs, dance and laughter?

Self-confidently the writer assures the group that the 'world order is not so stupid and dirty, and yes, it will hardly change'.[113]

This identification of a generational gap that ran right across youth – and which was not necessarily linked to actual age – was a topos visible in many memoirs and other subjective accounts. While ideological conviction, with its polar points of sentimental revolutionary zeal and hedonistic, modern pleasure seeking, was a strong divider, the experience of war was also often cited to explain the growing rifts and sudden changes taking place in youth culture. Liudmilla Alexeyeva recalls in her memoirs the gap between the ideologically rigid *frontoviki* at the history faculty of Moscow and the younger or female students, whose critical questioning of ideology and reality made them more likely to distance themselves from certain aspects of regime policy.[114] Similarly the secretary of the Komsomol organization at the Leningrad Electro Technical Institute in the 1950s, B. Firsov, saw a clear contradiction between the values which had been introduced during the war, and the 'primitive' values of the *stiliagi*. He explains the sudden changes taking place at the institute during his tenure with the fact that people of his time were still deeply influenced by the spirit of the *frontoviki* generation, which then 'dominated the human landscape of the institute', while the following generation had lost this spiritual connection. For him and his contemporaries the *stiliagi*, who appeared in growing number in his institutes, were the antithesis of war-related virtues such loyalty, sacrifice and obedient collectivity.[115] Implicit in his statement – even though in all likelihood completely outside his imagination – is the fact that indeed the glorification of war participation excluded the young post-war generation from participating in an all-dominant identity. Both the young radical revolutionary ideologues and the hedonistic, fashion-conscious *stiliagi* were expressions of this lack of identity for a generation that was virtually squeezed between the dominant myths of war and Thaw and located on the watershed between two distinct periods in Soviet history. The author Vladimir Britanishskii, himself representative of a younger generation that identified with Krushchev's reforms, aptly called his predecessors the *mezhpokolenie* – the generation that was in between and neither here nor there.[116] In short: a generation that was on a permanent search of itself and its place in the system they lived in.

Conclusion

Contradictory and diverse as young people's views and actions were, they all indicate that the place and role of the youthful individual in the Soviet post-war collective was hotly debated and, depending on the standpoint of the analyst, either open to renegotiation or in need of correction. The very concept of Sovietness, painfully constructed during the 1920s and 1930s, had been shaken and put into question by the upheavals, sacrifices and consequences of war. Having missed the age of revolutionary iconoclasm and unable to match the glories of action at the front or with the partisans, the young post-war generation was in need of an identity of its own. A decline in institutional control and the influx of Western goods and ideas facilitated the rise of individualism in youth culture and behaviour – a tendency that caused an outspoken backlash towards more radical collectivism among another section of the same generation. However, young people increasingly considered it their right to express and distinguish themselves through personal markers. While rarely consciously directed against the norms of the Soviet state, these actions nonetheless signified an attempt to extend the public notion of Sovietness and integrate apolitical and private pursuits into an acceptable Soviet lifestyle.

Such an elevation of the apolitical and private elements of life necessarily had important repercussions on the power relationship between the individual and the collective. If private, and thus individual, desires and lifestyle options were respected as integral to general Sovietness, then the system inevitably had to curtail the totality of its demands on the time, thoughts and actions of the Soviet individual. In the atmosphere of the early Cold War this was not an option. It was thus not surprising that any form of individualism was soon bedevilled as selling out to the West. Yet, rather than killing the nascent trend, the suppression of private and apolitical pursuits caused a split between youth and authorities and between the youth cultures they advocated and practised. Ultimately, the desire of the Soviet subject to align his or her personal self with the collective demands gave way to a compartmentalization of life and a hollowing out of the official structures. The obscure and opaque nature of late Stalinist campaigns – the concept of a *kosmopolit* was far more elusive than that of an enemy of the people – failed to capture youth and unite them under a major idea.

When Stalin died the system had lost much of its control over what it meant to be a young citizen of the Soviet Union. In the following years official and unofficial youth cultures grew apart, separating Soviet life more and more into private and public spheres. The early years of the Thaw featured both an intensification of youth's identification with individual style and a renewed attempt by the regime to regain control over the spaces and spheres inhabited by youth. The three years between Stalin's death and Khrushchev's secret speech saw the debate over what it meant to be a young Soviet and how individual and collective rights and needs break into the open with a vengeance unexpected by the authorities. There was nothing new in the questions asked and the behaviour displayed. Yet, freed of the hush-hush nature of debates under Stalin and encouraged by a press that for the first time acknowledged burning problems, young people reinvented themselves and their position within society with enthusiasm. The routes chosen were manifold. There was an explosion of *stiliachestvo*, which found new notoriety through numerous press articles, as well as a plethora of groups concerned with the renewal of the revolutionary ideal. There were youngsters who believed in their right to consume modern art, and youngsters who campaigned for a cleaner and leaner Komsomol. Khrushchev's answer to his growing youth problem was an attempt to recapture that space, which was the interface between individual and collective culture. He tried to regain control over the youth clubs, cleaning them of hooligans, *stiliagi* and unwanted music. He wanted to make the Komsomol more interesting by establishing self-rule that included the creation of so-called patrols to fight non-conformist elements. Official culture attempted to embrace youth's desire to travel both close to home and far away, to stage self-fabricated drama and sing self-composed songs. Yet, in the end the forge between system and young individual, which had started after the war, proved too advanced to recapture. Official culture – no matter if it called for youth to explore the new spaces of the Virgin Land, asked young people to build a new railway through the Siberian wilderness or subscribed to the toleration of rock and pop – always found itself a step behind the newest trend.

Notes

1 Vladimir Gusarov, *Moi Papa ubil Mikhoelsa*, Frankfurt am Main: Passev, 1978, p. 90.
2 See: Mark Edele, 'Strange young Men in Stalin's Moscow: The Birth and Life of the Stiliagi, 1945–1953', *Jahrbücher für die Geschichte Osteuropas* 50 (2002), pp. 37–61. Juliane Fürst, 'Stalin's Last Generation: Youth, State and Komsomol 1945–53', PhD thesis, University of London 2003, p. 87–89.
3 See: Jeffrey Jones, '"People without a Definite Occupation": The Illegal Economy

and "Speculators" Rostov-on-the-Don', in Donald Raleigh, *Provincial Landscapes: Local Dimensions of Soviet Power, 1917–1953*, Pittsburgh: University of Pittsburgh Press, 2001, pp. 236–254. Vladimir Kozlov, *Massovye Besporiadki v SSSR pri Krushcheve i Brezhneve*, Novosibirsk: Sibirskii Khronograf, 1999.

4 See for example: Elena Zubkova, *Obshchestvo i reformy 1945–64*, Moscow: Rossiia Molodaia, 1993. Elena Zubkova, *Poslevoennoe sovetskoe obshchestvo*, Moscow: Rosspen, 2000.

5 There was a widespread rumour that kolkhozes were to be abolished after the war (Zubkova, *Obshchestvo*, 5, 61–69); on the issue of religion there were also expectations that the tolerant wartime policies were to continue, see Daniel Peris, ' "God is Now on Our Side": The Religious Revival on Unoccupied Soviet Territory during World War II', *Kritika* 1 (1), Winter 2000, 97–116.

6 See: Juliane Fürst, 'Stalin's Last Generation: Youth, State and Komsomol 1945–53', PhD thesis, University of London 2003. Anna Krylova, 'Healers of the Wounded Soul: the Crisis of Private Life in Soviet Literature 1944–46, *Journal of Modern History* 73 (June 2001), 307–331. Don Filtzer, *Soviet Workers and Late Stalinism: Labour and the Restoration of the Stalinist System after World War 2*, Cambridge: Cambridge University Press, 2002.

7 Interview Maria Ivanovna, born 1932, Volgograd, 27.9.2001.

8 Interview Tat'iana Lavrenova, Ustran', *Riazan'skaia oblast'* 13.1.2001.

9 Ibid. Interview Maria Zabotina, Spassk, *Riazan'skaia oblast'* 19.11.2000.

10 Mirra Ginsburg (transl.), *The Diary of Nina Kosterina*, Chicago: Camelot, 1970, pp. 24, 28, 30, 34, 39, 57.

11 Aleksei Kozlov, *Kozel na sakse i tak vsiu zhizn'*, Moscow; Vagrius, 1998, p. 68; on street dances see Ginsburg, *Kosterina*, pp. 68–69. Interview Victor Kosmodam'ianskii, born 1934, Moscow, 7.10.2000.

12 See Anne Gorsuch, *Youth in Revolutionary Russia: Enthusiasts, Bohemians, Delinquents*, Indianapolis: Indiana University Press, 2000, pp. 116–138.

13 Anna Krylova, 'Soviet Modernity in Life and Fiction: The Generation of the "New Soviet Person" in the 1930s', PhD thesis, Johns Hopkins University 2000, p. 154.

14 Rossiiskii Gosudarstvennyi Arkhiv Sotaial'no-Politicheskoi Istorii (RGASPI) M-f. 1, op. 23, d. 1304, ll. 107–108.

15 Krylova, 'Modernity', p. 156.

16 RGASPI M-f. 1, op. 8, d. 1, ll. 102–112.

17 Ibid., l. 103.

18 Tsentr Dokumentatsii Noveishei Istorii Volgogradskoi Oblasti (TsDNIVO) f. 113, op. 35, d. 28, l. 73.

19 Nonna Mordiukova, *Ne plach' Kazachka!*, Moscow: Olimp, 1998, pp. 183–184.

20 RGASPI M-f. 1, op. 2, d. 285, ll. 463–464.

21 RGASPI M-f. 1, op. 6, d. 468, l. 149.

22 RGASPI M-f. 1, op. 4, d. 1398, l. 192.

23 Alexander Clifford and Jenny Nicholson, *The Sickle and the Stars*, London: Peter Davies, 1948, p. 214.

24 RGASPI M-f. 1, op. 46, d. 154, ll. 4–5.

25 RGASPI M-f. 1, op. 46, d. 76, l. 163.

26 Frederick Starr, *Red and Hot: The Fate of Jazz in the Soviet Union*, New York 1994, p. 241. Vasilii Aksenov, *In Search of Melancholy Baby*, New York: Random House, 1985, p. 13.

27 RGASPI M-f. 1, op. 2, d. 285, l. 464.

28 G. Gornostaev, 'Kontrabandnye fokstroty', *Komsomol'skaia pravda* 29.6.1947, p. 6. TsDAHOU f. 7, op. 3, d. 1482, l. 79.

29 'Ne takoi klub nam nuzhen!', *Komsomol'skaia pravda* 9.8.1946, p. 2.

30 TsGASPb (Tsentralnyi Gosudarstvennyi Arkhiv St. Peterburga) f. 6276, op. 271, d. 1047, ll. 1–53.

228 *Juliane Fürst*

31 Tantseval'naia likhoradka used in Timofeev, Iu. 'Tantsoval'nyi vikhr', *Komsomol'skaia pravda* 8.5.1946, p. 2.
32 Ibid.
33 Mikahil Gorbachev, *Memoirs*, London: Doubleday, 1995, p. 42.
34 Peter Kenez, *Cinema and Soviet Society*, Cambridge: Cambridge University Press, 1992, p. 209.
35 Aksenov, *Melancholy*, p. 17; Interview Chernov; RGASPI f. 17, op. 132, d. 92, l. 13.
36 Aksenov, pp. 17–18.
37 Kenez, *Cinema*, p. 214.
38 RGASPI M-f. 1, op. 6, d. 468, l. 48; RGASPI M-f. 1, op. 32, d. 450, l. 149.
39 RGASPI M-f. 1, op. 6, d. 467, l. 70.
40 Gosudarstwvennyi Arkhiv Autonomnoi Respubliki Krym (GAARK) P-f. 147, op. 1, d. 449, l. 27.
41 Interview Chernov/Chernova; Richard Stites, *Russian Popular Culture: Entertainment and Society since 1900*, Cambridge: Cambridge University Press, 1992, p. 125.
42 RGASPI M-f, 1, op. 46, d. 154, l. 13; Stites, *Entertainment*, p. 126.
43 RGASPI M-f. 1, op. 6, d. 468, l. 11.
44 RGASPI f. 17, op. 131, d. 49, l. 119.
45 Maia Turobskaia, 'Kubanskie kazaki', *Rossiiskii Illiusion*, Moscow: Materik, 2003, 261–266.
46 There are no viewer numbers for the trophy film and they would not have been counted in the official statistics.
47 Okeev Tolomush in *Domashnaia sinematika*, Moscow: Dubl'D, 2002, p. 219.
48 See Juliane Fürst, '"The Arrival of Spring?" Changes and Continuities in Soviet Youth Culture and Policy between Stalin and Khrushchev', forthcoming in Polly Jones (ed.), *Problems of Destalinisation*, London: Routledge, 2006.
49 Kozlov, *Kozel*, pp. 70–71.
50 Ilia Ehrenburg, 'Fal'shivyi golos', *Kul'tura i zhizn'*, 10.4.1947, p. 4.
51 Andrei Konchalovskii, *Nizkie Istiny*, Moscow: Sovershenno sekretno, 2001, p. 76.
52 Kozlov, *Kozel*, p. 90; Interview Victor Kosmodam'ianskyi, Moscow, November 2000.
53 Gosudarstvennyi Arkhiv Riazan'skoi Oblasti (GARO) P-f.3, op. 4, d. 71, ll. 153–154; Interview Nina Chernova and Nikolai Chernov, Volgograd, 5.10.2001.
54 See for example 'Davaite tantsevat' po-drugomu', *Smena*, 30.6.1953, and the subsequent discussion on the topic by cultural workers under the aegis of the Leningrad trade unions: TsGASPb f. 6276, op. 271, d. 1047, ll. 1–53.
55 TsGASPb f. 6276, op. 271, d. 825, l. 39.
56 Gosudarstvennyi Arkhiv Rossiskoi Federatsii (GARF) R-f. 5707, op. 1, d. 15, l. 279; Iurii Bondarev, *Silence: A Novel of Post-War Russia*, London: Chapman & Hall, 1965, p. 16.
57 Tsentral'nyi Derzhavnyi Arkhiv Hromadskikh Ob'iednan Ukrainy (TsDAHOU) f. 7, op. 3, d. 494, l. 11.
58 Interview Chernov/Chernova.
59 GARO P-f. 3, op. 3, d. 351, l. 85.
60 'Zapiski zakroishchitsy', *Krokodil* no. 14, 1948, 4.
61 RGASPI M-f. 1, op. 46, d. 175, ll. 91–92.
62 M. Menshikov, 'Zolotaia koronka', *Komsomol'skaia pravda* 18.5.1946, p. 3.
63 S. Gorbusov, 'Vecher v Gigante', *Komsomol'skaia pravda* 16.4.1946, p. 2.
64 Iu. Ganf, 'Na vse sto', *Krokodil* 10.7.1946, 5. Translation by Mark Edele.
65 RGASPI M-f. 1, op. 46, d. 175, ll. 91–92.
66 Kozlov, *Kozel*, p. 76; Interview Kosmodam'ianskii.
67 TsGAGSP f. 6276, op. 271, d. 1047, l. 18.
68 Interview Kosmodam'ianski.

69 Kozlov, *Kozel*, p. 79; Interview Kosmodam'ianskii.
70 Kozlov, *Kozel*, p. 83.
71 Ibid., p. 79.
72 Menshikov, p. 3.
73 D. Beliaev, 'Stiliagi', *Krokodil* no. 7, 10.3.1949, 10.
74 Don, Dallas, *Dateline Moscow*, Melbourne: Heinemann, 1952, p. 116.
75 Kozlov, *Kozel*, p. 79.
76 RGASPI M-f.1, op. 46, d. 175, l. 91.
77 Kozlov, *Kozel*, p. 78. Beliaev, 'Stiliaga', p. 10.
78 Starr, *Red*, p. 237.
79 Volodimir Slavkin, *Pamiatnik neizvestnomu stiliage*, Moscow: Art Rezh Teatre, 1996, p. 5.
80 For inspiration see RGASPI M-f. 1, op. 46, d. 175, l. 91. For Soviet fashion and Western couture see Larissa Zakharova, 'Soviet Fashion and the Transfers of Western Clothing Practices to the USSR under Khrushchev', paper presented at the 7th World Congress of the International Council for Central and East European Studies (ICEES), Berlin 25–30 July 2005.
81 Kozlov, *Kozel*, p. 90. Interview Kosmodam'ianskii.
82 Volodimir Slavkin claimed that most *stiliagi* were neither politically engaged nor aware of the political and ideological implications of their behaviour. Slavkin, *Pamiatnik*, p. 8.
83 RGASPI M-f. 1, op. 46, d. 175, ll. 91–92.
84 RGASPI M-f. 1, op. 46, d. 175, ll. 89–90.
85 RGASPI M-f. 1, op. 46, d. 175, ll. 91–92.
86 Starr, *Red*, p. 241.
87 Volodimir Slavkin, one of Russia's foremost playwrights, created a memorial to the culture of *stiliachestvo* with his play *Vsrozlaia Doch' Molodogo Cheloveka/The Adult Daughter of a Young Man*. Himself never a *stiliaga* Slavkin pays tribute to his contemporaries, the generation of the 1960s, whose creative beginnings he sees in the phenomenon of the *stiliagi*.
88 Slavkin, *Pamiatnik*, p. 8.
89 Interview Kosmodam'ianskii.
90 Konchalovskii, *Nizkie*, p. 79.
91 Ibid., p. 94.
92 Ibid., p. 90.
93 'Soprotivlenie na nevskom prospekte', *Pchela* 1997, no. 11, pp. 29–30. Interview Boris Pustinstev, born 1935, St. Petersburg, September 2004.
94 Aksenov, *Melancholy*, p. 12.
95 Aksenov's mother was Evgenia Ginzburg.
96 Aksenov, *Melancholy*, p. 18.
97 TsDNIVO f. 113, op. 35, d. 28, l. 73.
98 GARO P-f. 366, op. 3, d. 288, l. 90.
99 RGASPI M-f. 1, op. 2, d. 285, l. 457.
100 RGASPI M-f. 1, op. 46, d. 159, l. 14.
101 TsGADSp f. 6276, op. 247, d. 1047, ll. 1–60, here ll. 9, 16.
102 RGASPI M-f. 1, op. 3, d. 553, l. 74.
103 Kenez, *Cinema*, p. 213.
104 TsDAHOU f. 7, op. 3, d. 494, l. 30.
105 GAARK f. 147, op. 1, d. 449, l. 19; RGASPI M-f. 1, op. 6, d. 468, l. 91.
106 Kozlov, *Kozel*, p. 77.
107 Gorbusov, *Gigant*, p. 3.
108 Gorbusov, *Gigant*, p. 2.
109 Beliaev, 'Stiliaga', p. 10.
110 TsDAHOU f. 7, op. 3, d. 494, l. 33 obo; Shelepin, Report to the Komsomol Con-

gress; Alan Kassof, *The Soviet Youth Program*, Cambridge, Mass.: Harvard University Press, 1965, pp. 154–161.

111 For a more detailed discussion of such young people, whose purist understanding of communist ideology propelled them to engage in political opposition, see Juliane Fürst, 'Prisoners of the Soviet Self? Political Youth Opposition in late Stalinism, *Europe-Asia Studies* 54, no. 3, May 2002, pp. 353–376.

112 RGASPI f. 17, op. 132, d. 196, ll. 9–15.

113 RGASPI f. 17, op. 132, d. 196, ll. 18–20.

114 Liudmilla Alexeyeva, *The Thaw Generation: Coming of Age in the Post-Stalin Era*, Boston: Little, Brown and Company, 1990, pp. 56–57.

115 Interview B. M. Firsov, generously supplied by Lev Lur'e and St. Petersburg State Television.

116 Vladimir Britanishskii, 'Studencheskoe Peticheskoe Dvizhenie v Leningrade v Nachale Ottepeli', *Novoe Literaturnoe Obozrenie* no. 14, 1996, pp. 1–18.

Part V

Post-war spaces
Reconstructing a new world

11 "Where should we resettle the comrades next?"

The adjudication of housing claims and the construction of the post-war order[1]

Rebecca Manley

In the late summer of 1944, historian Saul Borovoi, a Jewish Odessite who had spent the bulk of the war in Samarkand, began the long journey home. The journey to his native city lasted two months; the journey home, however, remained incomplete. Borovoi's apartment was occupied. In his memoirs, he recalls how he received a letter while in Samarkand from his old neighbour in Odessa, who "wrote that our furniture had been preserved and that she was waiting for us. Later, evidently having better understood the situation, she realized that she had rushed with the invitation and she did everything not to let us back into our old apartment."[2] Borovoi was not only unable to reclaim his old apartment, he was not returned all of his belongings. His neighbour, who had occupied the apartment, "kept part of my books, my stamp collection, and many domestic objects for herself."[3]

While Borovoi ultimately reconciled himself to the unfortunate turn of events, the loss of an apartment was no small matter. Housing in the Soviet Union was the item in greatest shortage in an economy defined by shortage.[4] This had been true before the war, and it was doubly so at its conclusion. For those, like Borovoi, who had spent the war "in evacuation," pre-war housing assumed a position of particular importance, for it constituted a crucial component of the re-evacuation. Return was contingent upon official authorization, which, in turn, depended upon proof of a place to live.[5] While Borovoi had circumvented these restrictions, most evacuees were not so fortunate. As one Odessite explained in an attempt to secure the return of his family in the summer of 1946, he had been demobilized in December 1945, but "given that my apartment was destroyed and I did not have living space, I was not able to secure a summons for my family."[6] Clearly, housing was not simply a matter of shelter. For many Soviet citizens, it was also the space in which the "return to normalcy" transpired.[7] At the end of the war, pre-war living quarters emerged as elements of stability in a society that had been subjected to strain and suffering. They were repositories of cherished items, precious links to a pre-war world. They were also objects of contention, as Borovoi was made painfully aware.

Indeed, battles over housing became one of the primary sites of conflict in the period of post-war reconstruction.

The backdrop to conflicts over housing was the extraordinary death, destruction and displacement caused by the war. In Odessa, local authorities estimated that one-third of the city's housing had been destroyed or damaged during the war.[8] Even when housing had survived intact, it could not always be reclaimed. As one state prosecution official from the Dnepropetrovsk region wrote of returning evacuees: "Their living space has been occupied, in part on the orders of the housing division, after the liberation of the city, and in part during the occupation."[9] In some cases, it had been occupied on no orders at all. In the city of Kharkov, the dislocation was so great that, according to the *oblast'* prosecutor, there was not a single person living in the apartment in which he or she had resided before the war.[10] In these circumstances, it was inevitable that many evacuees would return to find their living quarters already inhabited. Indeed, almost half of all complaints regarding housing submitted to the Supreme Soviet in 1945 concerned "the failure to return former living space to the previous owner."[11] Borovoi was obviously no exception.

Borovoi's experience raises a host of questions for the historian. First, was Borovoi legally entitled to his former apartment? Had he lodged a claim requesting the return of his apartment, how would his claim have been adjudicated? Finally, if Borovoi was entitled to the return of his former apartment, how to explain his failure to retrieve it? Borovoi himself attributed his difficulties, at least in part, to the emergence of officially sanctioned and popularly endorsed anti-Semitism. His supposition would seem to support a growing body of literature which increasingly locates the origins of official anti-Semitism not in the anti-cosmopolitan campaigns of the late 1940s, but in the latter years of the war itself.[12] To what extent, however, were the problems in reclaiming housing a specifically Jewish issue? More broadly, what do official directives regarding housing tell us about the construction of the post-war polity?

The adjudication of post-war housing claims compelled judicial and state authorities to rank competing claims. As such, it emerged as a principal site in which post-war hierarchies were articulated. In debates over housing rights and in the practice of resolving housing disputes, we can observe the emergence of new categories of entitlement as well as exclusion. In addition, housing disputes provide a window onto popular conceptions of entitlement at the end of the war. Housing emerged as a contested terrain in which individuals and groups fought not only over scarce material resources, but over who won the war, and the extent to which the war would determine the post-war order. Such conceptions were, I would suggest, crucial determinants of the way housing disputes were resolved in practice and are key to any analysis of the emergence of the post-war order.

To begin with housing rights: did Borovoi have a right to his apartment? Neither Borovoi nor the majority of other returnees actually owned their pre-war apartments.[13] Nonetheless, they had, at least in theory, certain rights as tenants. To be sure, according to a law from 1937, an absence of over six months from one's apartment could result in a loss of occupancy rights, as could failure to pay rent for three months. That, however, was in ordinary times, and officials in the prosecutor's office were well aware of the problems inherent in applying these regulations in times of war. Central archives contain dozens of letters from state prosecutors and judicial officials in recently liberated territories asking how and if the existing legislation should be applied.[14] "When should the six-month period begin," queried one prosecution official, "from the day of evacuation or from the day the place was liberated from the Germans? How should it be regarded if the people in question have not paid their rent? Can section 30 of the law of 17 October 1937 be applied?"[15] In a clear indication that the official himself had his doubts about the applicability of these laws to the current situation, he concluded with one final question: "Is there no plan to issue a new law in relation to the housing rights of evacuees?"[16]

To the presumed dismay of prosecution and judicial officials, whose repeated queries testify to their continued struggles to make sense of the complex situation (which had raised, in the words of the chairman of an *oblast'* court, "extraordinary difficulties"),[17] no such law was ever issued. This is not to say, however, that the housing rights of evacuees were not defended and upheld. When the Council of People's Commissars drafted a proposal whereby those returning to Leningrad would be accorded housing from the general stock (rather than provided with their previous living space), prosecution officials objected: "One cannot agree with this proposal. It cannot be considered right that tenants who were evacuated from Leningrad and who in good conscience fulfilled all of their responsibilities should be deprived of their right to their living space."[18] An only slightly less extreme measure proposed by the State Defence Committee with respect to Moscow met with a no less categorical response. The draft resolution proposed that evacuees who had failed to return to Moscow by 1 January 1944 should lose their claim to their pre-war housing. Once again, prosecution officials responded that "one cannot agree with this proposal," arguing that

> we cannot permit a situation in which tenants who left Moscow as part of the evacuation, and who have fulfilled all of their obligations ... can be deprived of their rights to their living space only because they were unable, for one reason or another, to return to their place of permanent residence in Moscow by 1 January 1944.[19]

Clearly, in the minds of central prosecution officials, the fact that evacuees had left for reasons not of their own choosing, coupled with the unspoken difficulties attendant upon return, mitigated against the application of the six-month rule. Should there be any doubts, however, the prosecution official added one final, and in his view decisive, argument:

> If one takes into consideration the fact that in the main this affects the families of those serving in the army (judicial practice has shown that the

principal group of people who have not yet returned to Moscow from evacuation are the families of service people), then it becomes clear that this
point of the draft cannot be accepted.[20]

This exchange is of interest on several counts. First, it suggests that the rights
of re-evacuees to their former apartments were susceptible to administrative
annulment. Proposals similar to those put forward in Moscow and Leningrad
were advanced in a host of formerly occupied cities as well. In Odessa, authorities proposed an expiration date on tenants' rights, and in Voronezh, local
authorities, "on their own initiative," simply abolished all rights to previously
occupied living space.[21] While such proposals appear to have foundered in each
case on the objections of the state prosecution (in Voronezh the city soviet's
decision was not approved by legal organs), they underscore the feeble foundations on which such rights rested. In addition, they point to a tendency towards
spatial differentiation, which the war had done little to diminish. Universal laws,
applicable to the entire space of the Soviet Union, were mediated by a series of
individual resolutions applicable to specific cities, thereby establishing spatially
differentiated rights for re-evacuees.

The rights of re-evacuees were further differentiated with respect to service
on the frontlines. The response of the state prosecution in the exchange over
housing rights in Moscow is of particular interest in this regard: Ultimately, the
prosecutor chose to defend the rights of returnees not as "evacuees," but as the
"families of service people." The choice of words was not accidental. It reflected
the fact that the rights of re-evacuees relied not only on the 1937 legislation, but
on a variety of other directives as well, issued during the war and in its immediate aftermath, which established differential treatment of specific groups in the
evaluation of their claims to occupancy. The complex of additional directives
established an effective hierarchy of entitlement whereby certain groups of evacuees were endowed with stronger claims to their pre-war housing than others.

The emergent hierarchy reflected both wartime service and pre-war status.
First, as the comments of the prosecutor suggest, the rights of service people and
their families emerged as a powerful category of entitlement. Those serving in
the Red Army and their families were guaranteed the right to return to their
former living space by a government resolution issued on 5 August 1941.[22] For
this group, and this group alone, moreover, the requirement to pay the rent, a
key part of the 1937 legislation, was waived. Not all wartime service, however,
was recognized as equal. While workers shipped to factories in Siberia and the
Urals clearly felt that they too had made sacrifices for the war, they were effectively excluded from the post-war hierarchy of entitlement. In February 1942 a
government resolution deprived workers and employees evacuated with factories and enterprises of their right to their former housing.[23] The resolution struck
workers as manifestly unfair. As one worker put it in a letter excerpted by the
censors: "The workers have given all their strength to defeat the enemy, and
they want to return home, to their own people, their own homes. And now it
turns out we have been deceived. They've shipped us out of Leningrad, and they

want to leave us in Siberia. In this case we workers should say that our government has betrayed us and our work."[24] Similar sentiments were expressed by
workers from Moscow, who likewise viewed the policy as a betrayal.[25] Not even
all officials were convinced of its correctness. In the city of Kalinin, for
example, the head of the people's court wrote in 1944 to the prosecutor's office
in Moscow that the rejection of workers' claims to their former housing was, in
his opinion, "essentially incorrect" as there was no other housing to be given
these people.[26] His query, however, was met with a definitive answer: evacuated
workers had "lost their right to the living space they occupied before the evacuation."[27] Only if the worker happened to have lived with someone currently
serving in the army before the war would he or she have a chance of reclaiming
the apartment: such, at least, was the way a Moscow prosecutor interpreted the
existing legislation.[28]

Wartime behaviour emerged as a basis for exclusion in a more punitive sense
as well – those who had voluntarily departed with the Germans and now sought
the return of their former living space, for example, were explicitly deprived of
their rights.[29] Moreover, people who had occupied the apartments of evacuees on
"their own initiative" (*samovol'no*), either with the sanction of the occupying
powers or with no sanction at all, were subject to administrative eviction and
were not entitled to the protections afforded by a legal process.[30] While the war
thus created new categories of entitlement and exclusion, behaviour during the
war was but one axis of the new post-war hierarchies. The role of the war in
reshaping the polity was, in other words, by no means absolute.[31] Pre-war elites
saw their privileges ensconced and strengthened in legislation that secured their
return to their previous apartments. Such privileges were extended not only to
the party and state apparatus, but also to the cultural and scientific elite. In some
cases, such groups were not only guaranteed the return of their pre-war housing,
but were assured of living space, from which they could not be evicted, even if
their own housing had been destroyed (a right that was not extended, at least in
theory, to the families of service people). Thus party authorities in Odessa, for
example, appealed to state prosecution officials to exempt certain categories of
citizens from expulsions, namely

> the Soviet and party *aktiv*, war invalids, and scholars, who returned to
> Odessa in conjunction with a decree of the Central Committee or *oblast'*
> authorities. In practice this is what happens: some comrades arrive, receive
> living space, and then a family that used to live in the apartment returns
> from evacuation, and the comrades are resettled in another apartment, and
> another family returns. Again resettlement – where should we resettle the
> comrades next![32]

Of course, the hierarchy of entitlement established by central and regional
directives did not seamlessly translate into a corresponding hierarchy on the
ground. The gulf between theory and practice was in part a product of bureaucratic corruption and bribery, of which there is ample evidence in the archives.[33]

It was also a product, however, of competing conceptions of entitlement. The way housing claims were assessed in practice provides insight into the way legitimate claims were popularly constructed and into how authorities and the population at large attempted to "make sense of the war."[34] As the cases of correspondence between central and regional prosecution officials suggest, the adjudication of housing claims was a source of considerable conflict. Both within and between the government and judicial organs charged with resolving housing disputes, officials articulated different and often incompatible visions of how to resolve the matter at hand. The division was more complex than a simple one between municipal authorities anxious to solve the housing crisis and judicial authorities anxious to uphold the letter of the law. At play were different, often competing, conceptions of rights and of entitlement. Correspondence among prosecutors, the dictates of local officials, and the petitions of individuals all suggest that the claims of the families of servicemen enjoyed widespread support. Appeals for the return of housing rarely failed to mention service on the front lines. In the consideration of municipal government and even prosecution officials, moreover, service appears to have constituted a much stronger ground for the reclamation of living space than the fact of evacuation. It was typical, for instance, that in a case in which the rights of a returning evacuee came into conflict with the desires of local industry, the rights of the former were upheld not with reference to the fact that the living space was rightfully his, but rather with reference to his connection to the army. "Citizen Reznikov," the prosecutor claimed in one such case, "is the father of a *frontovik* and in conjunction with the *Sovnarkom* decision of 5 August 1941, he has an indisputable right to his former living space, regardless of whether or not he has documents permitting him to enter Odessa."[35]

More striking still, the rights of the families of service people were widely considered to trump the rights of other groups. This category did, of course, have special rights, as laid out in the August 1941 resolution. The resolution, however, while it entitled the families of service people to the return of their own living space, did not entitle them to the pre-war living space of others. Not, at least, in theory. In practice, however, the rights of evacuees to the return of their former living space were frequently abrogated in the name of the rights of the families of servicemen. While the letter of the law was quite clear, the practice appears to have been widespread. Consider the following case: one re-evacuee who returned to Odessa attempted to evict residents who had settled in her apartment "willfully." Despite the fact that the eviction order was upheld by the courts, however, the other woman was nonetheless given an order for the apartment "as she is also the family of a serviceman."[36] Another woman returning from evacuation and seeking to reclaim her former living space sought to strengthen her case by pointing out that her husband was on the front. It turned out, however, that the man in question was in fact her ex-husband (they had divorced shortly before the war). Did her divorce thus deprive her of the right to her apartment? The military official writing on behalf of those about to be evicted certainly thought so.[37] While the archives contain no trace of how the conflict was actually resolved, it would not be at all surprising if she was indeed denied her claim. The correspondence of regional judicial officials suggests that, in practice, the families of service people were often endowed with rights for which there was in fact no legal foundation. The chairman of the Kursk *oblast'* court, for instance, was convinced that the law prevented evacuees from reclaiming their former housing if it had been occupied by the families of service people with official sanction.[38] While the deputy prosecutor of the Soviet Union informed the latter in no uncertain terms that he had misinterpreted the existing legislation, and risked violating the "legal interests" of evacuees, such assumptions appear to have been widespread.[39]

In practice, then, it would seem that the rights of the families of service people were substantially extended by officials on the ground, encroaching upon the less well-defined, but nonetheless recognized rights of others. This is not to say that all families of service people were successful in reclaiming their former apartments.[40] The housing crisis was extreme, the operation of the housing and judicial organs slow, and even an invalid, whose rights to housing were in theory very well protected, found himself compelled to live on a balcony for four months in Odessa.[41] As party officials themselves acknowledged in a meeting of the City's Party Bureau, there were serious "deficiencies" in the way local authorities were handling the housing problem, as a result of which even

> war invalids, the families of service people, and party, Soviet and economic workers who have come from the east lose a tremendous amount of time searching for an apartment and only receive apartments after lengthy appeals and pressure on the part of leading party and Soviet authorities.[42]

It is to say, however, that there appears to have been a consensus among officials (and certain sectors of the population as well) that the sacrifices of those who served on the front had conferred upon them and their families an entitlement not enjoyed by other groups.

Evacuees enjoyed no such entitlement. While many of the officials adjudicating post-war housing claims had themselves spent the war "in evacuation," evacuees tended to be stigmatized for having "sat out the war" in the rear. Indeed, there was no hero's welcome for those returning from Central Asia or Siberia. One Leningrader who had remained in the city later described the resentment towards evacuees. "There were cases," he recalled,

> in which individual workers treated those who arrived in a hostile fashion, called them deserters, accused them of saving their skins in the rear, at the same time as those who remained at the factory had lived through tremendous difficulties. It had to be explained. After all, these people had not been at a resort, they had also had a tough time.[43]

The notion that evacuees had "sat it out in the rear" undermined the legitimacy of their housing claims in the popular imagination. Those who had been evacuated, or had simply departed on their own, were viewed by some as having forfeited their rights to their previous apartments. It was those who remained (those who remained reasoned) who had suffered through the war and either occupation or bombardment, and were thus entitled to the housing they currently inhabited.[44]

Hostility towards re-evacuees was particularly pronounced with regard to the Jewish population. In some regions, local officials refused to help returning Jews on the grounds that "we didn't summon you here. . . ."[45] In other cases, Jewish re-evacuees were told to "go back to where they came from."[46] The stigma attached to Jews was nowhere more marked and nowhere as enduring as in the emergence and flowering of a new wartime myth – that of the "Tashkent front," where Jews were widely said to have served. The essentials of the myth are clearly contained in the comments made by a war veteran to a Jew on a city bus in 1948: "all you Jews are cunning and didn't fight in the war, you were all sitting in Tashkent while we fought."[47] The alleged non-participation of Jews, symbolized by their supposed flight to the rear, became a staple of post-war humour. Jokes circulated that "the Jews have taken the cities of Alma Ata and Tashkent by storm" and that "the Jews took Tashkent without a fight."[48] The "Tashkent front" quickly became a byword for Jewish cowardice and non-participation in the war, underscoring the centrality of the wartime experience, and the experience of evacuation, in giving shape to post-war anti-Semitism in the Soviet Union.

This brings us back to Saul Borovoi. In legal terms, Borovoi's claim to his apartment was sound. To be sure, Borovoi had not served in the army – nor had any of his immediate family members. As such, he could not enjoy the benefits of the blanket resolution issued in August 1941 promising all servicemen and their families the return of their pre-war abodes. However, as an evacuee who had not been evacuated with a factory he was nonetheless legally entitled to his pre-war living space. His memoirs suggest that he did not seek legal recourse – he lacked the strength, and as a Doctor of Science he was ultimately offered another apartment. Moreover, the building had been taken over by the steamship-line, and he was under the impression that "it was practically impossible to receive an order for expulsion."[49] Had he sought legal recourse, it is hard to say what the decision might have been. While central prosecution officials upheld the rights of evacuees to return to their previous apartments, even if the latter had been resettled by the families of service people, local practitioners, as I have suggested, did not, or not always. Many appealed for the return of their living space and were refused. Such was the fate of one woman who returned to Odessa with her daughter from evacuation in Central Asia and whose request for the return of her pre-war apartment was rejected, despite the fact that she was legally entitled to its return both as an evacuee and as the family of someone serving in the Red Army. Whether she ultimately managed to reclaim the living space is unclear, but it was only the fact that her son served in the military that led district authorities to take up her case once again.[50] Borovoi had no such recourse.

Either way, a legal ruling and an order authorizing one to reclaim one's apartment were only half the battle. Had Borovoi sought legal recourse and been granted his request, it is not clear that he would have retrieved his former apartment. Even when requests for the return of housing were granted, they were not always fulfilled. As one man explained in a letter to the Jewish Anti-Fascist Committee penned almost a year after the end of the war, the order for his living space, which he had received from the city soviet of his native city of Kharkov, "is thus far just a piece of paper, for the city soviet does nothing to help the realization of the order."[51] A similar case was reported by a Jewish resident of Kiev, whose claim to his former abode, a room in a communal apartment, was bolstered by his service on the front lines. Appealing to Erenburg in desperation, he informed him that "despite a whole slew of resolutions from the prosecution office of the Kaganovich district of Kiev and resolutions of the Soviet of the same district about freeing up the room for my family, the room has still not been vacated."[52] According to the letter-writer, the individual who had occupied the room, a former neighbour, "acts from the outside (*deistvuet so storony*), imperceptibly, and all the decisions concerning the vacation of the room for my family come to nothing."[53] As a result, his wife and children were living in Moscow without housing, and his sister-in-law, who had gone ahead to Kiev, had no alternative but to sleep in the apartment's communal kitchen.

As these cases suggest, the files of judicial and municipal authorities provide only a partial picture of how the post-war order was constructed. All too often, the adjudication of housing claims transpired on the contested terrain itself, in the apartments, hallways and buildings that were the object of contention. It was in these spaces, among neighbours, rather than in the offices of prosecution or municipal officials that, de facto, decisions were made. Consider the case of Borovoi, who seems to have never even filed a claim with local officials. His housing situation was effectively decided on the ground: his neighbour refused to return his living space, and he was not up to the fight.

I would suggest that we could understand Borovoi's failure in several ways. First, it bears emphasizing that almost all evacuees had trouble reclaiming their apartments. Even those at the top of the emergent hierarchy of entitlement had no guarantee of success. Borovoi, however, found himself at a particular disadvantage. First, as someone with no connections to the front, his claim lacked popular legitimacy, if not a legal basis. Second, Borovoi was a Jew. The difficulties encountered by Jews in reclaiming their housing were widely reported by contemporaries, as some of the examples cited above suggest. In the spring of 1944, the Chairman and Executive Secretary of the Jewish Anti-Fascist Committee wrote to Molotov asserting that Jews "are encountering obstacles in their re-evacuation to their native areas." The letter went on to note that

even those few who manage, through various ways, to get to their native towns where their grandfathers and great-grandfathers had lived, find that their homes have been occupied since the German occupation. Thus, those returning are left without a roof over their heads.[54]

The representatives of the Committee were clearly of the opinion that Jews were encountering exceptional obstacles in their efforts to reclaim their housing.

Evidence from letters sent to the Committee seems to corroborate their suspicion. Across the formerly occupied territories, there were reports of Jews being barred from reclaiming their prior abodes. As one Jewish Odessite wrote to Erenburg:

> I was in evacuation for three years, I returned to my native city recently. The things and furniture in my apartment had been stolen, my apartment was occupied. I have two sons who are officers defending the motherland, and for seven days I had to sleep in the front entrance before a neighbour felt sorry for me and let me into his apartment. The bureaucrats in the housing division have still not given me an order for an apartment.[55]

While the difficulties encountered by Jews are incontrovertible, the sources of their problems are more difficult to identify. Were the "bureaucrats in the housing division" simply slow, perhaps corrupt, or was this a case of anti-Semitism? How should we explain the large number of Jews who had difficulties reclaiming their apartments?

In part, the prominence of Jews among supplicants seeking to reclaim former housing reflects their over-representation among the returning population as a whole. Jews had constituted a full quarter of the evacuee population, and accordingly constituted a substantial proportion of the re-evacuees.[56] This is not to say, however, that anti-Semitism was not a factor. Housing emerged at once as both a site and an important source of post-war anti-Semitism. One Soviet Jew directly attributed the "unfriendly" reception accorded to re-evacuated Jews to the fact that "disputes arise over the return of apartments and demands for the return of plundered property when it is found...."[57] Another Soviet Jew, a resident of Odessa, went so far as to claim that the anti-Semitism in his native city

> doesn't particularly worry me, as I qualify it exclusively as a phenomenon of love for Jewish property, and, insofar as it is in practice already stolen, there is reason to believe that the lovers of such property will soon understand that there is no basis for the manifestation of hateful feelings towards the Jewish nation.[58]

"Love for Jewish property" undoubtedly exacerbated anti-Semitism in the post-war Soviet Union. Ultimately, however, the source of Soviet Jewry's problems in the post-war period lay not in their property, but rather in their perceived non-participation in the war.

Was discrimination against Jews a new part of the post-war hierarchy? Did Jews constitute a new category of exclusion at the same time as new categories of entitlement were being created? While the obstacles encountered by Jews do not seem to have been the result of a systematic policy of exclusion elaborated by central or even regional authorities, they were shaped, at least in part, by central policies. The period of re-evacuation corresponded with the onset of a series of initiatives divesting Jews of their responsibilities in a variety of fields, namely in the arts, in academia, and in party and government organizations.[59] Borovoi himself had been a victim of such a policy. When the members of the university were re-evacuated from Uzbekistan, he was not among the list of those authorized to return.[60] Given these policies, it is hardly surprising that there were rumours in Odessa that Jews would not be permitted to return to the city.[61] State-sponsored discrimination, in other words, fed and reinforced the notion that Jews would be excluded from the post-war polity. It both confirmed the local population in its conviction that Jews were not entitled to reclaim their housing, and emboldened them to take the steps required to assure they would not be able to do so.

To the extent that the adjudication of housing claims was based not only on the existing legal framework, but on assessments of wartime service and entitlement, Jews were clearly at a disadvantage. Widely perceived as having "sat it out in the rear," their claims were more likely to be rejected by local officials or inactively pursued. The individuals who had taken over their apartments, moreover, were more likely to feel emboldened to remain there, and to feel justified in doing so. This is not to say that discrimination against Jews was systematic. There were many Jews who succeeded in reclaiming their apartments, and many non-Jews who did not. Ultimately, the degree to which Jews were singled out for exclusion, and the regional variations in this process, remain to be determined. What the evidence presented here does make clear, however, is that official resolutions, decrees and laws offer only a limited window onto the processes of post-war reconstruction. To understand fully how post-war disputes played out in practice it is necessary to examine not only the dictates of judicial and municipal authorities, but what actually happened on the ground, what was decided by the inaction of local authorities and the actions of ordinary citizens.

Notes

1 This article is based on research undertaken in state and party archives in Moscow and Odessa. It forms part of a larger project on the evacuation of Soviet civilians during the Second World War, which was the subject of my dissertation, "The Evacuation and Return of Soviet Civilians, 1941–1946," PhD dissertation, University of California, Berkeley, 2004. I am grateful for the helpful comments of Juliane Fürst, Andrew Jainchill and participants at the University of Chicago workshop on *Property Issues in the Postwar Soviet Union and Eastern Europe* and at the Berlin Congress of the International Council for Central and East European Studies.
2 S. Ia. Borovoi, *Vospominaniia*, Moscow: Evreiskii Universitet v Moskve, 1993, p. 285.
3 Ibid., p. 288.
4 On the pre-war housing crisis see Sheila Fitzpatrick, *Everyday Stalinism. Ordinary Life in Extraordinary Times: Soviet Russia in the 1930s*, New York: Oxford University Press, 1999, pp. 46–50.
5 Like the evacuation itself, the return of evacuees was, at least in theory, a highly regulated process. Wartime restrictions on freedom of movement that made train travel dependent on the permission of the police remained in effect until the summer of

1946. See M. N. Potemkina, *Evakuatsia v gody Velikoi Otechestvennoi voiny na Ural: liudi i sud'bi*, Magnitogorsk: Magnitogorskii gosudarstvennyi universitet, 2002, pp. 182–183.

6 Gosudarstvennyi Arkhiv Odesskoi Oblasti (hereafter GAOO), f. R1234, op. 7, d. 363, l. 222.

7 On the incomplete nature of this process and the problematic notion of "normalcy" in a Soviet context, see Sheila Fitzpatrick, "Postwar Soviet Society: The 'Return to Normalcy,' 1945–1953," in *The Impact of World War II on the Soviet Union*, ed. Susan J. Linz, Totowa, N.J.: Rowman & Allanheld, 1985, p. 150.

8 GAOO, f. R1234, op. 7, d. 288, l. 45. The rates for other cities in formerly occupied territory were even higher. Belorussian authorities estimated that in the cities of Minks, Vitebsk, Gomel' and Mogilev only 23 per cent of the housing stock remained intact. Rossiisskii Gosudarstvennyi Arkhiv Sotsialno-Politicheskoi Istorii (hereafter RGASPI), f. 17, op. 88, d. 718, l. 20. For a discussion of the longevity of the post-war housing crisis, see Elena Zubkova, *Russia after the War: Hopes, Illusions, and Disappointments, 1945–1957*, trans. Hugh Ragsdale, Armonk, N.Y.: M. E. Sharpe, 1998, pp. 102–103.

9 Gosudarstvennyi Arkhiv Rossisskoi Federatsii (hereafter GARF), f. 8131, op. 21, d. 21, l. 10.

10 GARF, f. 8131, op. 21, d. 23, l. 154.

11 The precise percentage was 45.2 per cent. Also note that the number of housing related complaints had reportedly doubled in 1945. GARF, f. 7253, op. 65, d. 579, ll. 127, 129.

12 See G. B. Kostyrchenko, *Tainaia politika Stalina: Vlast' i antisemitizm*, Moscow: Mezhdunarodnye otnoshenie, 2001, and M. Altshuler, ed. and trans., "Antisemitism in Ukraine Toward the End of Second World War," *Jews in Eastern Europe*, no. 3 (1993): 40–81.

13 While there was private ownership of housing, most urban housing was owned by city soviets, and by a host of other organizations including enterprises, trade unions and ministries. In Odessa, for example, 87.6 per cent of the pre-war housing stock was in the socially owned sector. Of that, the vast majority, 94.7 per cent, was owned by the local soviets. GAOO, f. R1234, op. 7, d. 288, l. 45. On the various forms of housing tenure and their evolution, see G. Andrusz, *Housing and Urban Development in the USSR*, London: Macmillan in association with the Centre for Russian and East European Studies, University of Birmingham, 1984, pp. 29–110.

14 See, for example, GARF, f. 8131, op. 21, d. 21, ll. 10, 92.

15 GARF, f. 8131, op. 19, d. 59, l. 11.

16 Ibid.

17 GARF, f. 8131, op. 21, d. 21, l. 104.

18 Ibid., l. 77.

19 Ibid., l. 36.

20 Ibid.

21 See, respectively, GAOO, f. P9, op. 3a, d. 123, l. 81 and GARF, f. 8131, op. 21, d. 21, l. 92.

22 GARF, f. 5446, op. 1, d. 195, ll. 88–89.

23 The resolution, entitled "on freeing up the living space of local soviets and enterprises previously occupied by workers and employees, evacuated to the East," was reprinted in *Pravda*, no. 48, 17 February 1942.

24 RGASPI, f. 17, op. 117, d. 530, ll. 56–57. Cited in Zubkova, *Russia after the War*, 37.

25 See RGASPI, f. 17, op. 88, d. 137, ll. 70–71.

26 GARF, f. 8131, op. 21, d. 21, l. 50.

27 Ibid., l. 51.

28 GARF, f. 8131, op. 19, d. 59, l. 46.

29 See, for example, the unequivocal instructions to this effect issued by the central state

prosecution office to prosecution officials in the Dnepropetrovsk *oblast'*. GARF, f. 8131, op. 21, d. 21, l. 11.

30 See, for example, GARF, f. 8131, op. 19, d. 59, l. 26. Similar procedures were applied in Moscow. See GARF, f. 8131, op. 21, d. 21, l. 3.

31 On the war's partial erasure of previous stigmas, see Amir Weiner, *Making Sense of War: the Second World War and the Fate of the Bolshevik Revolution*, Princeton, N.J.: Princeton University Press, 2001 and Mark Edele, "A 'Generation of Victors?' Soviet Second World War Veterans from demobilization to organization, 1941–1956," PhD dissertation, University of Chicago, 2004, pp. 108–111.

32 GAOO, f. P9, op. 3a, d. 123, l. 81. In Odessa, certain categories of citizens were further exempt from the post-war reduction in the amount of living space each individual was entitled to (and hence were entitled to the same amount of living space that they occupied before the war). The categories listed in the resolution were "academics, scholars, cultural figures, leading engineering and technical workers, directors of enterprises and organizations and other workers who are entitled to supplementary living space . . ." GAOO, f. R6105, op. 1, d. 3, l. 8.

33 In Odessa, for instance, 18 members of the city housing bureau were arrested in the fall of 1945 on charges of corruption. The individuals in question had allegedly accepted bribes, in the words of an NKVD report on the topic, in return for "the illegal signing of orders for the occupation of living space and the illegal settlement of apartments." GAOO, f. P9, op. 3a, d. 123, l. 77.

34 The phrase is from Weiner, *Making Sense of War.*

35 GAOO, f. P17, op. 4, d. 140, l. 47.

36 GAOO, f. P9, op. 3, d. 400, l. 52.

37 GAOO, f. P22, op. 9, d. 310, l. 418.

38 GARF, f. 8131, op. 21, d. 21, l. 104.

39 Ibid., l. 106.

40 Mark Edele concludes that "while as a rule the eviction process worked to the advantage of veterans, there were cases when they had to fight for years for their housing. Such cases occurred if the apartment or room which the veteran had inhabited before the war was occupied either by an individual with competing legal rights, with superior connections, or by an institution which could claim to be more important than an individual veteran." Edele, "A 'Generation of Victors?' Soviet Second World War Veterans from demobilization to organization, 1941–1956," p. 216.

41 GAOO, f. P17, op. 4, d. 140, l. 5.

42 RGASPI, f. 17, op. 44, d. 1784, l. 167. A similar point was made in a report on complaints submitted to the Supreme Soviet in 1945. Noting that the supplicants included "service people and their families returning from evacuation, invalids of the patriotic war, and more recently demobilized service people," the report blamed the situation on "the fact that local organs of power fulfill the resolution of the government on this question in an insufficiently energetic manner." GARF, f. 7253, op. 65, d. 579, l. 129.

43 Quoted in Potemkina, *Evakuatsia v gody Velikoi Otechestvennoi voiny na Ural*, p. 195.

44 The sense of entitlement felt by those who had remained was not necessarily endorsed by authorities. In formerly occupied territories, those who had remained and had failed to engage in partisan activity were frequently suspected of collaboration. Party members were particularly vulnerable, and many were expelled from the party on the basis of their wartime conduct. See Weiner, *Making sense of war*, 84–85.

45 These were reportedly the words with which the Chairman of a District Party Executive Committee greeted Jewish re-evacuees in the Kalinindorf region. Mordechai Altshuler *et al.*, eds., *Sovetskie evrei pishut Il'e Erenburgu, 1943–1966*, Jerusalem: Prima-Press, 1993, p. 176.

46 This is how one Soviet Jew described the reception accorded evacuees in the Ukraine in a letter to Il'ia Erenburg penned in 1947. Altshuler *et al.*, *Sovetskie evrei pishut Il'e Erenburgu*, p. 267.

47 Ibid., p. 293.
48 The first joke was reported to Il'ia Erenburg in 1947 by a Soviet Jew evacuated to Alma Ata. See Altshuler *et al.*, *Sovetskie evrei pishut Il'e Erenburgu*, p. 276. For the second, which a Jewish survivor of the Odessa ghetto recalls hearing upon his return to the city, see Leonid M. Dusman, *Pomni! Ne povtori!* Odessa: Druk, 2001, p. 63. For other references to the "Tashkent front" see see Weiner, *Making Sense of War*, pp. 115, 192. See also Nikolai Mitrokhin, "Etnonatsionalisticheskaia mifologiia v sovetskom partiino-gosudarstvennom apparate," *Otechestvennye zapiski*, no. 3, http://magazines.russ.ru/oz/2002/3/2002_03_25.html (accessed 19 March 2003).
49 Borovoi, *Vospominaniia*, pp. 288–289.
50 GAOO, f. P22, op. 9, d. 310, ll. 220, 221, 241.
51 GARF, f. 8114, op. 1, d. 1056, l. 73.
52 Altshuler *et al.*, *Sovetskie evrei pishut Il'e Erenburgu*, p. 150.
53 Ibid., pp. 149–150.
54 Shimon Redlich, *War, Holocaust and Stalinism: A Documented Study of the Jewish Anti-Fascist Committee in the USSR*, Luxembourg, United States: Harwood Academic Publishers, 1995, p. 243.
55 Altshuler *et al.*, *Sovetskie evrei pishut Il'e Erenburgu*, p. 148.
56 According to data compiled by Soviet authorities as of 12 December 1941, 26.94 per cent of evacuees were listed as Jews. Rossiisskii Gosudarstvennyi Arkhiv Ekonomiki (hereafter RGAE), f. 1562, op. 20, d. 249, ll. 67–68. To some extent, the high percentage of Jews was a function of a general over-representation of Jews among those sectors of the population designated for evacuation by the Soviet state, namely the political, managerial and cultural elite. It was also a function, however, of the nature of the war, a war in which Jews had been singled out by the enemy, thus increasing their proclivity to depart. Note also that while Jews were over-represented among the evacuee population, the official statistics undoubtedly exaggerate the phenomenon, as official statistics during the evacuation generally took much better account of the urban population than the rural one, and the vast majority of Soviet Jews (86.9 per cent) lived in urban areas on the eve of the war. On the population distribution of Jews before the war, see Mordechai Altshuler, *Soviet Jewry on the eve of the Holocaust a social and demographic profile*, Jerusalem: Centre for Research of East European Jewry, Hebrew University of Jerusalem, Yad Vashem, 1998, p. 34.
57 Redlich, *War, Holocaust and Stalinism*, p. 226.
58 Altshuler *et al.*, *Sovetskie evrei pishut Il'e Erenburgu*, p. 140.
59 On the removal of Jews from key posts see Kostyrchenko, *Tainaia politika Stalina*, pp. 258–273.
60 Borovoi, *Vospominaniia*, pp. 284–285.
61 Borovoi recalls how an older woman at a station outside of Odessa, seeing him on the train, cried out that "in Odessa there are no residence permits for Jews." Ibid. p. 285.

12 The Moscow Gorky Street in late Stalinism

Space, history and *Lebenswelten*[1]

Monica Rüthers

The concepts of city and society are closely knitted together. A town entails intense communication, actors concentrate here, different forms of public interests prevail, important social decisions are made. Marginalized dissident groupings articulate themselves at central urban scenes, thereby converting these into sites of negotiation and conflict.[2] In this chapter we shall explore a central urban site in Moscow in relation to its evaluation as communicational space. What do interactions at this site disclose about relationships between individuals and the late Stalinist system?

The case of Gorky Street was chosen for study because of its location at the very heart of Moscow. This central public space was subject to various transformations before and after the war. Especially during late Stalinism, its central position did not only entail the official functions suggested by its closeness to the Kremlin. As will be shown, public space was put to very different uses regarding the time of day or night, and it was perceived in very different ways. Who were the actors on Gorky Street after the war? What do the uses and experiences of public spaces on Gorky Street reveal about living in late Stalinism?

Space as social construction

Space is a social construction, formed by people, artefacts and actions.[3] Space, analogue to the analytical concept gender, does not exist by itself, but is composed of social and material components that interact dynamically. Space is not monolithic, but diversified, a conglomeration of many spaces. Generally the idea of space is that of a uniform totality and its fragmentation is conceived as problematic. Physical and abstract spaces are not differentiated. At a given site, various spaces can exist. A single space can entail several sites which are united by their significance. For example, a separate space existed for the Soviet *nomenklatura*, the sites of which were apartments, governmental dachas in closed recreation areas, a system of segregated shops and vacation camps. Social spaces are not necessarily defined by territory, rather they gain existence through symbolic connection, such as the *blat* networks of underground economy.[4] Here economic spaces were created to facilitate the exchange of services and the acquisition of scarce goods.

The relational quality of space becomes clear when the concept of space is linked to the concept of *Lebenswelt*. If space is created by the relationship of objects, an individual cannot simply be *inside* a space. We create our space by our relationship to sites, other individuals, actions and objects. The definition of space comprehends space as plural. Accordingly, space exists as a general entity no more than history does. It seems useful to approach the theory of space through the acting individual. The definition of space combines all aspects of the worlds of living (*Lebenswelten*). It connects obscure micro-happenings to overlapping historical processes. Social goods are distributed unequally. Not everybody has the same possibilities to create or alter spaces. This is the context, which ties the individuals back to the structures where they live. Structures define spaces for action. They draw limits. But structures are ever-changing frames made by people and their actions. This play of structured–structuring life is an important part of what *Lebenswelt* means.[5] Both concepts, space and *Lebenswelt*, are social constructionist and actor-oriented. They focus on the actor and see no dichotomy between micro- and macro-history. The interferences between individual and system are investigated in case studies. Those can be single or collective biographies of chosen parts of a city.

The concept of public in Soviet-type societies remains a much-discussed subject in current research.[6] In Soviet everyday life citizens had various points of contact with officials of one kind or another. These ranged from higher bureaucrats and superiors at the workplace to ladies behind the counter at ticket offices in the metro and in public libraries and saleswomen at the *gastronom*. In context with the definition of space as described above, public spaces appear as communication spaces. The order of various public levels at specific places can be explored empirically through the media of communication, urban scheming, the formal language of architecture, rules of admission to official spaces, the calendar of public holidays, day and night, parades at official festivities, shop window decorations, clothing, behaviour, symbolic actions, horizontal social control, militia control, idlers, protesting masses and rhetoric declamatory demonstrations. Acts of communication in public spaces signify an intertwined coexistence of formal and informal messages.

Gorky Street as a place of multiple spaces

Ulitsa Tverskaya was the old road leading from the Kremlin to the capital, St Petersburg. Not only was this road central, giving access to Red Square, it was the centre itself. Before the revolution it had already been one of the main streets of commerce in Moscow.[7] After the revolution of 1917, several attempts were made to occupy this street space symbolically. In line with Lenin's monumental propaganda, monuments of socialist heroes were erected.[8] The yearly October Revolution anniversaries show decorated facades, processions with *living pictures* and other theatrical elements.[9] In the 1920s new edifices in the constructivist style were built. In form and function these had signal character. In 1927 on the 10th anniversary of the Great October the Central Telegraph Office was inaugurated, a symbol of technical development and communicational prowess. On Strastnaya Square, later renamed Pushkin Square, the main office buildings of the Izvestiya were constructed. These buildings advertised the nature of the new rulers and displayed their claim to power and influence over public spaces. During the Stalin period, these claims were intensified. The General Plan for the Reconstruction of Moscow of the year 1935 and therewith the claim on space covered all of Moscow. It extended under the earth with the metro and reached into the sky with the Palace of Soviets. Pre-revolutionary symbolic meanings were to be erased: *merchant Moscow*, the *city of 40 times 40 cupolas*, the *big village* was to become the capital of communism.

The main axes laid out in the General Plan reflected the visions of seventeenth-century urban schemers.[10] Tverskaya was the dominating ceremonial axis and represented the higher order of Soviet power.[11] Here the mass parades were to be held, leading to Red Square. This was also the reason why the buildings on Manezh Square were torn down. Voskresenskie Vorota, a tower portal which gave access to Red Square, was destroyed at the same time. This process began with the renaming of Ulitsa Gor'kogo in 1933.[12] Reconstruction of the street started in 1937, when the houses on Gorky Street 4, 6 and 8 were built, called Corpus A and B and planned by the architect A. G. Mordvinov.[13] Between Soviet Square and Okhotny Ryad there now stood only two buildings instead of 20 different houses. To level out the sharp bend in the former Tverskaya, the axis became funnel-shaped, widening from 40 to 60 metres (Figure 12.1).

The facades of the edifices had nothing to do with the buildings' purpose, since hotels, offices and private living space all showed the same exterior. The shops on the ground floor, however, were pompously furnished and decorated, thereby living up to the grand facades. Shop windows in the capital had an important show effect. On Gorky Street there were several notable establishments: the two delicatessen shops Eliseev and Filippov, the exclusive gift shop Podarki with clothes and porcelain, a jewellery store, the largest Moscovian furrier, an international book shop, the Hotel Lyuks (Luxury), which served as living quarters of the Comintern, the Grand Hotel National, the Central Telegraph Office and the famous Georgian restaurant Aragvi. Many of these establishments had existed prior to the reconstruction and quite a few are still in existence nowadays. The interior of Eliseev delicatessen was taken out and installed afresh in the new building after Gorky Street was widened.

Discourse on urban renewal

Urban renewal was an issue that evoked much discussion, being itself object and subject of the discourse at the same time. In the revolutionary master narrative, the new grew right out of the destruction of the old. Thus the destruction of the old Moscow stood for the demolition of the ancient social order and lay therefore at the grain of the establishment of the new society. The violent interventions into the urban substance severely influenced the citizens' everyday lives. Reconstruction of the main axes was accompanied by massive propaganda. On 1

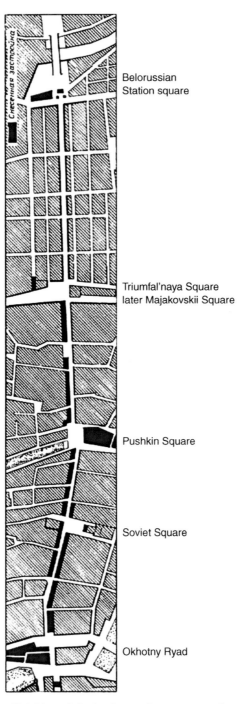

Torn down buildings

Спесимая застройка

Belorussian
Station square

Triumfal'naya Square
later Majakovskii Square

Pushkin Square

Soviet Square

Okhotny Ryad

Figure 12.1 Plan of Gorky Street after reconstruction showing demolished buildings, *Sovetskaya arkhitektura 1917–1957. Zhilishchno-grazhdanskoe
stroitel'stvo, kurortnoe stroitel'stvo, sel'skoe stroitel'stvo,
gidrotekhnicheskoe stroitel'stvo, tipovoe proektirovanie, stroitel'naya
industriia*, Moscow 1957.

May in 1935, the fourth exhibition of schemes and architecture on Gorky Street took place.[14] Citizens were requested to deposit their written opinions on the project in a special letterbox.

> One stands on an estrade in front of the gigantic model showing Moscow in the future year 1945, a Moscow that compares to the present-day city as does the latter to the town of the Tsar, which then was a large village. The model is electrified, and ever new blue, green and red lights show future streets, light up the Metro routes and the highways for automobiles, displaying the schemes for future living areas as well as traffic solutions for the big city.[15]

The General Plan of 1935 was part of an international discourse on the ideal city.[16] The future development of the city of Moscow was considered to represent a picture of society in general. Discourse on the plans of the city of the future was multifunctional. It legitimized power and created identity. Axes and squares of the New Moscow emphasized the radial structure of the centre as a sun beaming into all directions, thereby underlining governmental and authoritative functions. The gigantic Palace of Soviets was to become the centre point. In all, 160 international groups of architects took part in the competition, but the projection was too big and never became reality. The Second World War interrupted its construction. After the war, architects worked on plans for a downsized version, and a further competition was held for a new location in the vicinity of Moscow State University on the Sparrow Hills. In 1960, the whole idea was abandoned and the big hole in the ground with prepared foundations was turned into an open-air swimming pool. The vision of a compact, well-ordered city took European ideas as model, expressing clarity, comfort and pomposity at the same time. It stood in direct contrast to the wide countryside and also to the topos of the *large village* Moscow. Gorky Street lies in the centre of the area comprised in the Plan for the Reconstruction of Moscow as capital. Hence most procedures are well documented. Articles on the projects and photographs reporting on the progress on the construction sites or showing new buildings appeared on a regular basis in technical journals.[17] Especially the reconstruction of Gorky Street was much commented upon and permitted remarkable demonstrations of power on the part of the Kremlin regime. Houses were displaced. For instance, a splendid turn-of-the-century trading centre was moved backward on rails to broaden the street. Then the new building was erected in front of it, and so it finally came to stand in the backyard of Gorky Street 4. Photographs of the demolition and of spectacular displacements document the violent disruption, which took place in 1937–1938 only a few metres from the Hall of Columns, where show trials were being staged at the same time. The *cleansing* of the city was proceeding on all levels and designed to be of high visibility.

At the outbreak of the Second World War complete reconstruction was not yet achieved, although several basic operations had already taken place.[18] After the war further important decisions were made in regard to Gorky Street. Urban planning policies were discussed on a new basis. The municipal building council and its chairman, G. M. Popov, who was also chairman of the city council, Mosso-viet, proposed the construction of modern, comfortable condominiums of a high standard. In contrast the departmental heads of several ministries and municipal offices emphasized the lack of living space with a view to removing citizens from barracks or overcrowded communal apartments in which they lived to mass build-ing apartments with limited comfort. The conflict about the building opposite the Central Telegraph Office was exemplary. The planners debated on the high costs with the ministry responsible for financing the works. The architects argued that the granite for the enfacement of the building would not only embellish the most important urban axis but would also commemorate victory over the Germans. The granite, originally acquired for the invaders' victory monument, was left behind when the German troops fled the country. Popov turned to Stalin for intervention. The latter declared that nothing cheap was to be built in the capital. Each building should radiate the spirit of the communist epoch for future generations. This state-ment allowed for the highest planning standards to be applied and for the best archi-tects to be consulted for each individual project in Moscow.[19]

A break in the building policy was imminent at the end of the war. Already in 1943 A. Burov suggested at a plenum of the architects' union to let the great plans rest for five years. Instead low, prefabricated houses of American design were to be built. This proposition could not have been made a few years earlier or a few years later.[20] Nonetheless, the proposal was not realized in the Stalinist post-war years. Beria, who was chief of the Ministry of Inner Affairs (MVD) at the Lubyanka and supervised concentration camps with millions of hard labourers, promised to solve the labour problems. He had in mind the successful employment of prisoners as labourers after the White Sea Canal and also several towns and fac-tories had been constructed. But the municipality declined the offer. The assump-tion that a large number of prisoners of war and other captives were to be employed on Moscow building sites did not happen according to Taranov's docu-mentation. Archival documents until now only confirm the employment of German prisoners of war on construction sites in the centre of Moscow, namely on Gorky Street. Moscovites were to see the defeated Germans doing hard labour.

Image and representation: a view of the city

The widened Gorky Street between Okhotny Riad and Soviet Square became an icon of the New Moscow. After 1935, well-known architectural photographers regularly produced picture series. The city and the interiors as well were put into the scene.[21] The architecture, in itself already a stage, became a scene in the scene. Several different subjects of interest were touched upon: the scarce living space, the socialist city, New Moscow and the New Man. Photographs and paintings of the ideal townscape constituted an important part of the new urban space and were widely noted. Coffee-table books such as *Sovietskaya Arkhitek-tura za 30 let RSFSR* by V. A. Svarikov, edited 1950 in Moscow as well as in Leipzig (in German), featured large photographs of the communist achievements (Figure 12.2).[22] The pictures were heavily retouched, not to say painted over,

Figure 12.2 Sovetskaja Architektura za 30 let RSFSR, double page with figures 12 and 13, view of Gorky Street from Pushkin Square to Okhotny Ryad.

and tried to convey an aura of eternity. Those books reveal the outstanding role of Gorky Street. The popular architectural journal *Arkhitekturnaya Gazeta* used to publish large illustrations of recently completed projects. Such images constituted an important part of Soviet visual culture. They created an image of the new urban spaces as a monumental unity and creator of identity.

Ignored: the backyards

Illustrations of street space mostly depicted its unified appearance. Facades were an important subject in representative publications. However, analysis disrupts this logic. The complete living space must be explored, including not only the surface facades, but also the throughways to the courtyards and the rear sides of buildings. All of them constituted spaces of everyday life by day as well as by night, on festive days as well as on normal workdays.

Buildings, which formed a unified block, show a sharp separation between outer and inner usage. This can also be observed in Berlin or Vienna. Vivid accounts by several observers tell of the rural character of the back parts of the buildings on Gorky Street. Jelena Bonner who stayed at the Hotel Lyuks in the 1930s reports on passageways through a system of courtyards called Bakhrushchenka behind the hotel, stretching out as far as Bolshaya Dmitrovka. These areas were considered dangerous and unsafe because of the gangs of neglected juveniles who controlled them.[23] Walter Benjamin gives a similar account in 1921 as does Paul Thorez in 1962:

> Behind the facades of Gorky Street, the two worlds (village and town, MR) lie only a few steps apart. If you are heading towards Mossoviet, it suffices to take the first turn left into Ogarev or Stankevich Street. Or just to pass through the hall of one of the buildings, thereby gaining access to the backyards that form a kind of parallel street to Gorky. On the one side the large buildings, on the other, in irregular formation, gardens, barracks, and low houses with narrow windows and overhanging roofs. And then, behind this, right up to Herzen Street you will find a maze of *pereulki* and pathways in which you can easily lose your way. The borderline between these two worlds lies in the apartments in the large buildings. There the representative rooms face Gorky Street, whereas the kitchen windows give a view of the extensive village behind, and in the background other high buildings and skyscrapers. Very often did I stroll from one backyard (*dvor*) to the next. There the walls lack colonnades or reliefs; higher up, square windows open out in the naked brick walls. All have a barred ledge, which serves as pantry in wintertime. In summer, housewives sit there, bending forward to watch their children playing in the yard. They converse at such a loud pitch from one story to the next that I could not but be fully informed on the gossip of this district. And below lies the wonderful life of the *dvors*.[24]

Such accounts show that the courtyards were spaces of horizontal social control, where in contrast to the anonymous and urban front, rural familiarity reigned, and everyone knew everything about each other.[25] Here the very young and the old lived, here hours of leisure were spent. The courtyards are a favoured motif of nostalgic childhood memories, lyrics and paintings.[26] Along with the official image of street space and the new city there was the daily experience of these spaces of the backyards with their own flow of life. In the everyday life, two seemingly contradictory spaces existed: the anonymous and monumental in the front, and the retreat into the back as a space of self-determination (*eigen-sinn*).[27]

Urban experience and street life

The Hotel Lyuks became famous in 1921 during the 3rd World Congress of Comintern. Here the foreign visitors were accommodated en bloc and well segregated. The Lyuks became exceptional, and a *propusk* system (passport control) turned it into a ghetto. Comintern had its own global political space. In Moscow, its sites were the Lyuks or the dachas in the fashionable suburb Kuntsevo. The noble hotel on Gorky Street cultivated hierarchic structures; disgraced persons were assigned rooms in the back building, where the lesser guests had resided in former days. In the famine years of the 1930s, the hotel became a point of economical attraction.

> Prostitution flourished along Tverskaya and around the corner into the side alley, where the smell of boiled cabbage and other tempting dishes wafted through the boarded-up windows of the *stolovaya*. It was a normal issue through the weeks of the Congress. Moscow hungered severely – no wonder then that prostitution, openly and secretly, besieged and conquered the foreigners' hotel.[28]

Spaces came together where people gathered. Impersonated on site by its economically privileged members, the global political space of the Comintern came in touch with the local economic space of hunger and prostitution.

Crowds also gathered in front of the Eliseev shop windows, now a *torgsin* store,[29] to gloat over the appetizing but unattainable stacks of goods displayed.[30] This display of scarce edibles in the pompous delicatessen revealed a certain policy, namely the 'secret of the caviar sandwich'.[31] New culture politics covered the needs of the new middle class in the Stalin period: during the first Five-Year-Plan 1928–1932, a new Soviet-educated hegemony moved up into positions vacant owing to continuous purging. This created a demand for social acknowledgment and values. In 1934, a historic turnabout can be detected, when the Communist Party condemned all self-inflicted asceticism. Street festivals, music, dance, make-up and a cultured dress-sense were now part of the programme just as conservative family policy and reflection on popularized classic culture. At the same time, the principle of equal salaries for all was abandoned and privileges were granted. The morale of the new middle class legitimated social injustice. In 1935 the end of rationing was titled 'arrival in abundance'. Commercial fairs exhibited goods that were out of reach in everyday life, such as washing machines, cars or cameras. Advertisement became a part of everyday Soviet culture. The act of shopping became an educational process. The store became the classroom where exemplary saleswomen instructed clients on the issues of hygiene and good taste.[32] Moscow's shop windows, especially those on Gorky Street, were the showpiece of the Soviet Union. They emanated a sense of luxury and plenitude, despite the fact that the displayed goods were available only for a privileged few. The showcases fulfilled the same function as did the ostentatious underground halls of the metro: they proved true the claims of propaganda. Their spatial significance reached way beyond the public on the streets.

Ego-documents, but also poetry and novels are an important source when it comes to exploring cognitive maps and significance of locations and spaces. Especially in relation to cognitive and topographic systems of orientation, where actions are placed onto locations, they constituted a ground of shared experience between the reader and the author through their authenticity. The actual location evoked can have various functions in a text. The simplest is verification of the action; furthermore composition of a certain ambiance that transports time and location, since the reader shares significant knowledge of the place with the author. Texts convey urban images when a city as system of social spaces and significations establishes locations for literary actions. In his autobiographic novel *Kommunalka*, Boris Jampolski describes nightly wanderings through Gorky Street. The action is taking place in the winter of 1953, a few days or weeks before Stalin's death. The narrator lives in a communal apartment on Arbat. He is a returnee from war and imprisonment and experiences all the apprehensions and fears of the days of the alleged doctor's conspiracy. On each corner a guard is standing. Everybody spies on everybody in the apartment where he lives. Jampolski's hero feels constantly shadowed. Night and day we see the restless returnee prowling through post-war Moscow. He is completely isolated. He shuns contact with his family or friends for fear of endangering them. The book, published in 1988 after the author's death in 1972, conveys an atmosphere of depression.

> I passed (. . .) by Pushkin, still standing there where literature lovers had placed him (the common people never found their way to him). Then I passed by the house with the woman in stone on its little turret, went on past the Eliseev store aglow in the glitter of its mercantile lustre, past the former Hotel Lyuks where the last Comintern functionaries still lived, left the Mordvinov houses, looking like old chests of drawers, behind me and I passed on down grey and deserted Gorky Street towards Okhotny Ryad.[33]
>
> The Café National's large windows shimmered through the snow with a friendly light. It was still open at three o'clock in the morning. This had been kept on from the time of commercial restaurants, when Stalin had suddenly given permission for restaurants to be open in the wee hours (. . . .) In the café the light was dim, and waltzing pairs glided past me through the

light reflections thrown by a rotating mirrored globe suspended from the ceiling. (...) Everything was as always, unchanged. Nightlife in the café continued uninterrupted. (...) every evening, in winter around ten or eleven, in summer at midnight, the same people, each knowing each other, meet here, after strolling through the city centre from the Dolgoruky monument to the Hotel Moskva and back on the other side past the Aragvi, the Cocktail Hall and Café Morozhennoe, later called Kosmos; whoever entered the café alone or in company of a girlfriend or with a girl picked up under the lanterns by the Telegraph Office, was called to the small tables in the various nooks. And by three o'clock in the morning, the occupants of those tables had so thoroughly intermingled, that no one knew to whom he or she belonged.[34]

The rules of space and time were invalidated by nightlife. Spatial contrasts and social boundaries dissolved in the darkness and in the effects of artificial lighting. Everything seemed to be rotating. The depicted nightlife and its anonymous, semi-public atmosphere gave Jampolski's hero a place of rest. For the slightly younger generation the same nightlife offered a possibility of creating a subculture. After the end of the Second World War, extravagantly dressed young men appeared on Moscow's streets, meeting regularly on Gorky Street, now nicknamed *brodvei*. They spent their evening in the notorious Kokteil Kholl, set up in 1940, or at the restaurant Aragvi. In the Soviet press the denomination *bezdelnichestvo* (loafing) appeared, the juveniles were called *stiliagi*. However, they called themselves *chuvaki*.

The Kokteil Kholl: meeting place of the *stiliagi*

'From afar shouts, discords, joyful cries and groans were heard' (Figure 12.3).[35] An especially vivid account of the customers and of the cocktails served here is to be found in *Die Stunde Moskaus* by Hélène and Pierre Lazareff. The cocktails served in high glasses with drinking straws, rare at that time, bore fantasy names like *Wallbreaker for the beginning*, (Armenian brandy with apricot, peach and plum liqueur) or *In Flight* (raspberry liqueur, plum liqueur and benedictine).

The *stiliagi* were a post-war appearance in several Soviet cities. After a wartime childhood, they were out to enjoy life: dancing, listening to music and going to the cinema to watch foreign films captured during the war.[36] The anxieties endured by Jampolski's hero were non-existent for them. They had no personal remembrances of the pre-war terror under Stalin. They were no war heroes, not having been old enough to be called to the front. The *frontoviki*, older by a few years only, swaggered around in looted Western apparel, war medals on their breasts. The younger men had to find their own symbols of successful manliness. The post-war semi-public sphere of restaurants and cafés brought about social interaction and interchange of information, where the juveniles could show off their fashionable trophies. Foreigners living in Moscow also frequented the bars and restaurants, thereby constituting an important

Figure 12.3 The Kokteil Kholl in *Sovetskaja Arkhitektura za 30 let RSFSR*, Moscow
1950, p. 46.

intersection of information on Western lifestyles. The *stiliagi* preferred to be
apolitical and expressed their attitude by their dress. Critical towards wartime
heroism, they rejected the Soviet way of life and sought its opposite. They wore
tight pants, shoes with crepe soles several centimetres thick and gaudy che-
quered jackets. Their jargon was intermixed with English expressions. On the
street and in the tramway they were subjected to derision and jostling by their
co-citizens.[37] There existed a favourite bet in Leningrad and Moscow: if it were
possible to pass down *brodvei* without being jostled.[38] Their lifestyle brought to
mind the dandyism favoured by oppositional Russian aristocrats, including
Pushkin, around 1830, when such behaviour had been the fashion of the time.

A dandy cultivates melancholy and *Weltschmerz* at the same time, he is remote and unapproachable and superciliously bored. Their disappointment over the lack of political reforms was implicit in their attitudes and actions. Some experienced this consciously and felt despondency over the impossibility of bringing about a change through their own doing.[39]

The *stiliagi*'s dress was not only the means, but the end in itself and as such very suspect to the authorities.

Magazines sneered at the *stiliagi*, caricaturists made fun of them and described them as sluggards and criminals (Figure 12.4). *Komsomolskaia*

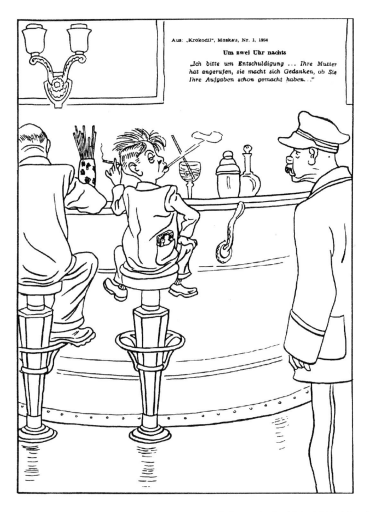

Figure 12.4 Caricature of the Kokteil Kholl from *Krokodil* 1, 1954 (*Ost-Probleme* 8, 1954, 324).

Pravda on 22 February 1952 commented on dudes with garish ties and presented them as liars and loafers.[40] The young man who served as a negative example led an unobtrusive and inoffensive life during daytime. In the evenings, after changing clothes and personality, he turned from being a good socialist into a pleasure-seeking idler. The newspaper campaigns were intensified after Stalin's death. The same newspaper on 19 November 1953 described the career of three juveniles whose regular visits to Kokteil Kholl in a logical succession of events finally led to murder. The Kokteil Kholl was finally closed down during the anti-alcohol campaign in 1954.[41] By 1960 it had become an ice-cream parlour and, at the same time, a legend. The Swedish exchange-student Harald Hamrin, citing a chance acquaintance, gives the following account:

> There used to be a cocktail bar here with a real bar with high chairs, candle-light music and decent drinks. All the rich teddy boys gathered there and sat around – sons of ministers and vice-ministers or university professors. Some drove up in red sportscars, breaks screeching when they pulled up at the curb. Sometimes the rows and fights among the crowd grew so violent that police had to interfere. About a year ago, the bar was the scene of a tragedy. One of the teddy-boys was shot and killed by the son of a minister. Police arrived immediately and there was a big fuss. Only days later the bar was closed down. Since then, Moscow's wealthy teddy-boys are homeless.[42]

After their meeting place was closed, they moved to private quarters. In 1955 stories about sons of members of the Moscow party elite who had allegedly held orgies in their parents' apartments and *dachas*, and even entertained an amateur brothel, caused a stir.[43] The press campaign waging against the *stiliagi* developed into criticism of the young poets, calling them *tuneiadtsy* (bohemians) and *nihilists*. It only died out in the mid-1960s. The name *brodvei* was given over to Prospekt Kalinina, the city's new and shining boulevard of the 1960s that created yet another vision of the city. Both Soviet and foreign commentators saw youth culture as an imitation of Western lifestyle. This imitation was looked upon favourably by Western observers, but less so by the Soviets who considered it an evil intrusion. Cold War reasoning overlooked the influence of Soviet everyday living that could very well have been the cause of emerging alternative youth cultures. Another factor was cultural contact brought about by war and the occupation of Germany. Alternative cultures can also be seen as a means of adaptation to urban surroundings. The Soviet urban population had mostly moved to the city during the 1920s and 1930s. Youth cultures were a part of urban survival techniques. The young generation may also have shared a feeling of having no say in political matters and no direction over choices for their future. It is significant that the *stiliagi* did not choose some backyard for a meeting place, but occupied the main streets of Soviet towns, renaming them *brodvei*. They chose to hang out near the centre of official power, visible to everyone. This was a communicative act. New ways of expressing oneself and the growing number of youths attracted to alternative subcultures were part of the emergence of a Soviet consumer society in the 1950s.

Considering the contrasting urban experiences of Jampolski's protagonist on the one side, and the *stiliagi* on the other, it becomes clear that there was a disruption in the urban experience. Gorky Street became the site of new public spaces on several levels. Social conflict broke out, causing scandal and the closing down of the Kokteil Kholl. Using in their turn a tool usually employed by the official power, *stiliagi* renamed Gorky Street *brodvei* and thereby laid claim to it. Their slang was full of English expressions and created an alternative space. By means of language, clothing and attitude, they constituted their own social cultural space on the map of Moscow.

The city as visual communication of power

The city itself became a communicative space and the creation of media. Displacing houses and erecting monumental new buildings were acts of communication, signifying a show of strength. At the same time, they invited citizens to identify with this display of power, and gave rise to a new collective identity. The widening of the axis and the erection of large buildings of neo-classical design resulted in much greater dimensions and a new visual axis to Red Square. Urban space was given new layers of meaning: the sites designed for mass parades and festivities made it into the space of rulers and rituals. Urban schemes and architecture were part of the uppermost level of communication, together with mass media, monumental propaganda and red banners featuring political slogans, ritual and pomp.

'Public political rituals showed patterns of symbolic representations of different aspects of this political culture. Parades, party conventions and similar events are held regularly to confirm the legitimacy of a regime and the positions of its elites'.[44] The roles of citizen and leader, the slogans, pictures and choreographies met and confirmed mutual expectations, so that people and leaders could predict each other's actions. One and the same place could be sacred on holidays and profane on common weekdays. The mute consensus concerning cooperation and acceptance by citizens was typical for the interaction between individual and Soviet system. Citizens came to Gorky Street in their best clothes to spend their time at leisure. They went window-shopping and strolled up and down Gorky Street between Okhotnyi Riad and Pushkin Square, all the while showing off to themselves and to the world the achievements of socialism. They thereby took possession of the public space as one of leisure, consumption and fashion, a space of urban lifestyle offering identity. They negotiated this cultural public space as one of civil society, while on official holidays it fell under the rule of the government.

Gorky Street spaces and their scopes

Gorky Street became a place where actors created several different spaces. One of them was communicating messages from the official public sphere. Other spaces functioned as stages, where different roles were played out and confirmed. Yet other spaces provided room for alternative publics. The uses of the ceremonial axis on holidays and weekdays, by day and by night were strictly separate. The common space was turned into a public space at distinct moments. Nightlife tended to lift the rules, which were in force during daytime. It offered anonymous spaces of escape and more liberties for self-representation. The limits of these liberties were contested when members of juvenile subcultures met to hang out on Gorky Street just like their conformist elders, thus imitating a popular pastime, while at the same time probing the limit of its possibilities. Those social spaces were not territorial, but grew out of symbolic connections: the publishing of pictures, especially of photographs of the New Moscow served to communicate 'space' as the surroundings of New Man.[45]

The social, symbolic and communicative spaces of Gorky Street were of very different meanings, lifespans and reach. Seen on a macro-level, their meaning for the modernization of society was most important. Gorky Street was part of an urban system of axes that spread out from the centre to the periphery like the beams of a sun. Economic spaces also functioned on several levels, as for instance the above-mentioned prostitution or the careful arrangements of consumer goods in the shop windows on Gorky Street indicated. On an everyday level their effect was exclusive and hierarchic. At the same time, shop windows functioned as advisers for urban lifestyle and as a vision of abundance of consumer goods available in the future. They also gave visiting foreigners proof of Soviet success. One place thus became the site of multiple communicative spaces.

Notions of space

The investigation of Gorky Street brings to attention complex layers of social utopia, representation, discourse and urban experiences. The monumental character of the new axis speaks not only of pomposity and demonstrations of power, but also of different forms of constructing identity: people could take part in this splendid new centre and identify with the power that built it. Yet they could also, at least during weekdays or nights, come here to escape, to meet non-conformist peers, to deal in black-market goods or to prostitute themselves. The modernization of the axis turned it into a showcase of the Soviet Union and made it more hierarchical. Global, national and local spaces and meanings intersected. The exhibitions of plans and models and the depiction of urban spaces formed notions of what a city should look like as well as notions of space itself. The pictures of Gorky Street, published many times in the press, became more important than the place itself. There is an interesting parallel to the all-union agricultural exhibition (VSKhV) opened in 1939 and remodelled and renamed into VDNKh between 1950 and 1954. This exhibition substituted the reality of a dying Soviet agriculture with pictures of a thriving economy.[46] In a perfect Garden of Eden, the agricultural exhibition displayed an ideal picture of the countryside made of plaster and bronze, while collectivization destroyed nearly all of the real farming areas. The contrast between representation and reality in this case equalled the opposition of the carefully designed facades of Gorky Street and its backyards. The idea behind creating perfect spaces within limits had much to do with notions of space. Soviet authorities created islands of the future – socialist microcosms – in the chaotic present through the construction of Gorky Street, the Park Kultury i Otdykha imena Gorkogo (1928), the Metro (1935), the above-mentioned agricultural exhibition and Moscow State University (1953). These limited sites can be defined as 'other places' in the meaning of Foucault's heterotopias.[47] They are segregated places with a special purpose. They may be cemeteries, motels, libraries, museums, barracks, fitness centres or hospitals; all of them are places outside the current day-to-day life – places of crisis and transition, of exclusion or inclusion, places out of time. They serve different purposes. Heterotopias of compensation denounce the rest of the world as anything but perfect, but at the same time offer their model order for recreation and edification. In the Soviet Union those places were sites, where the *other spaces* of society were constituted and where society could recover from itself. Soviet heterotopias were pictures of common ideas of a perfect place, not unlike the Persian gardens depicted on delicate silk rugs. They were the opposite of utopia, because they existed, however small, in reality. They were not the sites of everyday socialist life, but artificial perfect places inside an imperfect society. They were proof that the future was to come – a future one could already visit. Those heterotopias proved the power of the authorities to make the future happen. By providing social spaces belonging to all, the Soviet state convincingly demonstrated care for its subjects. On a symbolic level, heterotopias were places of encounter and mutual understanding between the people and their leaders. These were places of beauty and perfection.

The islands of compensation floated in a master narrative of revolution and socialism, which consisted of social realist epic novels, films, monumental paintings and general plans for cities. Disrupting the old to create the new, building socialism, and the advent of abundance were elements of this narrative. During the years of the Cold War, the focus shifted to foreign policy. A rhetoric of catching up with and overtaking capitalism prevailed. The scheme of building small parts representing the whole proved effective in the long term and was used for exportation: the Palats Kultury was given to the people of Warsaw, the Stalinallee to the people of Berlin, and near Krakow the working-class settlement of Nowa Huta was erected. The recognizable orders of these ensembles were meant to communicate the respective social order and to establish a corporate identity. Communism appeared as a network of perfect microcosms. Stalin's dictum of communism in one country had turned into communism in some selected places.

The rituals held on some of the islands in their turn constituted spaces of power. The choreographic order of the bodies of sportsmen, soldiers and leaders on top of the mausoleum on Red Square was repeated over and over. Like the picture of the new urban spaces in photographs and paintings, the festivities were a picture in the picture, a double heterotopia. On weekdays between holiday festivities, other spaces were created. Prostitution, idling and nightlife showed different arrangements of people, objects and actions. After the war, a generational gap in the perception, experience and organization of leisure time and nightlife added contrasting spaces.

The visual strategies in Stalin's time aimed at creating the illusion of perfect communism through the means of socialist realism. The art of socialist realism provided the illusion of a monolithic space that ignored the less-than-perfect zones between the perfect spots.[48] It guided the perception of the public. People were supposed to learn to blend out imperfect surroundings and create an ideal synthetic space out of islands of perfection. Photographs and paintings provided a guide to this. Movies of the 1930s featured the reconstruction of Moscow employing special effects. They visualized the miraculous synthetic process and construed a virtual New Moscow. After the construction of the Moskva–Volga Canal, Moscow featured in the Film *Volga-Volga* (1938) as Venice of the north. In *The New Moscow* by Aleksandr Medvedkin (1939), landmarks of the old Moscow like the Strastnoy Monastery on Pushkin Square, the Cathedral of Christ the Saviour or the Sukharev tower tumbled down in clouds of dust. The author had inserted real takes of the sites. But only in the movie were the old buildings replaced overnight by new ones.[49] However jagged real city life appeared, however violent and hostile it was in the eyes of its habitants, in film and propaganda it was represented as a whole and pure. A study of the reconstruction of Moscow consciously has to disrupt the arranged logic of places and pictures. The seemingly perfect microcosms were – as the *other place* – only part of the complex total of the *Lebenswelten* of Moscovites. They have to be understood as constantly referring to elements, temporarily invisible, yet intimately connected to people's individual experience of their space.

Conclusion

By the reconstruction of Gorky Street in the late 1930s, one of the most important parts of the Moscow General Plan was acted out: the narrow, winding Tverskaya became a large ceremonial axis leading to Red Square. Still it stayed an important shopping mile and even preserved its most fashionable and luxurious shops, hotels and restaurants from before the revolution. There was always a difference in uses and regulations during weekdays and holidays. In Stalin's time, Gorky Street became a heterotopy as an icon of the New Moscow in pictures widely published, and it was fenced off to serve representational purposes as a ceremonial axis on official holidays. Yet even the important and – one should suppose – tightly controlled central space of Gorky Street stayed the place of thriving nightlife and prostitution before, during and after Stalinism. The most striking feature of post-war nightlife was the generational gap experienced by the returnees and the *stiliagi*. The Gorky Street of late Stalinism had multiple aspects: there were the problems of the demobilized returnees, unemployment and taverns, and the campaigns against 'cosmopolitism'. The younger post-war youth on the other side, lacking the glamour of war heroes, but also the memory of terror and purges, tried to carve out an identity for themselves. All of this took place right at centre-stage Moscow. Since the time of its reconstruction, an important function of Gorky Street has been to convey a sense of 'normality'. It served as an after-work meeting point, where citizens could take a stroll and look at nicely decorated shop windows. The front side became a showcase of new Soviet Moscow, whereas the rear, the courtyards and backstreets, belonged to Old Moscow. However, the Stalinist system created a corporate identity through its building policy, especially as a victor. At the time of post-war patriotism, monumental high-rises in 'Russian' style were planned and the building of representative edifices on Gorky Street resumed. To this day, it is seen as a 'beautiful' avenue that Moscovites identify with and are proud of.

Notes

1 This chapter is part of a more comprehensive study on the nature of social spaces in Moscow by the author. It was made possible by a generous grant from the Swiss National Fund.
2 On the subject of city and public sphere see A. V. Wendland, A. R. Hofmann, 'Stadt und Öffentlichkeit: Auf der Suche nach einem neuen Konzept in der Geschichte Ostmitteleuropas. Eine Einführung', in A. R. Hofmann, A. V. Wendland (eds), *Stadt und Öffentlichkeit in Ostmitteleuropa 1900–1939. Beiträge zur Entstehung moderner Urbanität zwischen Berlin, Charkiv, Tallinn und Triest*, Stuttgart: Steiner, 2002, pp. 9–26.
3 The concept of space I refer to was developed by M. Loew, *Raumsoziologie*, Frankfurt/M.: Suhrkamp, 2001.
4 A. V. Ledeneva, *Russia's Economy of Favours. Blat, Networking and Informal Exchange* (Cambridge Russian, Soviet and Post-Soviet Studies 102), Cambridge/UK: Cambridge UP, 1998. *Blat* means 'connections'.
5 On the concept of *Lebenswelten* see H. Haumann, 'Lebensweltlich orientierte Geschichtsschreibung in den Jüdischen Studien: Das Basler Beispiel', in K. Hödl (ed.), *Jüdische Studien. Reflexionen zu Theorie und Praxis eines wissenschaftlichen Feldes*, Innsbruck: Studienverlag, 2003 (Schriften des Centrums für Jüdische Studien Band 4), pp. 105–122.
6 G. T. Rittersporn, M. Rolf, J. C. Behrends (eds), *Zwischen partei–staatlicher Selbstinszenierung und kirchlichen Gegenwelten: Sphären von Öffentlichkeit in Gesellschaften sowjetischen Typs / Between the Great Show of the Party–State and Religious Counter–Cultures: Public Spheres in Soviet–Type Societies*, Frankfurt/M.: Lang, 2003; G. Hausmann, 'Öffentlichkeit', in T. M. Bohn, D. Neutatz (eds), *Studienhandbuch Ostliches Europa*, vol. 2, Geschichte des Russischen Reiches und der Sowjetunion, Cologne: Boehlau, 2002, pp. 260–266.
7 W. Giljarowski, *Kaschemmen, Klubs und Künstlerklausen. Sittenbilder aus dem alten Moskau*, Berlin: Ruetten & Loening, 1988 (*Moskva i moskviči*, Moscow 1934), pp. 262–263.
8 J. E. Bowlt, 'Russian Sculpture and Lenin's plan of monumental propaganda', in H. A. Milton, L. Nochlin (eds), *Art and Architecture in the Service of Politics*, Cambridge/Mass.: MIT Press, 1978, pp. 182–192.
9 See also H. Haumann, A. Guski, 'Revolution und Fotografie', *O.10 – Iwan Puni und Fotografien der Russischen Revolution*. Museum Tinguely, 12 April 2003 to 28 September 2003, Berne: Benteli, 2003, pp. 101–130; V. Tolstoy, I. Bibikova, C. Cooke (eds), *Street Art of the Revolution. Festivals and Celebrations in Russia 1918–1933*, London: Thames and Hudson, 1990.

10 S. Kostof, *The City Shaped. Urban Patterns and Meanings through History*, London: Thames and Hudson, 1991, p. 271.

11 Kostof, *City*, p. 271 ff.

12 *Architektura SSSR* 4, 1938, 17.

13 *Architektura SSSR* 11, 1938, 3–13.

14 *Pervomajskaja (4–ja) vystavka planirovki i architektury na ul. Gor'kogo. Pervomajskaja komisija MGK VKP(d) i mossoveta orgkomitet sojuza sovetskich architektov*, Moscow 1935.

15 L. Feuchtwanger, *Moskau 1937. Ein Reisebericht für meine Freunde*, 2nd edn, Berlin: Aufbau, 1993, S. 24.

16 For a discussion of the international context see P. Hall, *Cities of Tomorrow. An Intellectual History of Urban Planning and Design in the Twentieth Century*, Oxford: Basil Blackwell, 1988. For discussion of the Soviet context see H. Bodenschatz, C. Post (eds), *Städtebau im Schatten Stalins. Die internationale Suche nach der sozialistischen Stadt in der Sowjetunion 1929–1935*, Berlin: Verlagshaus Braun, 2003; A. de Magistris, *La costruzione della città totalitaria. Il piano di Mosca e il dibattito sulla città Sovietica tra gli anni venti e cinquanta*, Milano: CittàStudiEdizioni, 1995; P. Noever (ed.), *Tyrannei des Schönen. Architektur der Stalin–Zeit*, Munich: Prestel, 1994. E. Pistorius (ed.), *Der Architektenstreit nach der Revolution. Zeitgenössische Texte, Russland 1925–1932*, Basle: Birkhäuser, 1992; J. H. Bater, *The Soviet City. Ideal and Reality*, London: Arnold 1980; R. A. French, F. E. Ian Hamilton (eds), *The Socialist City. Spatial Structure and Urban Policy*, Chichester: John Wiley, 1979; N. A. Miljutin, *Sozgorod: die Planung der neuen Stadt*, [1930], Basle: Birkhäuser, 1992; H. Altrichter, '"Living the Revolution". Stadt und Stadtplanung in Stalins Russland', in W. Hartwig (ed.), *Utopie und politische Herrschaft im Europa der Zwischenkriegszeit* (Schriften des Historischen Kollegs, Kolloquien 56), Munich: R. Oldenbourg, 2003, pp. 57–75; A. R. French, *Plans, Pragmatism and People. The Legacy of Soviet Planning for Today's Cities*, Pittsburgh: Univ. of Pittsburg Press, 1995 (Pitt series in Russian and East European studies), pp. 9–68.

17 *Architektura SSSR* 8, 1934, 22–27; *Architektura SSSR* 2, 1937, 42–45; *Architektura SSSR* 4, 1938, 14–30; *Architektura SSSR* 1, 1938, 3–13.

18 A. Latour, *Mosca 1890–1991*, Bologna: Zanichelli, 1992, S. 236.

19 E. Taranov, 'Gorod Kommunizma. Idei liderov 50–60–ch godov i ich voploshchenie', in: *Moskovskij Archiv. Istoriko–kraevedcheskii Al'manach*, Vypusk I, Moscow: Mosgorarkhiv, 1996, pp. 372–390, see p. 372.

20 de Magistris, *Costruzione*, pp. 129–130.

21 Tsentral'nyi moskovskii arkhiv dokumentov na spetsial'nych nositelyakh (TsMADSN) has over 150 photographs on *grodostroitel'stvo* of Ulitsa Tverskaya/Gor'kogo.

22 On Soviet architecture as a model for the GDR see S. Hain, '"Von der Geschichte beauftragt, Zeichen zu setzen". Zum Monumentalitätsverständnis in der DDR am Beispiel der Gestaltung der Hauptstadt Berlin', in R. Schneider, W. Wang (eds), *Moderne Architektur in Deutschland 1900 bis 2000. Macht und Monument*. Ostfildern–Ruit: Cantz, 2001, pp. 189–220.

23 J. Bonner, *Mütter und Töchter. Erinnerungen an meine Jugend 1923 bis 1945*, Munich: Piper, 1992, pp. 154–155.

24 W. Benjamin, *Moskauer Tagebuch*. Frankfurt/M.: Suhrkamp, 1980, p. 99. Citation: P. Thorez, *Moskau*, Lausanne: Editions Rencontre 1964, pp. 28–29.

25 B. Chasanow, 'Moskau als Zeichensystem', *Merkur. Deutsche Zeitschrift für europäisches Denken* 2, 1988, 85–98, see pp. 87–88, 91.

26 V. Semenova, 'Ravenstvo v nishchete. Simvolicheskaya znachenie "kommunalok"', *Sudby lyudei. Rossiia XX. vek. Biografii semei kak obekt stsciologicheskogo issledovaniia*. Moscow: RAN, 1996, pp. 373–389, see pp. 382–384. Compare also the paintings of Aleksandr Pavlovich Vasil'ev: *Moi dvor*. 1976, tempera, in possession of

the artist, and N. I. Osenev: *Moscow in winter*, oil on canvas, 1967, in possession of the artist, *Obraz tvoi, Moskva*. *Moskva v russkij zhivopisi*, Moscow 1982.

27 On *eigen-sinn* see A. Lüdtke, 'Geschichte und Eigensinn', in Berliner Geschichtswerkstatt (ed.), *Alltagskultur, Subjektivität und Geschichte: zur Theorie und Praxis von Alltagsgeschichte*, Munster: Westfälisches Dampfboot, 1994, pp. 139–153.

28 R. von Mayenburg, *Hotel Lux*, Munich: Bertelsmann, 1981, p. 56.

29 The purpose of the *torgsin*-shops was to deal with foreign tourists and to procure Western currency (*vsesojuznoe obedinenie dlja torgovli s inostrancami*). Soviet citizens could exchange valuables for vouchers.

30 S. Fitzpatrick, *Everyday Stalinism. Ordinary Life in Extraordinary Times: Soviet Russia in the 1930s*, Oxford: Oxford UP, 1999, S. 57.

31 J. Gronow, 'Kitsch and Luxury in the Soviet Union', in J. Gronow, *The Sociology of Taste*, London: Routledge, 1997, pp. 49–70.

32 J. Hessler, 'Cultured Trade. The Stalinist Turn Towards Consumerism', in S. Fitzpatrick (ed.), *Stalinism. New Directions*, London: Routledge, 2000, pp. 182–209, see p. 188; Fitzpatrick, *Everyday Stalinism*, pp. 90–91; C. Kelly, V. Volkov, 'Directed Desires: Kul'turnost' and Consumption', in C. Kelly, D. Shepherd (eds), *Constructing Russian Culture in the Age of Revolution: 1881–1940*, Oxford: Oxford UP, 1998, pp. 291–329.

33 B. Jampolski, *Kommunalka. Ein Moskauer Roman*, Leipzig: Reclam, 1991, S. 221.

34 Jampolski, *Kommunalka*, pp. 247–249.

35 H. Lazareff, P. Lazareff, *Die Stunde Moskaus. Russland wie es wirklich ist*, 2nd edn, Dusseldorf: K. Rauch, 1955, pp. 38–39, 204–206.

36 E. Yu. Zubkova, *Obshchestvo i reformy 1945–1964*, Moscow: Rossiya molodaya, 1993, p. 138 sees the *stiliagi* merely as an especially visible part of a general movement toward self-determination among the youth after the war; on *stiliagi* see also M. Edele, 'Strange young men in Stalin's Moscow: The birth and life of the stiliagi, 1945–1963', *Jahrbücher für Geschichte Osteuropas* 50, 2002, 37–61; M. Hindus, *Haus ohne Dach. Russland nach viereinhalb Jahrzehnten Revolution*, Wiesbaden: Brockhaus, 1962, pp. 374–383.

37 On fashion and attitude of the *stiliagi* see also E. Crankshaw, *Russia without Stalin: The Emerging Pattern*, London: Michael Joseph, 1956, pp. 114–116; see also A. Kassof, 'Youth vs. The Regime: Conflict in Values', *Problems of Communism* 3, 1957, 15–23.

38 R. Kirsanova, 'Stiljagi. Zapadnaja moda v SSSR 40–50–ch godov', *Rodina. Rossijskij istoričeskij ill. žurnal*, Moscow, 8, 1998, 72–75, see p. 74.

39 H. Haumann, *Geschichte Russlands*, 2nd edn, Zurich: Deutscher Taschenbuch Verlag (dtv), 2003, p. 234.

40 *Ost–Probleme* 29, 1952, pp. 950–952.

41 Crankshaw, *Russia without Stalin*, pp. 116–117.

42 H. Hamrin, *Zwei Semester Moskau*, Frankfurt/M.: Fischer, 1962, pp. 72–73.

43 *Sovetskaja Kul'tura*, 10 January 1955, cited in Crankshaw, p. 128. Kassof speaks of orgies, referring to *Komsomol'skaja Pravda*, 15 August 1956, p. 2.

44 G. D. Hannah, 'Legale und dissidente Formen politischer Kommunikation in der Sowjetunion nach Stalin', *Osteuropa* 7, 1976, 491–510, see p. 505.

45 On the dominating role of Moscow as sacred place of political power see also K. Gestwa, 'Sowjetische Landschaft als Panorama von Macht und Ohnmacht. Historische Spurensuche auf den Grossbauten des Kommunismus und in dörflicher Idylle', *Historische Anthropologie* 1, 2003, 72–100.

46 G. N. Jakovleva, 'Massenbewusstsein und Dritte Realität', in G. Gorzka (ed.), *Kultur im Stalinismus: sowjetische Kultur und Kunst der 1930er bis 1950er Jahre*, Bremen: Edition Temmen, 1994, pp. 147–152. M. Ryklin, *Räume des Jubels. Totalitarismus und Differenz*, Frankfurt/M.: Suhrkamp, 2003, pp. 134–148.

47 M. Foucault, 'Andere Räume', in M. Wentz (ed.), *Stadt–Räume*, Frankfurt/M.: Campus, 1991, pp. 65–72.

48 E. Degot, 'Zwischen Massenreproduktion und Einzigartigkeit: Offizielle und inoffizielle Kunst in der UdSSR', in P. Choroschilow *et al.* (eds), *Berlin–Moskau, Moskau–Berlin 1950–2000*, Berlin: Nicolai, 2003, pp. 133–137; C. Cooke, 'Beauty as a Route to "the Radiant Future": Responses of Soviet Architecture', *Journal of Design History* 2, 1997, Special Issue: *Design, Stalin and the Thaw*. Guest Editor: Susan E. Reid, pp. 137–160; Ryklin, *Räume*, pp. 87–133.

49 See also O. Bulgakowa, 'Film–Phantasien im Wettbewerb', *Berlin–Moskau, 1900–1950*, pp. 361–365.

Conclusion: Late Stalinism in historical perspective

Sheila Fitzpatrick

Time was when "Stalinism" or "Soviet totalitarianism" was a seamless whole extending from the late 1920s until 1953. In the composite picture of Stalinism built up in Western scholarship in the 1950s and 1960s, phenomena of the post-war years like monumental architecture, preoccupation with ideology and ideological correctness, anti-Semitism and the full flowering of the Stalin cult were prominent, partly because they belonged to the recent past, though notable pre-war episodes like collectivization and the Great Purges were also featured. The history of the Soviet Union was still the province of political scientists rather than historians, which meant that change over time – especially evolutionary rather than systemic change – was of little interest. A progression was noted from "the totalitarian embryo" to "totalitarianism full-blown," in Fainsod's often-quoted phrase;[1] and in the 1970s an argument developed about whether it was the revolution and Lenin that launched Russia on this fateful path or Stalin's "great break" at the end of the 1920s.[2] In the decades after Stalin's death, political scientists were preoccupied with the issue of fundamental systemic change: had Soviet totalitarianism died along with Stalin, or was it still alive? The idea that the Soviet political and social system might be *evolving* was implicitly ruled out for many years (partly because of the premise that totalitarian systems, once in place, were immutable and could be destroyed only by external pressure); and even when Jerry Hough raised the issue of change in the mid-1970s,[3] it was with reference to the shift in the 1930s from transformative assault to the more conservative policies that Timasheff labelled the "Great Retreat," not change from the pre-war to the early post-war period.

Thanks to the Cold War and the birth of Sovietology that accompanied it, Western scholars (particularly American) were watching the Soviet Union much more attentively in the post-war years than they had ever done before the war. The temporary cracking open of a closed society that the war produced led to a considerable increase in knowledge of the Soviet Union, notably through the Harvard Interview Project and various refugee or defector testimonies like Victor Kravchenko's (though the reliability of the latter was hotly contested). But these sources were informative primarily on the pre-war situation. What was happening inside the Soviet Union in the post-war years was much harder to find out, unless one were to accept the picture disseminated to sympathizers in the outside world by Soviet propaganda agencies: censorship of the press and other publications was even more stifling than before the war, so the Kremlinological skills of reading between the lines and decoding hidden messages in the press, however derided by later scholarly generations, were in fact indispensable.[4] The post-war years remained opaque even after historians had taken over the writing of Soviet history in the 1970s and 1980s: it was too hard to write without archives (which were available to Western Soviet historians only for earlier periods), too recent to be fully legitimate as a field of study in the conservative Russian history profession, and in any case less interesting to the first cohorts of historians than the revolution and the tumultuous 1930s.

Now, with archives for the post-war period open and historians eager to explore them, all that has changed. After a decade of concentration on the 1930s in the immediate aftermath of the Soviet Union's collapse, young historians are turning eagerly to the late 1940s – not to mention the 1950s, 1960s, and even 1970s – for topics. They are lucky, since Soviet archives turned out to be richer and more informative for the post-war years than they are for the 1930s. This was a surprise, given the poverty of post-war published sources. My impression is that the greater richness of post-war archives is not only a matter of 1930s documents being damaged, mislaid and destroyed during the war, but also a result of improved bureaucratic record keeping – and, perhaps, bureaucratic functioning in general – in the post-war 1940s compared with the 1930s. In the new scholarship, which this volume represents, cultural and socio-cultural interests predominate. Reflecting the current disposition of the international historical profession as a whole, politics is comparatively neglected, except by Russians,[5] as is the economy.[6] Post-1991 Russian scholarship has been notable for its strength in the historical sociology of science,[7] as well as the study of political history.

The overall picture of the post-war Soviet Union that emerges from this volume is grim. Peasants struggle to survive and pay ever higher taxes in money and kind, urban living standards are terrible and officials corrupt. The country is full of lost children, physically and psychologically damaged veterans, returning evacuees who find their apartments occupied and possessions stolen, single mothers struggling on inadequate state benefits. Famine reappears in 1947. There are fears of a new war. In the midst of this devastation, somewhat incongruously, we find a culture of stylishness emerging among privileged urban youth, and new forays into architectural monumentalism in the country's flagship city, Moscow. Though chaos and "self-supply" reigned in everyday life, the Soviet police state was at its zenith in the late Stalin period. The Gulag had more prisoners than ever before, their numbers increased by contingents of returned Soviet POWs suspected of disloyalty, enemy POWs who were kept for many years after the war because of their labour value, and "counter-revolutionaries" from the Baltics and Western Ukraine, as well as a constant influx of non-political prisoners like those convicted under the harsh labour discipline laws introduced in 1940.[8] The security forces were also responsible for the (still uncounted) number of persons living as administrative exiles in Siberia, the

Urals and Central Asia, including entire ethnic groups like the Chechens who had been deported east during the war.[9]

The war's impact was everywhere. Huge wartime population losses had left the country with labour shortages as well as a substantial imbalance between men and women in the working age groups. In the great swathe of territory occupied by the Germans during the war, there had been wholesale destruction of factories, dams, roads, railroads and dwellings. The Soviet Union faced economic reconstruction tasks almost on a level with those of the defeated powers, Germany and Japan, but without an equivalent infusion of outside capital. By most reckonings, the Soviet economic system had stood up well to the challenge of war.[10] While the "first" economy had dealt with the business of mobilizing and deploying resources on the macro-scale, however, the "second" (informal) economy spread its tentacles yet more widely than before the war, particularly in the sphere of consumption, and official corruption and bribery flourished. In the countryside, peasants, who had encroached on collective fields for their own individual production during the war, hoped that, now the war was won, the unpopular kolkhoz would be disbanded.

The Soviet Union had been devastated by the war, but it had also emerged as a victor. The products of victory were perhaps even more important in the long term than the devastation. In the inter-war period, the Soviet Union had been a pariah nation, accepted grudgingly if at all as a great power. After the Second World War, by contrast, it was recognized not only as a great power but as one (admittedly, the lesser) of two "superpowers." Its own territories had been substantially enlarged in the West, and in addition it had acquired a buffer zone in the form of a Sovietized Eastern Europe, a region where its new de facto mastery was quasi-imperial. For all the historic Soviet mismanagement of international communism, the war had changed the balance here, too, in ways that in the late 1940s looked very favourable for the Soviet superpower. Communist parties in Western Europe had so much popular support that in France and Italy it seemed possible that they might come to power by democratic means. To the East, Asia was in the process of throwing off colonial domination, and a new Communist power – initially the Soviet Union's ally – was emerging in China. The Soviet Union's superpower status, coupled with what seemed a rising tide of communist and anti-imperial revolution in the post-war world, generated great anxiety in the United States and Europe; Soviet intentions were construed simultaneously as imperialist and revolutionary, no clear distinction being made between the two. The tension developing between the two superpowers, exacerbated by fears about the atomic bomb as the Soviet Union joined the United States as a nuclear power in 1949, laid the ground for the Cold War – a political and propaganda confrontation which seemed to have the potential of becoming military and nuclear at any crisis.

Soviet and American Cold War fears were mutual; the difference was that in the Soviet case the post-war fear of the US was an updated continuation of the fear of Western capitalist attack that had dominated the first two decades after the revolution. Nevertheless, victory in the Great Patriotic War had significantly changed Soviet elite and popular self-perception: it was a national triumph that had been achieved under Soviet Communist leadership, and that fact added significantly to the regime's legitimacy in the eyes of the population.[11] In the new propaganda theme of the 1940s, "Soviet patriotism," love of the native land (typically, Russia) was combined with love of the Soviet project ("building communism").[12] The war became the cornerstone of Soviet patriotism, the demonstration of national virtue and proof of the link between the Communist Party and the people.[13] As the cult of the war flourished, it increasingly encroached on the Revolution's status as the nation's founding myth.[14]

Psychologically, Soviet patriotism was an easier creed to embrace than Marxist–Leninist internationalism. Another change that simplified life for many was the regime's move away from earlier policies of class stigmatization,[15] which meant that those with "alien" class origins were under less pressure to conceal them than in the pre-war period and could more easily feel themselves to be full Soviet citizens. Utopianism – or at least the kind of utopianism built on uncertainty about belonging that was characteristic of the 1930s – was on the wane, too. It was no longer necessary for a Stepan Podlubnyi to agonize over his self-imposed task of becoming Soviet, squeezing out the alien ("kulak") parts of his soul; indeed, even the real Podlubnyi had stopped bothering about such things.[16] For the majority of post-war citizens, being Soviet was easy, not something that had to be achieved through internal struggle: "Sovietness" was a natural attribute of all those who had fought in the war, either actually at the front or metaphorically on the "home front," and survived.

There was, nevertheless, a substantial minority of people for whom "Sovietness" had become a problem. This was not for the old reasons of class (except in the newly incorporated Western territories) but rather because of circumstances connected with the war: suspected collaboration with the Germans, either as a POW or resident of occupied territory, or being deported as a member of a "traitor" nationality like the Chechens or Volga Germans. Close contact with foreigners was also a problem, to the point that in the late 1940s the Supreme Soviet passed a law forbidding marriages between Soviet citizens and foreigners.[17] The strong element of xenophobia in post-war "Soviet patriotism" was expressed in the "anti-cosmopolitan" campaign. "Anti-cosmopolitanism" was shorthand for anti-foreignism, on the one hand, and anti-Semitism, on the other.[18] Official condonement and even encouragement of popular anti-Semitism was a distinguishing feature of the post-war period, something that might be called a policy innovation were it not for the fact that the late Stalinist regime put out such contradictory signals about what its policy regarding Jews and anti-Semitism actually was. The rise of popular and official anti-Semitism in the Soviet Union in the post-war years is at first sight a paradox, given that anti-Semitism was a core aspect – much reviled in the Soviet media of the 1930s – of the ideology of the Soviet Union's wartime enemy, Nazi Germany. It appears that among its breeding grounds during the war were those regions in the rear where war invalids and other Soviet Army veterans (perhaps exposed to Nazi propaganda at the front) encountered Jewish evacuees whom they perceived as arrogant shirkers.[19] Jews, it was believed by many, had sat out the war in Tashkent while Russians were dying at the front; this was a symbol not only of their alienness to the Soviet people but also of their privilege in Soviet society (good connections, elite jobs).[20]

Official anti-Semitism was no doubt influenced to some extent by the upsurge of popular anti-Semitism in the 1940s, but there were other factors at work as well. During the war, the leadership had allowed the formation in 1942 of a "voluntary" body, the Jewish Anti-fascist Committee (JAC), which was in many respects an anomalous institution in Soviet terms. Founded as one of five anti-fascist committees for conducting propaganda in support of the Soviet war effort among a particular constituency in the West,[21] it appears to have been by far the most successful in collecting financial contributions, medicine and warm clothing for the war effort from foreigners, particularly American Jews.[22] In addition, it was the only one of the five to attract substantial interest from a domestic constituency (Soviet Jews) and to assume a kind of representative function – critics called it the "commissariat of Jewish affairs"[23] – that was virtually without precedent in the Soviet period. The combination of foreign connections and aspirations to representation was clearly a dangerous one,[24] all the more as JAC's chief policy interests turned out to be protesting domestic anti-Semitism and advocating the creation of a Soviet Jewish republic in the Crimea. But the danger was magnified by the rapturous reception that Golda Meir, emissary of the new state of Israel, received from Soviet Jews when she visited Moscow in September 1948. JAC was disbanded in November 1948 and its leaders arrested and put on secret trial for nationalism, ties with American intelligence and other criminal anti-Soviet activity, in 1952. After initially supporting the establishment of the state of Israel, Soviet diplomacy switched to a sharply "anti-Zionist" position once it became clear that Israel would be part of the US camp, not the Soviet one, in the bipolar Cold War world. Official Soviet anti-Semitism reached its height with the announcement at the beginning of 1953 of the "Doctors' plot," whose alleged aim was a terrorist conspiracy to murder Soviet leaders at the behest of foreign intelligence, and whose medical co-conspirators were almost all Jewish.

Official anti-Semitism was a dirty secret that the regime tried to hide.[25] But it was not the only dirty secret. Wartime deportations of nationalities from the Caucasus and elsewhere, the arrest of many POWs on their return from captivity, wartime episodes like the Katyn murders, and repression of popular resistance in the newly acquired West Ukraine and Baltic states were all on the list. Of course the Soviet Union was not alone among the victorious Allies in having done things during the war it wanted to forget. But, apart from the question of magnitude, there was also the complicating factor that the regime was already carrying a big dirty secret from the pre-war period, namely the mass executions and extra-legal dispatch to the Gulag of victims of the Great Purges of the late 1930s. While Purge victims, for the most part, continued to languish in the Gulag, and public discussion of the Great Purges was taboo, there seems to have been a tacit consensus within the leadership that no mass blood-letting of this type – particularly no mass shedding of elite blood – was to happen again.[26] For the intelligentsia, the Great Purges were a painful but repressed topic; over time, the memories of suffering and losses associated with the Purges became intertwined with those associated with the war, so that, depending on context, either (but rarely both simultaneously) could be invoked as the cause of a great national and personal grief.[27] This made it possible for Shostakovich's Seventh Symphony – lauded from the time of its first performance in Kuibyshev in 1942 as a lament for Soviet wartime suffering – to be reinterpreted decades later, perhaps even by the composer himself, as a lament for the suffering produced by the Great Purges.[28]

In many ways, it was a new world that the Soviet Union was entering after the war. But this was not how things were presented by the regime, and probably not how they were experienced by the millions of Soviet citizens who were running around desperately trying to recover the shattered pieces of their pre-war lives. The official rubric under which post-war policies were presented was "reconstruction": that is, recreating as exactly and as quickly as possible the pre-war situation that had been destroyed. That meant not just reconstructing ruined cities and railroads but also – to the disappointment of those who had hoped for institutional reform and liberalization – strengthening (and, if necessary, recreating) collective farms and re-imposing discipline in the cultural sphere.[29] Troublemakers in the kolkhoz were liable to be exiled.[30] In literature, the arts and scholarship, the punishments meted out in connection with the disciplinary campaigns in 1946–1948 known collectively as the *Zhdanovshchina* were lighter than they might have been in the 1930s, involving mainly professional shunning and loss of status and perks rather than arrests and executions, but they nevertheless made it abundantly clear that liberalization and innovation were off the agenda. The K–R affair,[31] or more precisely the closed letter from the Central Committee to local party organizations that this affair provoked,[32] added the message that everyone, but particularly intellectuals, should beware of "cosmopolitanism" and "low bowing" to foreign influences, in other words, any excessive contact with or admiration for the West or Western culture. Press and book censorship in the immediate post-war years was as heavy as it had ever been; indeed, it was even more difficult for careful Aesopian readers to glean information than it had been in the 1930s.

Yet this restorationist facade was in many ways misleading. As two Russian scholars have recently pointed out, the conventional historiographical characterization of the late Stalin period as an "era of reconstruction" is superficial and leaves a great deal out of the picture.[33] Economic and institutional reconstruction was going on, to be sure, but in a radically changed geopolitical and psychological context. Even the economic reconstruction was "not so much conservation as a reorientation of production within the frame of the military-industrial complex," as Pyzhikov and Danilov write.[34] Mindful of what they saw as a real likelihood of military attack in the early post-war years, the Soviet leaders focused single-mindedly on building up the country's military and defence-industry capacity and developing the atomic bomb, non-restorationist enterprises that involved both administrative and technological innovation.[35] In other spheres, innovation was not the order of the day, and Stalin personally functioned as an effective barrier to most kinds of policy change. At the same time, a spirit of reform was abroad, however cautiously: reform plans were discussed at length within the bureaucracy, even though officials knew that, while Stalin lived, the time was probably not ripe for their implementation.[36] (This, of course, explains why radical changes could be introduced so rapidly after Stalin's death.) Julie Hessler has drawn attention to one remarkably radical, though unrealized, reform originating in the Ministry of Finance and seriously discussed for several years, which would have legalized "second economy" enterprises in order to tax them.[37] Within the central bureaucracy, there were palpable changes compared with the 1930s. It was not only that central offices now had better typewriters and better-quality paper than before the war (trophy goods from Germany?); also visible was a new self-confidence, perhaps associated with the entry of a younger cohort into middle-level positions, a higher level of competence and a new valuation of expertise. The old dichotomy of "Red" and "expert" had apparently lost its importance for the younger generation. In many cases, specialized government ministries were not just attentive to expertise but run by (Communist) professionals in the given field.[38]

Bureaucracy and its evolution in the post-war years remains a comparatively unstudied and unproblematized topic, and Cynthia Hooper's and James Heinzen's articles in this volume are a welcome contribution. The picture that emerges from them is of a bureaucracy in which corruption, particularly bribe-taking, increased greatly during the war and, despite token efforts, was not effectively reined in afterwards. This was associated with a new (though unannounced) policy of respect for cadres that gave officials much greater security of tenure than before the war and usually protected them from criminal prosecution. There were experiments with new, milder forms of discipline for Communists and other elite members,[39] notably the "courts of honour," in which bureaucrats and professionals were subjected to scrutiny and rebuke (but not prison or Gulag sentences) by their peers for offences involving Western contacts and influences.[40] All this provides an important context for understanding the increased privileges and sense of entitlement of the group Djilas called "the New Class," whose unabashed emergence in the late Stalin period is captured so effectively in Vera Dunham's study of post-war literature.[41]

Another change of the post-war period was the heightened attention now paid within the Communist Party to "private life" questions.[42] Errant party members, guilty of adultery, excessive public drunkenness and other conduct unbecoming to a Communist, were often called into local or central party control commissions for talks and admonition; and the same thing occurred in the Komsomol. This was not novel in principle, as the party had always asserted the right to intervene in the private lives of its members; what was new was the pervasiveness of the practice of intervention. Party agitators and propagandists were now expected to engage actively with issues of private morality, and a brochure on the "moral make-up of Soviet man" was issued for their guidance.[43] The heightened scrutiny of private lives that developed in the immediate post-war years was to become so firmly entrenched in the post-Stalin period that later, after the 1991 collapse of the Soviet Union, many Russians remembered it as the central characteristic of Soviet "totalitarianism."[44] The practice of denunciation and its reception by central authorities provides an interesting insight into the evolution of Soviet political culture. Before the war, Communist officials and other elite members had been vulnerable to discipline from below via "abuse of power" (whistle-blowing) denunciation of superiors by subalterns, a practice that rose to its height during the Great Purges.[45] The old whistle-blowing denunciations continued to flow in the post-war period,[46] but, Cynthia Hooper argues in this volume, they were no longer encouraged, as they had been in the past, and sometimes actively discouraged. Denunciations based on class origin were also losing their former power. The tremendous stigma that "alien" class origin had carried in the 1930s was not forgotten, but it was much diminished. Ordinary people might still think to get even with a neighbour by denouncing them as the son of a kulak, and provincial authorities still sometimes responded as the writers hoped, but central authorities were much more inclined to dismiss such charges.[47] This is not to say that stigmatization was no longer an important part of Soviet life, but rather that the targets of stigmatization had changed, notably to wartime collaboration and Jewish ethnicity; and the effect of these changes was to make (non-Jewish) elite members generally less vulnerable.

Although Timasheff's "Great Retreat" paradigm has fallen out of favour with scholars in recent years,[48] it remains tempting to use it for the shift away from revolutionary ideology and political psychology to the "great power-and-proud-of-it" mindset of the post-war years. Some of the limitations of the paradigm are revealed, however, when we realize that, while bold revolutionary-transformation policies of the 1920s and early 1930s like class stigmatization and "proletarian promotion" were no longer practised in the Soviet Union, they were still part of the "Sovietization" policy package applied in the Baltic States and exported to Eastern Europe.[49] In other words, these policies had not been repudiated by the Soviet leadership, as an observer in the mid-1930s might have concluded; rather they were seen as essential policies *for the early phase of revolutionary transformation* that were no longer needed in the phase of maturity that the Soviet Union had entered. The appropriateness to Eastern Europe and the Baltics of the old transformation policies was never debated within the Soviet leadership, as far as we know, presumably because it appeared self-evident. There could scarcely be a clearer indication of the leaders' sense that the Soviet Union had ceased to be a state in revolution and become a state with a successful revolution behind it.

Another important sign of change in the post-war years was an emerging youth counter-culture, something that had been notably absent in the Soviet Union in the 1920s and 1930s,[50] when youth had been remarkably successfully mobilized behind regime projects, infused for that generation with a sense of adventure, challenge and danger, and despised the only real "counter-culture" of the period, the tradition- and religion-based culture of their parents. The *stiliagi*, young counter-cultural pioneers who embraced Western (or what they thought of as Western) forms of self-presentation calculated to distinguish them from the grey Soviet masses, were once thought to be part of the post-Stalin Thaw, but recent scholarship has shown them to be already highly visible in the late 1940s.[51] To be sure, this was only a beginning. The Soviet authorities neither took the *stiliagi* very seriously in the late Stalin period nor perceived it as a major problem that resentful lower-class youths entering manual employment in the hard times of the 1940s never subsequently bonded with the Soviet regime, as Donald Filtzer has argued.[52] Foreign radio broadcasts to the Soviet Union, a major source of counter-cultural encouragement, were still in their infancy.[53] Nevertheless, a path was set that would lead in the Khrushchev era to the regime's shocked discovery that "youth" as a group was no longer supportive of the Soviet project, not even in a reform version, but instead were looking westward to the fleshpots of capitalism.

At first glance, the changes of the late Stalin period pale in comparison with those that followed Stalin's death in the age of the Thaw, with its cultural liberalization and cautious opening to the West, dismantling of the Gulag, shift of priorities towards consumer goods, and provision of tens of million of new separate apartments for urban families, all sponsored by an overtly reform-minded regime whose ambitious projects ranged from renovation of agriculture and education to major bureaucratic restructuring. Yet this verdict may require qualification. The Khrushchev period was indeed the first great era of self-conscious Soviet reform, in many ways foreshadowing Gorbachev's perestroika 30 years later and, like perestroika, ending in failure. The Khrushchev reform effort differed from Gorbachev's, however, in one crucial respect: its outcome was not system collapse but conservative revanche. It has often been suggested that that status quo ante to which Brezhnev's regime sought to return was "Stalinism," but if so, it was clearly neither the transformational Stalinism of the First Five-Year Plan nor the Stalinism of mass purges and chronic insecurity of elites in the 1930s. A more plausible hypothesis might be that the referent for conservative revanche after Khrushchev's fall was late (post-war) Stalinism, the kind of Stalinism that is incomprehensible outside the new context of superpower status, quasi-imperial responsibilities, suspicion of Western influences and cultural liberalization and a revolution that had been relegated to history.

Late Stalinism was a watershed, Pyzhikov and Danilov argue, because it was at this period, and not earlier or later, that "the USSR acquired the status of a world superpower, with a sharply expressed militarized economy, a great-power ideology and the formation of a circle of top leadership which constituted the core of the leaders in subsequent decades."[54] We might add some social and psychological characteristics to this "watershed," notably a new acceptance of elite entitlement and security, a new sense of national pride and what it meant to be Soviet based on victory in the Second World War, and, as far as the intelligentsia was concerned, a lasting trauma that was as much a product of the state-sponsored anti-Semitism of the late Stalin period as of the *Zhdanovshchina*, and laid the ground for the later dissident movement. All in all, the configuration that emerged in the post-war period differed in essential respects from that of the 1930s and was to prove highly durable. One could argue that it was this configuration, combining military-technological innovation and assumption of the quasi-imperial superpower burden with rejection of anything smacking of liberal reform that sowed the seeds of the Soviet Union's ultimate downfall. By the same token, however, it was to set the stage for almost half a century and provided the template for Soviet "developed socialism" that was to last through the long Brezhnev era right up to 1991.

Notes

1 Merle Fainsod, *How Russia is Ruled*, rev. edn (Cambridge, Mass.: Harvard University Press, 1963), p. 59.
2 See Stephen F. Cohen, "Bolshevism and Stalinism," in Robert C. Tucker, ed., *Stalinism. Essays in Historical Interpretation* (New York: W. W. Norton & Co, 1977).
3 Jerry F. Hough, "The Cultural Revolution and Western Understanding of the Soviet System," in Sheila Fitzpatrick, ed., *Cultural Revolution in Russia, 1928–1931* (Bloomington, Ind.: Indiana UP, 1978), pp. 242–244.
4 "Kremlinology" was not only a specialty of the political scientists watching high politics. A similar technique of critical reading of Soviet published data was practised with considerable success by economists like Naum Jasny.
5 See especially Oleg Khlevniuk's work, e.g. Yoram Gorlizki and Oleg Khlevniuk, *Cold Peace Stalin and the Soviet Ruling Circle, 1945–1953* (Oxford: Oxford University Press, 2004) (Gorlizki is a young British scholar whose disciplinary location is a Department of Government); also G. V. Kostyrchenko, *Tainaia politika Stalina. Vlast' i antisemitizm* (Moscow: Mezhdunarodnye otnoshenii, 2001); A. A. Danilov and A. V. Pyzhikov, *Rozhdenie sverkhderzhavy. SSSR v pervye poslvoennye gody* (Moscow: Olma-Press, 2002). See also the document publications by Russians, e.g. V. A. Kozlov and S. V. Mironenko, eds, *"Osobaia papka" I. V. Stalina: Iz materialov Sekretariuata NKVD-MVD SSSR 1944–1953* (Moscow: Seriia "Katalogi," 1994); O. V. Khlevniuk *et al.*, eds, *Politbiuro TsK VKP(b) i Sovet ministrov SSSR, 1945–1953* (Moscow: ROSSPEN, 2002); idem, ed., *TsK VKP(b) i regional'nye partiinye komitety 1945–1953* (Moscow: ROSSPEN, 2004); A. S. Kiselev *et al.*, eds, *Moskva poslevoennaia 1945–1947. Arkhivnye dokumenty i materialy* (Moscow: Mosgorarkhiv, 2000); Joshua Rubenstein and Vladimir P. Naumov, *Stalin's Secret Pogrom. The Postwar Inquisition of the Jewish Anti-Fascist Committee*, trans. Laura Esther Wolfson (New Haven: Yale University Press, 2001).
6 Notable exceptions are Donald Filtzer, *Soviet Workers and Late Stalinism: Labour and the Restoration of the Stalinist System after World War II* (Cambridge: Cambridge University Press, 2002); Julie Hessler, *A Social History of Soviet Trade. Trade Policy, Retail Practices, and Consumption, 1917–1953* (Princeton, NJ: Princeton University Press, 2004); and P. Charles Hachten, "Property Relations and the Economic Organization of Soviet Russia, 1941–1948," PhD diss., University of Chicago, 2005.
7 See, for example, Nikolai Krementsov, *Stalinist Science* (Princeton, NJ: Princeton UP, 1997); idem, *The Cure. A Story of Cancer and Politics from the Annals of the Cold War* (Chicago: University of Chicago Press, 2002); Alexei Kojevnikov, "Games of Stalinist Democracy: Ideological Discussions in Soviet Sciences 1947–1952," in Sheila Fitzpatrick, ed., Stalinism: New Directions (London: Routledge, 2000).
8 Donald Filtzer, *Soviet Workers and Late Stalinism*, pp. 158–167.
9 On ethnic deportations, see Pavel Polian, *Ne po svoei vole: Istoriia i geografiia prinuditel'nykh migratsii v SSSR* (Moscow: "Memorial," 2001) and J. Otto Pohl, *Ethnic Cleansing in the USSR, 1937–1949* (Westport, Conn.: Greenwood Press, 1999).

10 See Mark Harrison, *Soviet Planning in Peace and War, 1938–1945* (Cambridge: Cambridge University Press, 1985) and idem, "Soviet Industrialisation and the Test of War," *History Workshop Journal* no. 29 (1990).

11 See Elena Zubkova, *Russia after the War. Hopes, Illusions, and Disappointments, 1945–1957*, trans. Hugh Ragsdale (Armonk, NY: M. E. Sharpe, 1998), p. 32: "The victory raised to an unprecedented height not only the international prestige of the Soviet Union but the authority of the regime inside the country as well. . . ."

12 Danilov and Pyzhikov, *Rozhdenie*, p. 173. For an interpretation that places more emphasis on continuities with the pre-war period, see David Brandenberger, *National Bolshevism. Stalinist Mass Culture and the Formation of Modern Russian National Identity 1931–1956* (Cambridge, Mass.: Harvard University Press, 2002), pp. 183–196.

13 The classic statement on this is Stalin's in his speech to Moscow voters on 9 February 1946: "War was not only a curse. It was at the same time a great school of experience and testing of all the forces of the people . . . War constituted something like an examination for our Soviet system, our state, our government, our Communist Party" – an examination which Stalin clearly thought had been triumphantly passed. I. V. Stalin, *Sochineniia* vol. XVI(3), ed. Robert H. McNeal (Stanford: Hoover Institution, 1967), p. 4.

14 On the myth of the war, see Nina Tumarkin, *The Living and the Dead: The Rise and Fall of the Cult of the War* (New York: Basic Books, 1994); Amir Weiner, *Making Sense of War. The Second World War and the Fate of the Bolshevik Revolution* (Princeton: Princeton University Press, 2001), pp. 18–21.

15 See below, but note that class discriminatory policies were still actively pursued in the newly incorporated territories of the Soviet Union and Sovietized Eastern Europe, as discussed below.

16 See the second instalment of Podlubnyi's diaries, 1941–1948, in Tsentr dokumentatsii "Narodnyi Arkhiv," f. 30 (Podlubnyi, Stepan Filippovich). For an insightful analysis of Podlubnyi's diaries of the 1930s, see Jochen Hellbeck, "Fashioning the Stalinist Soul: the Diary of Stepan Podlubnyi, 1931–9," in Sheila Fitzpatrick, ed., *Stalinism*, pp. 77–116.

17 Danilov and Pyzhikov, *Rozhdenie*, p. 178.

18 For a good discussion of the complexities of "anti-cosmopolitanism," which can by no means be reduced simply to anti-Semitism, see Kiril Tomoff, *Creative Union. The Professional Organization of Soviet Composers, 1939–1953* (forthcoming, Cornell University Press), ch. 6. In some instances, as Tomoff shows in his article "Uzbek's Music's Separate Path: Interpreting 'Anti-Cosmopolitanism' in Stalinist Central Asia, 1949–52," *The Russian Review* 63:2 (2004), "anti-cosmopolitanism" could also be a slogan for anti-Russianism (resistance of non-Russian nationalities to Russian cultural imperialism).

19 See G. V. Kostyrchenko, *Tainaia politika Stalina. Vlast' i antisemitizm* (Moscow: "Mezhdunarodnye otnosheniia," 2001), pp. 242–249.

20 On popular perception in the late Stalin period of Jews as a privileged group, see Sheila Fitzpatrick, "Vengeance and Ressentiment in the Russian Revolution," *French Historical Studies* 24:4 (2001), pp. 585–586.

21 The other anti-Fascist committees were for women, youth, scientists and Slavs. Joshua P. Rubinstein and Vladimir P. Naumov, eds, *Stalin's Secret Pogrom. The Postwar Inquisition of the Jewish Anti-Fascist Committee*, trans. Laura Esther Wolfson (New Haven: Yale University Press, 2001), p. 7.

22 Rubinstein and Naumov, *Stalin's Secret Pogrom*, p. 10.

23 G. Kostyrchenko, *V plenu u krasnogo faraona* (Moscow: "Mezhdunarodnye otnosheniia," 1994), p. 63.

24 q.v. Kostyrchenko, *V plenu*, p. 49, on Stalin's growing perception that both in the international and domestic arena JAC "was trying to occupy the position of a legitimate representative of Soviet Jewry. This was particularly painful for the Soviet system and strengthened its innate xenophobia."

25 See, for example, Khrennikov's report of how Stalin, at a Politburo meeting attended by Stalin Prize Committee representatives (i.e. intelligentsia leaders) in December 1952, made a great show of being shocked by hearing of anti-Semitism in the Central Committee and ordered that it be stopped forthwith, cited in Tomoff, *Creative Union*, ch. 9. On avoidance of outright anti-Semitism in the central press and the continued censure of overtly anti-Semitic remarks in the Communist Party, see Sheila Fitzpatrick, *Tear off the Masks! Identity and Imposture in Twentieth-Century Russia* (Princeton: Princeton University Press, 2005), pp. 290–298.

26 Gorlizki and Khlevniuk conclude that "although we have no incontrovertible evidence (and none is ever likely to surface), Stalin's actions in the postwar period suggest that he had been unnerved by the Great Purges and was extremely wary of embarking on a new round of bloodletting on such a scale," *Cold Peace*, p. 5.

27 See, for example, Catherine Merridale, *Night of Stone. Death and Memory in Twentieth-Century Russia* (London: Granta, 2000), pp. 214 and passim.

28 In Volkov's rendition, Shostakovich was "thinking of other enemies of humanity," not the Nazis, when he wrote the Symphony: Solomon Volkov, ed., *Testimony. The Memoirs of Dmitri Shostakovich*, trans Antonina W. Bouis (New York: Harper & Row, 1979), p. 155. For a critique of the "Great Purges" reading of the Seventh Symphony and its popularity with Russian intellectuals, see Richard Taruskin, *Defining Russia Musically: Historical and Hermeneutical Essays* (Princeton: Princeton University Press, 2000), p. 487.

29 On hopes of post-war liberalization, see Zubkova, *Russia after the War*, pp. 33, 59, and passim.

30 Ukaz Prezidiuma Verkhovnogo Soveta SSSR of 21 February 1948 "O vyselenii iz Ukrainskoi SSR lits, zlostno ukloniaiushchikhsia ot trudovoi deiatel'nosti v sel'skom khoziaistve i vedushchikh antiobshchestvennoi, paraziticheskoi obraz zhizni," followed by law of 2 June 1948, extending scope to other territories of USSR (excluding Baltics and Western Ukraine). Text of the law of 2 June in *Otechestvennye arkhivy*, 1993 no. 2, pp. 37–38.

31 Named for the cancer researchers Nina Kliueva and Grigorii Roskin. See Krementsov, *The Cure*, and V. D. Esakov and E. S. Lebina, *Delo KR. Sudy chesti v ideologii i praktike poslevoennogo stalinizma* (Moscow: Institut rossiiskoi istorii RAN, 2001).

32 The text was published by V. D. Esakov and E. S. Levina, "Delo 'KR' (Iz istorii gonenii na sovetskuiu intelligentsiiu)," in *Kentavr*, 1994 no. 2, pp. 54–69, and no. 3, pp. 96–118.

33 Danilov and Pyzhikov, *Rozhdenie sverkhderzhavy*, p. 296.

34 Ibid., p. 109. The term "military-industrial complex" was introduced in a Soviet context by N. Simonov in his book *Voenno-promyshlennyi kompleks v SSSR v 1920–1950-e gody* (Moscow: ROSSPEN, 1996).

35 On the creation of the atomic industry in the immediate post-war years, described as "a remarkable feat, especially for a country whose economy had been devastated by the war," see David Holloway, *Stalin and the Bomb. The Soviet Union and Atomic Energy, 1939–1956* (New Haven: Yale University Press, 1994), pp. 192–193 and passim.

36 Gorlizki and Khlevniuk, *Cold Peace*, pp. 61–62, 166–167.

37 Julie Hessler, "A Postwar Perestroika? Toward a History of Private Enterprise in the USSR," *Slavic Review* 57:3 (1998), 526–531.

38 On the role of qualified doctors in Soviet health administration, see Christopher Burton, "Medical Welfare during Late Stalinism. A Study of Doctors and the Soviet Health System, 1945–53," PhD diss., University of Chicago, 1999, pp. 89–93.

39 This mildness did not go unremarked. With regard to the K–R affair, as reported to the Central Committee in 1947, many expressed surprise that Kliueva and Roskin got off so lightly ("I expected that they would shoot them"). RGASPI. f. 17, op. 88, d. 819.

40 On courts of honour, see Esakov and Lebina, *Delo KR*, pp. 148–218, and Krementsov, *The Cure*, pp. 109–115. As Kuznetsov described the function of courts of honour in his introductory speech at the court held in the apparatus of the Central Committee on 29 September 1947, it was to educate cadres "in the spirit of Soviet patriotism, in recognition of their state and public duty" and eliminate "very harmful survivals of capitalist among some strata of our intelligentsia, survivals manifest in low-bowing and servility before things foreign and bourgeois reactionary culture." RGASPI, f. 17, op. 121, d. 616, l. 6. But not all courts of honour dealt with foreign-related offences: in those held in the ministry of the electrical and machine-tool industries in 1948, officials were accused of various forms of financial malfeasance ("anti-state and amoral actions"). RGASPI, f. 17, op. 122, d. 327, ll. 126–9.

41 Vera S. Dunham, *In Stalin's Time. Middleclass Values in Soviet Fiction* (Cambridge: Cambridge University Press, 1976).

42 This is discussed in Sheila Fitzpatrick, *Tear off the Masks!* ch. 12.

43 *O moral'nom oblike sovetskogo cheloveka: Sbornik materialov v pomoshch' propagandistam i agitatoram*, comp. A. S. Aleksandrov (Moscow: Moskovskii rabochii, 1948).

44 See the comment in a popular perestroika text that "the determining sign" of totalitarianism is that "Not one sphere of life remains non-transparent for the regime. All is illuminated by its searchlights and caught in their beams. All possibility for an individual to exit from the control of the state – be it through family, friendship, intimate relations, ... or personal tastes, opinions, and habits – is blocked," L. Gozman and A. Etkind, *Osmyslit' kul't Stalina* (Moscow: Progress, 1989), p. 339, quoted in V. Zamkovoi, *Totalitarizm. Sushchnost' i kontseptsii* (Moscow: Institute mezhdunarodnogo prava i ekonomiki, 1994), p. 2.

45 Fitzpatrick, "Signals from Below: Soviet Letters of Denunciation of the 1930s," in Sheila Fitzpatrick and Robert Gellately, *Accusatory Practices. Denunciation in Modern European History, 1789–1989* (Chicago: University of Chicago Press, 1997), pp. 85–120.

46 See Vladimir A. Kozlov, "Denunciation and its Functions in Soviet Governance: A Study of Denunciations and their Bureaucratic Handling from Soviet Police Archives, 1944–1953," in Fitzpatrick and Gellately, *Accusatory Practices*, pp. 121–152. Note that, in contrast to Hooper, Kozlov noticed no significant change of practice in the post-war period with regard to whistle-blowing and its reception in his sample of denunciations to the NKVD/MVD, and treats denunciation from below as a continuing threat to and constraint on officials.

47 For such a case, see, for example, RGANI, f. 6, op. 6, d. 1574, l. 3: this investigation by the Party Control Commission of a denunciation of a physician as son of a priest concluded that the allegations were true but "there are no grounds to show distrust" of the man.

48 Following the lead given in Stephen Kotkin's *Magnetic Mountain. Stalinism as a Civilization* (Berkeley: University of California Press, 1995), which treated the Stalinism of the 1930s as the real site of social and cultural transformation rather than an abandonment of revolution, David L. Hoffmann's attacked the Timasheff thesis in "Was There a 'Great Retreat' from Soviet Socialism? Stalinist Culture Reconsidered," *Kritika* 5:4 (2004), pp. 651–674; in the debate that followed, Matthew E. Lenoe was the only serious defender of Timasheff (ibid., pp. 721–730).

49 On post-war affirmative action programmes in East Germany, Czechoslovakia and Poland, see John Connelly, *Captive University. The Sovietization of East German, Czech, and Polish Higher Education, 1945–1956* (Chapel Hill: University of North Carolina Press, 2000).

50 In contrast to Germany, where the Westernizing counter-culture of the Edelweiss Pirates and groups of young devotees of swing had caused considerable concern to the Nazi regime in the 1930s. Detlev J. K. Peukert, *Inside Nazi Germany. Conformity,*

Opposition, and Racism in Everyday Life, trans Richard Deveson (New Haven: Yale University Press, 1987), pp. 154–169.

51 Mark Edele, "Strange Young Men in Stalin's Moscow: the Birth and Life of the Stiliagi, 1945–1953," *Jahrbücher für die Geschichte Osteuropas* 50 (2002). See also Juliane Fürst's article in this volume.

52 Filtzer, *Soviet Workers and Late Stalinism*, pp. 155–157.

53 The BBC's Russian Service was established in 1946, Voice of America in 1947: Robert D. English, *Russia and the Idea of the West. Gorbachev, Intellectuals and the end of the Cold War* (New York: Columbia UP, 2000), p. 63. Local party authorities were already criticizing those who listened to them in 1947 (see, for example, RGASPI, f. 17, op. 88, d. 819, ll. 129–30), suggesting that they quickly found an audience.

54 Danilov and Pyzhikov, *Rozhdenie sverkhderzhavy*, p. 296.

Index